AF846971

Conference Proceedings of the Society for Experimental Mechanics Series

Series Editor

Tom Proulx
Society for Experimental Mechanics, Inc.
Bethel, CT, USA

More information about this series at http://www.springer.com/series/8922

Randall Allemang
Editor

Special Topics in Structural Dynamics, Volume 6

Proceedings of the 33rd IMAC, A Conference and Exposition on Structural Dynamics, 2015

Springer

Editor
Randall Allemang
University of Cincinnati
Cincinnati, OH, USA

ISSN 2191-5644 ISSN 2191-5652 (electronic)
Conference Proceedings of the Society for Experimental Mechanics Series
ISBN 978-3-319-15047-5 ISBN 978-3-319-15048-2 (eBook)
DOI 10.1007/978-3-319-15048-2

Library of Congress Control Number: 2015931926

Springer Cham Heidelberg New York Dordrecht London
© The Society for Experimental Mechanics, Inc. 2015
This work is subject to copyright. All rights are reserved by the Publisher, whether the whole or part of the material is concerned, specifically the rights of translation, reprinting, reuse of illustrations, recitation, broadcasting, reproduction on microfilms or in any other physical way, and transmission or information storage and retrieval, electronic adaptation, computer software, or by similar or dissimilar methodology now known or hereafter developed.
The use of general descriptive names, registered names, trademarks, service marks, etc. in this publication does not imply, even in the absence of a specific statement, that such names are exempt from the relevant protective laws and regulations and therefore free for general use.
The publisher, the authors and the editors are safe to assume that the advice and information in this book are believed to be true and accurate at the date of publication. Neither the publisher nor the authors or the editors give a warranty, express or implied, with respect to the material contained herein or for any errors or omissions that may have been made.

Printed on acid-free paper

Springer International Publishing AG Switzerland is part of Springer Science+Business Media (www.springer.com)

Preface

Special Topics in Structural Dynamics represents one of ten volumes of technical papers presented at the 33rd IMAC, A Conference and Exposition on Structural Dynamics, 2015, organized by the Society for Experimental Mechanics, and held in Orlando, Florida, February 2–5, 2015. The full proceedings also include volumes on Nonlinear Dynamics; Dynamics of Civil Structures; Model Validation and Uncertainty Quantification; Dynamics of Coupled Structures; Sensors and Instrumentation; Structural Health Monitoring and Damage Detection; Experimental Techniques, Rotating Machinery and Acoustics; Shock and Vibration, Aircraft/Aerospace, and Energy Harvesting; and Topics in Modal Analysis.

Each collection presents early findings from experimental and computational investigations on an important area within structural dynamics. *Special Topics in Structural Dynamics* represents papers on enabling technologies for modal analysis measurements and applications of modal analysis in specific application areas. Topics in this volume include:

Analytical Methods
Biological Systems
Wind Turbine Applications

The organizers would like to thank the authors, presenters, session organizers, and session chairs for their participation in this track.

Lowell, MA, USA — C. Niezrecki

Contents

1 **Development of Reduced Order Models to Non-modeled Regions** 1
Kevin Truong and Peter Avitabile

2 **Prediction of Forced Response Using Expansion of Perturbed Reduced Order Models with Inexact Representation of System Modes** 13
Sergio E. Obando, Peter Avitabile, and Jason Foley

3 **Estimation of Rotational Frequency Response Functions** 35
T.A.N. Silva and N.M.M. Maia

4 **Estimation of Spatial Distribution of Disturbances** 49
Yalcin Bulut, Omer F. Usluogullari, and Ahmet Temugan

5 **Body Wise Time Integration of Multi Body Dynamic Systems** 55
Wolfgang Witteveen

6 **Structural Dynamic Modeling: Tales of Sin and Redemption** 63
Robert N. Coppolino

7 **Muscle Property Identification During Joint Motion Using the NL-LTP Method** 75
Michael W. Sracic

8 **On the Detectability of Femoral Neck Fractures with Vibration Measurements** 85
Wolfgang Witteveen, Carina Wagner, Patrick Jachs, Stefan Froschauer, and Harald Schöffl

9 **Static Calibration of Microelectromechanical Systems (MEMS) Accelerometers for In-Situ Wind Turbine Blade Condition Monitoring** 91
O.O. Esu, J.A. Flint, and S.J. Watson

10 **Predicting Full-Field Strain on a Wind Turbine for Arbitrary Excitation Using Displacements of Optical Targets Measured with Photogrammetry** 99
Javad Baqersad, Peyman Poozesh, Christopher Niezrecki, and Peter Avitabile

11 **Predicting the Vibration Response in Subcomponent Testing of Wind Turbine Blades** 115
Mohamad Eydani Asl, Christopher Niezrecki, James Sherwood, and Peter Avitabile

12 **Linear Modal Analysis of a Horizontal-Axis Wind Turbine Blade** 125
Gizem Acar and Brian F. Feeny

13 **Reduced-Order Modeling of Turbine Bladed Discs by 1D Elements** 133
Luigi Carassale, Mirko Maurici, and Laura Traversone

14 **Damping Estimation for Turbine Blades Under Non-stationary Rotation Speed** 145
Luigi Carassale, Michela Marrè-Brunenghi, and Stefano Patrone

15 **Finite Element Modeling of a 40 m Space Frame Wind Turbine Tower** 153
S.A. Smith, W.D. Zhu, and Y.F. Xu

16 Experimental Validation of Modal Parameters in Rotating Machinery 171
Bram Vervisch, Kurt Stockman, and Mia Loccufier

17 Estimation of Modal Damping for Structures with Localized Dissipation 179
M. Krifa, N. Bouhaddi, and S. Cogan

18 Design of UAV for Surveillance Purposes 193
F. Cheli, F. Ripamonti, and D. Vendramelli

19 An Innovative Solution for Carving Ski Based on Retractile Blades 201
F. Cheli, L. Colombo, and F. Ripamonti

Chapter 1
Development of Reduced Order Models to Non-modeled Regions

Kevin Truong and Peter Avitabile

Abstract Model reduction and model expansion techniques have been used in many structural dynamic modeling applications to map detailed FEM at the full set of DOF (NDOF) to an abbreviated set of active DOF (ADOF) while preserving the dynamic characteristics of interest. In order to perform model reduction and model expansion, the ADOF model must be a subset of the NDOF model. For complex structures with hollow spaces (i.e. wind turbine blades, airframes, hollow fuselage sections, etc.), model reduction to a "beam-like" neutral axis is not possible.

An approach for model reduction to a "line element style model with a neutral axis" is developed in this work. Several analytical models using hollow cantilever beams as academic structures are presented to illustrate the technique. This technique can be extremely useful for development of simplistic beam models from full 3D FEM that possess hollow regions where a neutral axis would exist. These simplistic beam models can be desirable for analyses such as flutter where computation is intensive for structures such as fuselage, aircraft wings, wind turbine blades, etc.

Keywords FEM • Reduced order modeling • Hollow beam • Correlation • Phantom nodes

Nomenclature

DOF	Degrees of freedom
NDOF	Full space DOF
ADOF	Active DOF
X_n	Physical space coordinate system
X_a	Reduced physical space coordinate
T	Transformation matrix
M_n	Mass matrix
M_a	Reduced mass matrix
K_n	Stiffness matrix
K_a	Reduced stiffness matrix
λ	Eigenvalue matrix
T_U	SEREP transformation matrix
$[U_a]^g$	Generalized inverse
U_n	Full space modal matrix (shape matrix)
U_a	Reduced space modal matrix
U	Modal matrix 1
u_i	*i*th vector from U
V	Modal matrix 2
v_j	*j*th vector from V

K. Truong (✉) • P. Avitabile
Structural Dynamics and Acoustic Systems Laboratory, University of Massachusetts Lowell, One University Avenue, Lowell, MA 01854, USA
e-mail: kevin_truong@student.uml.edu

© The Society for Experimental Mechanics, Inc. 2015

R. Allemang (ed.), *Special Topics in Structural Dynamics, Volume 6*, Conference Proceedings of the Society for Experimental Mechanics Series, DOI 10.1007/978-3-319-15048-2_1

1.1 Introduction

Finite element models are commonly developed from a CAD model which describes the structure in its actual geometric configuration. Many packages available enable the finite element mesh to be made from this geometry definition. However, there are many cases where the CAD generated models developed are very detailed with plate or solid elements to model local details such as holes, slots, rivets, grooves, etc.; thus, the CAD generated finite element models tend to become very large for no real reason other than it is easy to generate from the CAD representation. While there are many computation resources available at this point in time, a more effective modeling methodology is proposed for efficient analysis for use in different applications.

There are many instances where a simpler beam/line element model would enable much more efficient analyses to be performed to understand the overall system dynamics and perform perturbation studies to investigate a variety of effects. In fact, there are many applications where the analyst actually tries to develop a much reduced order line model from the complicated 3D model to allow for efficient dynamic response studies. One such case is shown in Fig. 1.1 for a generator where a complicated model is transformed to a simple beam type model with simple component connection elements.

Another very important problem is related to the structural dynamic problem for flutter analysis where simplistic beam type models are used for the aerodynamic analysis to address the flutter problem. This is done for systems such as airplane wing structures and wind turbine blades to name just a few of the many applications. In most cases, the models of these structures tend to be a much more detailed rib-stiffened plate structure for airplane wing models and rib-stiffened, multi-ply composite plate structure for wind turbine blades. Figure 1.2 shows some typical cross sections for these types of structures to illustrate the inherent problem of defining the simpler configuration. In this case, the complicated wind turbine blade made from a complicated composite layup must be converted to a simple beam cross section; the ability to accurately develop such a reduced model from a complicated cross section is, of course, not a trivial task.

In both cases, the models are very detailed and complicated and are not typically used for computationally intense analyses (such as flutter) because of the large size of the model and computational resources required to do so. These models are always transformed into a much simpler beam-type model, especially in the case of flutter analysis. However, the development of this simplistic model represented by line elements along a beam like neutral axis is not easily done in current finite element software packages.

This work proposes an approach to address this problem of development of a very accurate reduced order model to an equivalent neutral axis where no existing geometry exists. The reduction will utilize the SEREP model reduction scheme to exactly preserve the dynamic characteristics of interest. The technique allows for the development of a very accurate reduced order model to convert very complicated 3D shell and solid type models to equivalent reduced order line elements for further dynamic response studies including computationally intense flutter type analyses.

1.2 Theoretical Background

The development of this modeling technique is based on concepts related to model reduction and model expansion, which are summarized herein.

1.2.1 Model Reduction/Expansion Techniques

Model reduction is typically performed to reduce the size of a large analytical model to develop a more efficient model for further analytical studies. These techniques have been presented in earlier work cited in the references; only summarizing equations are presented below. Several model reduction methods have commonly been employed for analysis but the SEREP (System Equivalent Reduction-Expansion Process) technique is used for this work. For any reduction/expansion process, the relationship between the full set of degrees of freedom and a reduced set of degrees of freedom can be written as

$$\{X_n\} = [T]\{X_a\} \tag{1.1}$$

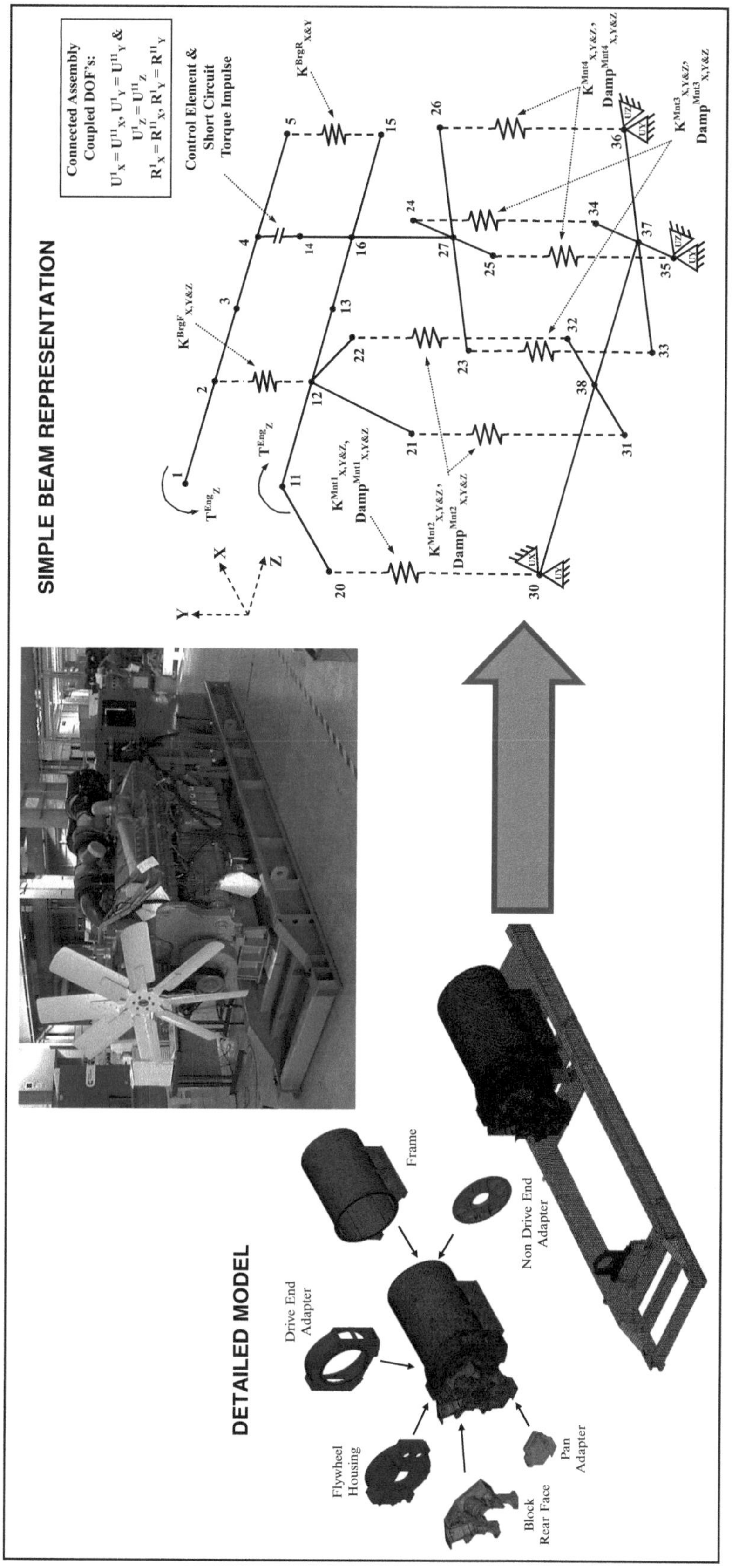

Fig. 1.1 Detailed model of genset reduced to simplistic beam element model

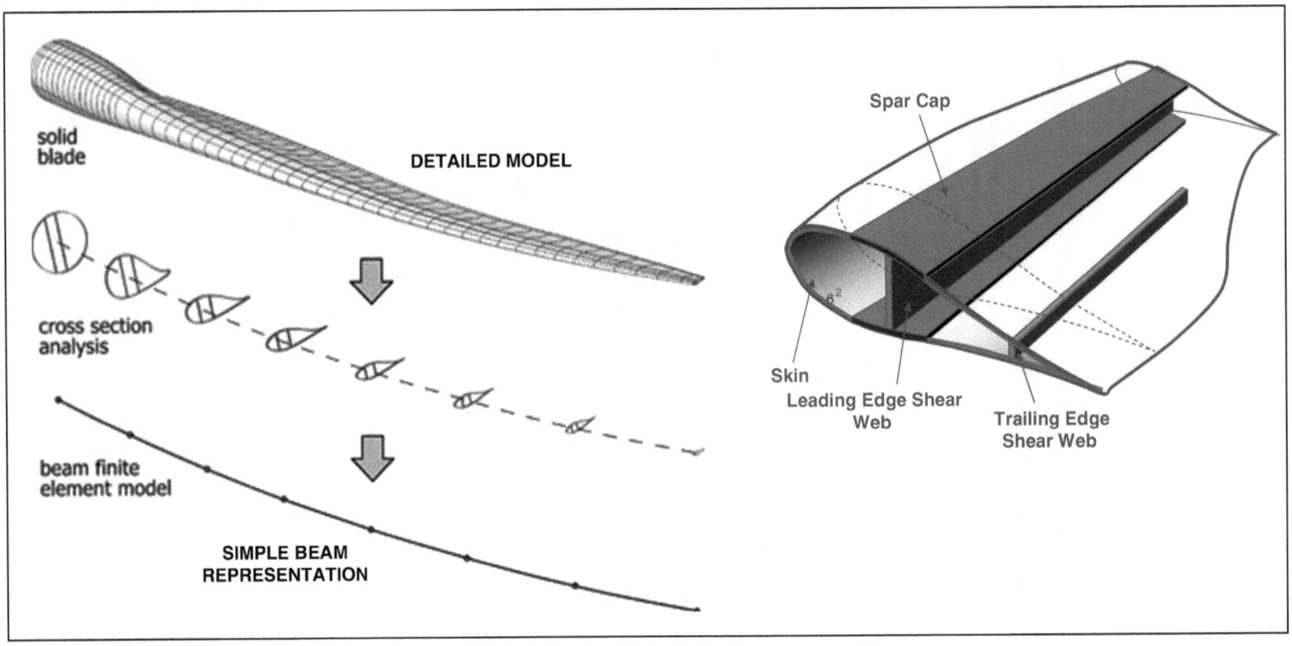

Fig. 1.2 Generation of a simplistic beam model from a detailed composite ply configuration

The transformation matrix [T] is used to project the full mass and stiffness matrices to a smaller size. The reduced matrices can be formulated as

$$[M_a] = [T]^T [M_n] [T] \tag{1.2}$$

$$[K_a] = [T]^T [K_n] [T] \tag{1.3}$$

For model reduction, it is important that the eigenvalues and eigenvectors of the original system are preserved as accurately as possible in the reduction process. If this is not maintained then the matrices are of questionable value. The eigensolution is then given by

$$[[K_a] - \lambda [M_a]] \{X_a\} = \{0\} \tag{1.4}$$

Because reduction schemes such as Guyan Condensation [1] and Improved Reduced System Technique [2] are based primarily on the stiffness of the system, the eigenvalues and eigenvectors will not be exactly reproduced in the reduced model. However, the System Equivalent Reduction Expansion Process (SEREP) [3] exactly preserves the eigenvalues and eigenvectors in the reduced model.

1.2.2 System Equivalent Reduction Expansion Process (SEREP)

For the specific work in this paper, the SEREP has been used to make the reduced order models. SEREP produces reduced matrices for mass and stiffness that yield the exact frequencies and mode shapes as those obtained from the eigensolution of the full size matrix. The SEREP transformation is formed as

$$[T_U] = [U_n] \, [U_a]^g \tag{1.5}$$

with

$$[U_a]^g = \left[\left(U_a^T U_a \right)^{-1} U_a^T \right] \tag{1.6}$$

1.2.3 *Modal Assurance Criterion (MAC)*

Modal Assurance Criterion is a correlation tool commonly used to compare mode shapes. MAC compares two vectors (u_i and v_j) and calculates a value from 0 to 1 that quantifies the degree of similarity between the vectors. The equation is

$$\mathrm{MAC}_{ij} = \frac{\left[\{u_i\}^T\{v_j\}\right]^2}{\left[\{u_i\}^T\{u_i\}\right]\left[\{v_j\}^T\{v_j\}\right]} \tag{1.7}$$

A MAC value of 1 signifies perfect correlation and 0 signifies no correlation.

1.2.4 *Pseudo Orthogonality Check (POC)*

Pseudo Orthogonality Check is a mass weighted orthogonality tool used to compare mode shapes and is given as

$$\mathrm{POC} = [V]^T[M][U] \tag{1.8}$$

The POC is mass weighted. If the shapes are scaled to unit modal mass, POC ranges from 0 to 1, similar to MAC.

1.3 Model Description

Five finite element models of a cantilever beam were used to develop and demonstrate the technique. Models with non-physical properties shall be referred to as the "phantom" model. The different models generated are:

- Model 1: Line model using 3D beam elements (BEAM3D)
- Model 2: Shell model using plate elements (QUAD4)
- Model 3: Shell model with "phantom" beam model constrained at the neutral axis (QUAD4 and BEAM3D)
- Model 4: Shell model with a "phantom" beam model located away from the neutral axis with same constraints as Model 3 (QUAD4 and BEAM3D)
- Model 5: Shell model with "phantom" beam model constrained at the neutral axis using a subset of nodes from Model 3 (QUAD4 and BEAM3D)

The properties used in the FEMs are shown in Table 1.1. These models were generated using the software, FEMtools [5].

Table 1.1 Cantilever beam dimensions and FEM characteristics

Property	Unit	Model	
		Line	Shell
Nodes	–	81	972
DOF	–	486	5832
Elastic modulus	[psi]	10E6	
Density	[lb/in**3]	0.098	
Poisson's ratio	–	0	
b	[in.]	4	
h	[in.]	2	
t (uniform)	[in.]	0.5	
Length	[in.]	80	

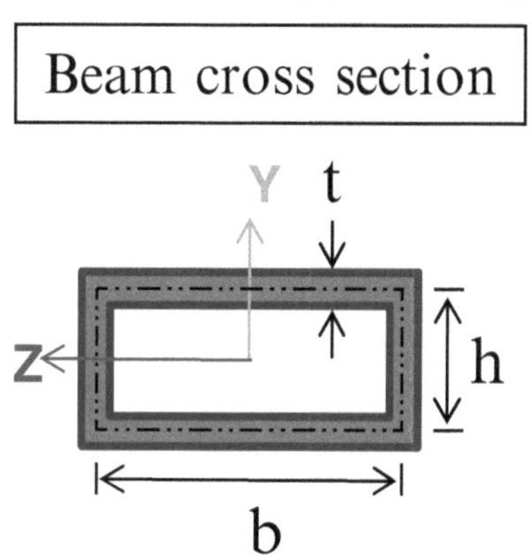

1.3.1 Model 1 (Line)

Model 1 uses simple beam elements to model the cantilever beam. Generally, the computation time for the solution is much quicker for models using line style elements than shell elements. Line models are common for beam structures with simple cross sections. However, for beam-like structures with complex geometries i.e. varying cross sections, rib stiffeners, etc., a FEM using beam elements would not be practical since the cross sectional properties for every element need to be specified.

1.3.2 Model 2 (Shell)

Model 2 uses linear plate elements to model the cantilever beam. Typically, FEMs of the structure are created from existing CAD models readily available. From these CAD models, a solid or surface mesh can quickly be generated on the structure and an eigensolution can be performed to obtain the structures frequencies and mode shapes. Because the solid/surface models contain many nodes, the computation time tends to be lengthy. Thus, an accurate reduced order model that captures the dynamics of interest, while making computation more efficient is desirable. For beam-like structures with hollow cross sections, a reduced order model at the neutral axis is not possible. This provides the motivation for Model 3.

1.3.3 Model 3 (Hybrid)

Model 3 is an extension of Model 2 with "phantom" beam elements constrained to the appropriate shell nodes. This modeling technique provides a set of DOFs to make a reduced order model at the neutral axis while obtaining the original frequencies and mode shapes of the shell model. A set of ADOF at the neutral axis were selected for the reduction process and expanded to the full space NDOF. Table 1.2 compares the reference mode shapes to the expanded mode shapes to illustrate the accuracy of the reduction/expansion process using SEREP. The "phantom" beam is constrained to the nodes on the surrounding four faces of the shell model. Note: To avoid numerical difficulties, small values for E and rho were used for the beam element properties.

1.3.4 Model 4 (Hybrid-Offset)

Model 4 is similar to Model 3 with the only difference of the location of the "phantom" beam; this model had the "phantom" beam offset from the actual neutral axis of the beam. For beam-like structures with complex geometries where the neutral axis does not coincide with the center, this model was develop to study the differences of the technique when the "phantom" beam is not modeled at the neutral axis.

1.3.5 Model 5 (Hybrid-1 Node)

The constraints in Model 5 differ that of Model 3. The "phantom" beam is constrained to only one face of the shell model.

The frequencies obtained from the models are shown in Table 1.3. Only bending modes were considered for reduction. Frequencies highlighted in blue and purple correspond to bending about the weaker and stronger axes, respectively. The frequencies of Model 1 differ from Model 2–5 due to formulations of the elements used in the models. Model 1 uses beam elements while Model 2 uses plate elements; and Models 3–5 uses a combination of plate and beam elements. Note that only the plate elements in Model 3, 4 and 5 have physical properties. The frequencies from Model 2 are used to validate the modeling technique used in Model 3, 4 and 5.

Table 1.2 Comparison of reference mode shapes to expanded mode shapes from model 3

U_{nm} - Reference modeshapes

DOF	Mode 1	Mode 2	Mode 3	. . .	Mode 10	Mode 11	Mode 12
1	0.00	–0.01	0.02	. . .	0.21	0.26	–0.34
2	0.00	0.00	0.01	. . .	0.27	–0.01	0.01
3	0.00	0.00	0.00	. . .	0.03	–0.21	0.33
4	0.00	0.00	0.00	. . .	–0.05	0.03	–0.05
5	0.00	0.00	0.00	. . .	–0.01	0.31	–0.47
6	0.00	0.00	0.03	. . .	0.41	0.00	0.00
7	0.00	–0.01	0.00	. . .	0.00	0.21	–0.26
8	0.00	0.00	0.01	. . .	0.28	0.00	0.00
9	0.00	0.00	0.00	. . .	0.00	–0.18	0.28
10	0.00	0.00	0.00	. . .	–0.01	0.00	0.00
11	0.00	0.00	0.00	. . .	0.00	0.31	–0.47
12	0.00	0.00	0.02	. . .	0.27	0.00	0.00
⋮	⋮	⋮	⋮	⋮	⋮	⋮	⋮
5821	0.00	0.00	0.00	. . .	0.00	0.00	0.00
5822	5.63	0.00	–5.34	. . .	–4.24	0.00	0.00
5823	0.00	5.62	0.00	. . .	0.00	–4.34	–4.03
5824	0.00	0.00	0.00	. . .	0.00	0.00	0.00
5825	0.00	–0.10	0.00	. . .	0.00	0.87	1.02
5826	0.10	0.00	–0.34	. . .	–1.04	0.00	0.00
5827	0.00	0.00	0.00	. . .	0.00	0.00	0.00
5828	5.72	0.00	–5.67	. . .	–5.31	0.00	0.00
5829	0.00	5.72	0.00	. . .	0.00	–5.23	–5.08
5830	0.00	0.00	0.00	. . .	0.00	0.00	0.00
5831	0.00	–0.10	0.00	. . .	0.00	0.88	1.03
5832	0.10	0.00	–0.34	. . .	–1.06	0.00	0.00

$Tu*U_a$ - Expanded fullspace modeshapes

DOF	Mode 1	Mode 2	Mode 3	. . .	Mode 10	Mode 11	Mode 12
1	0.00	–0.01	–0.02	. . .	–0.21	–0.26	0.34
2	0.00	0.00	–0.01	. . .	–0.27	0.01	–0.01
3	0.00	0.00	0.00	. . .	–0.03	0.21	–0.33
4	0.00	0.00	0.00	. . .	0.05	–0.03	0.05
5	0.00	0.00	0.00	. . .	0.01	–0.31	0.47
6	0.00	0.00	–0.03	. . .	–0.41	0.00	0.00
7	0.00	–0.01	0.00	. . .	0.00	–0.21	0.26
8	0.00	0.00	–0.01	. . .	–0.28	0.00	0.00
9	0.00	0.00	0.00	. . .	0.00	0.18	–0.28
10	0.00	0.00	0.00	. . .	0.01	0.00	0.00
11	0.00	0.00	0.00	. . .	0.00	–0.31	0.47
12	0.00	0.00	–0.02	. . .	–0.27	0.00	0.00
⋮	⋮	⋮	⋮	⋮	⋮	⋮	⋮
5821	0.00	0.00	0.00	. . .	0.00	0.00	0.00
5822	5.63	0.00	5.34	. . .	4.24	0.00	0.00
5823	0.00	5.62	0.00	. . .	0.00	4.34	4.03
5824	0.00	0.00	0.00	. . .	0.00	0.00	0.00
5825	0.00	–0.10	0.00	. . .	0.00	–0.87	–1.02
5826	0.10	0.00	0.34	. . .	1.04	0.00	0.00
5827	0.00	0.00	0.00	. . .	0.00	0.00	0.00
5828	5.72	0.00	5.67	. . .	5.31	0.00	0.00
5829	0.00	5.72	0.00	. . .	0.00	5.23	5.08
5830	0.00	0.00	0.00	. . .	0.00	0.00	0.00
5831	0.00	–0.10	0.00	. . .	0.00	–0.88	–1.03
5832	0.10	0.00	0.34	. . .	1.06	0.00	0.00

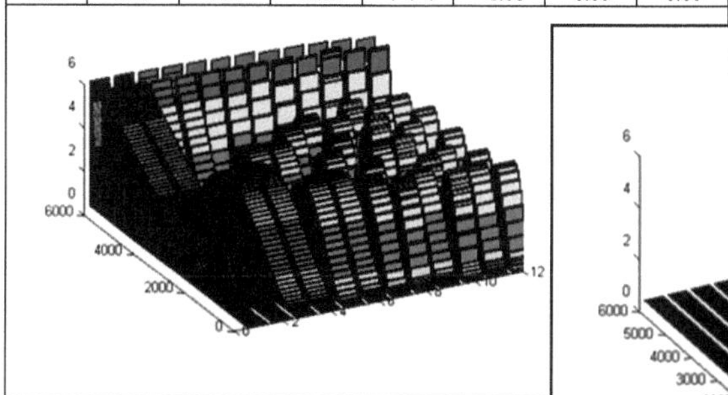

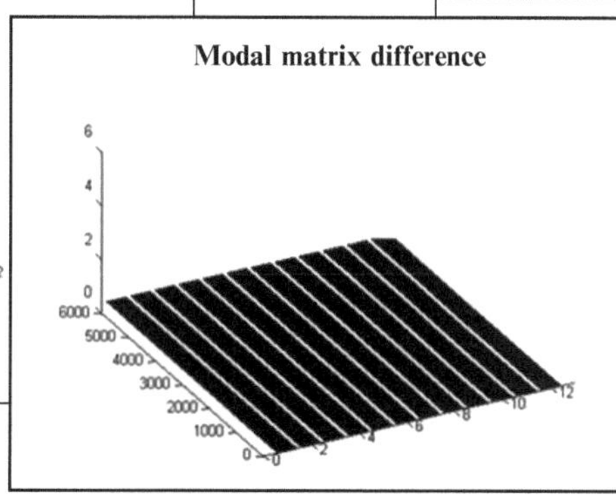

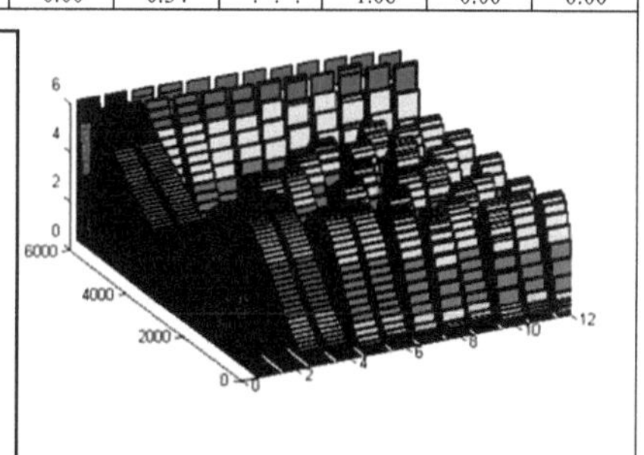

1.4 Cases Studied

Three cases were studied and presented in this section. Model reduction using SEREP is performed, and correlation tools such as the POC and MAC are applied to each of the cases. All reduction analyses were performed using Matlab [4] and correlation analyses were performed using FEMtools [5]. Note that only bending modes were considered for presenting the technique. The axial and torsion modes could have been preserved in the reduction process but provided no advantage in demonstrating the technique and therefore were not included or considered. The specified cases studied are:

- Case X: Correlate Model 1 to Model 3
- Case Y: Correlate Model 3 to Model 4
- Case Z: Correlate Model 3 to Model 5

In each of the cases, the models are reduced down to a set of 6 nodes with two transverse DOF at each node corresponding to the bending directions. The DOF selected and the modes preserved for SEREP are chosen consistently in each of the cases.

1.4.1 Case X

In Case X, the reduced line model (Model 1) is compared to the reduced hybrid model (Model 3) for similarities. First, the reduced mass matrices were studied and compared. Plots of the reduced mass matrices and the differences between the

Table 1.3 FEM natural frequencies for model 1–5

	Model 1	Model 2	Model 3	Model 4	Model 5
			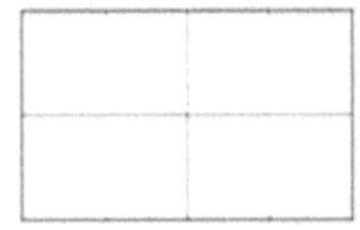	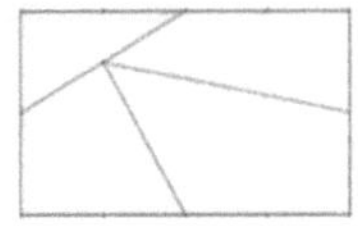	
Mode	Line beam	Shell beam	Hybrid beam	Hybrid beam (Offset)	Hybrid beam (1 Node)
	[Hz]	[Hz]	[Hz]	[Hz]	[Hz]
1	15.65	15.41	15.41	15.41	15.41
2	26.16	25.83	25.83	25.84	25.83
3	98.06	95.03	95.03	95.04	95.03
4	163.93	158.72	158.72	158.73	158.72
5	274.51	259.61	259.61	259.62	259.61
6	438.79	335.71	335.71	335.71	335.71
7	458.93	431.32	431.32	431.32	431.32
8	537.85	491.42	491.42	491.43	491.42
9	620.54	620.54	620.56	620.56	620.56
10	888.96	778.34	778.34	778.35	778.34
11	899.17	811.90	811.91	811.92	811.91
12	1,316.19	992.20	992.20	992.21	992.21
13	1,327.73	1,106.18	1,106.18	1,106.20	1,106.20
14	1,486.15	1,280.54	1,280.55	1,280.57	1,280.55
15	1,854.13	1,461.16	1,461.16	1,461.19	1,461.19
16	1,861.38	1,595.51	1,595.52	1,595.53	1,595.53
17	2,193.09	1,816.79	1,816.80	1,816.84	1,816.80
18	2,219.69	1,830.62	1,830.63	1,830.66	1,830.69
19	2,468.11	1,861.38	1,861.45	1,861.45	1,861.45
20	3,069.14	2,089.71	2,089.71	2,089.74	2,089.75

models are shown below in Fig. 1.3. There are some minor differences in the reduced mass matrices; this is to be expected because of the different formulations of the elements used in Model 1 and Model 3.

As an initial tool, the MAC was used to analyze similarities of the mode shapes in the models. POC was then used for vector correlation on a mass weighted basis. Two POC calculations were performed; one calculation using M_a from Model 1 and one calculation using M_a from Model 3. The MAC and POC values and plots for all cases are shown below in Table 1.4 and Fig. 1.4, respectively. Both high MAC and high POC values were obtained from the calculations. Because the DOF preserved for reduction were located at modally active nodes, the MAC calculation was dominated by the high mode shape values at these locations which accounts for the high MAC values. The POC shows that the mode shapes from Model 1 and Model 3 differ by a small scale factor, which results from the use of different element properties.

1.4.2 Case Y

In Case Y, the effects of modeling the "phantom" beam offset from the neutral axis was studied. The M_a from Model 3 and Model 4 have negligible differences as shown in Fig. 1.3. POC values of 1 indicate perfect model correlation shown in Table 1.4. Good model correlation is also indicated with the high MAC values. Identical results can be obtained independent of the "phantom" beam location. The nodes to which the "phantom" beam is constrained to, dictate the model behavior. The "phantom" beam does not have to be modeled at the neutral axis in order to use this technique. This is applicable for real-world structures with complicated geometries, where the neutral axis cannot be easily located.

1.4.3 Case Z

In Case Z, Model 3 is correlated to Model 5. Each model had a different set of constraints; Model 5 constraints are a subset of the constraints applied in Model 3. The M_a from both models have negligible differences shown in Fig. 1.3. High POC and MAC values from Table 1.4 show good correlation. Since only a handful of the lower order modes were preserved for reduction, very similar results were obtained from the two models. The effects of the higher order breathing type modes when using the technique requires further study and is beyond the scope of this initial investigation.

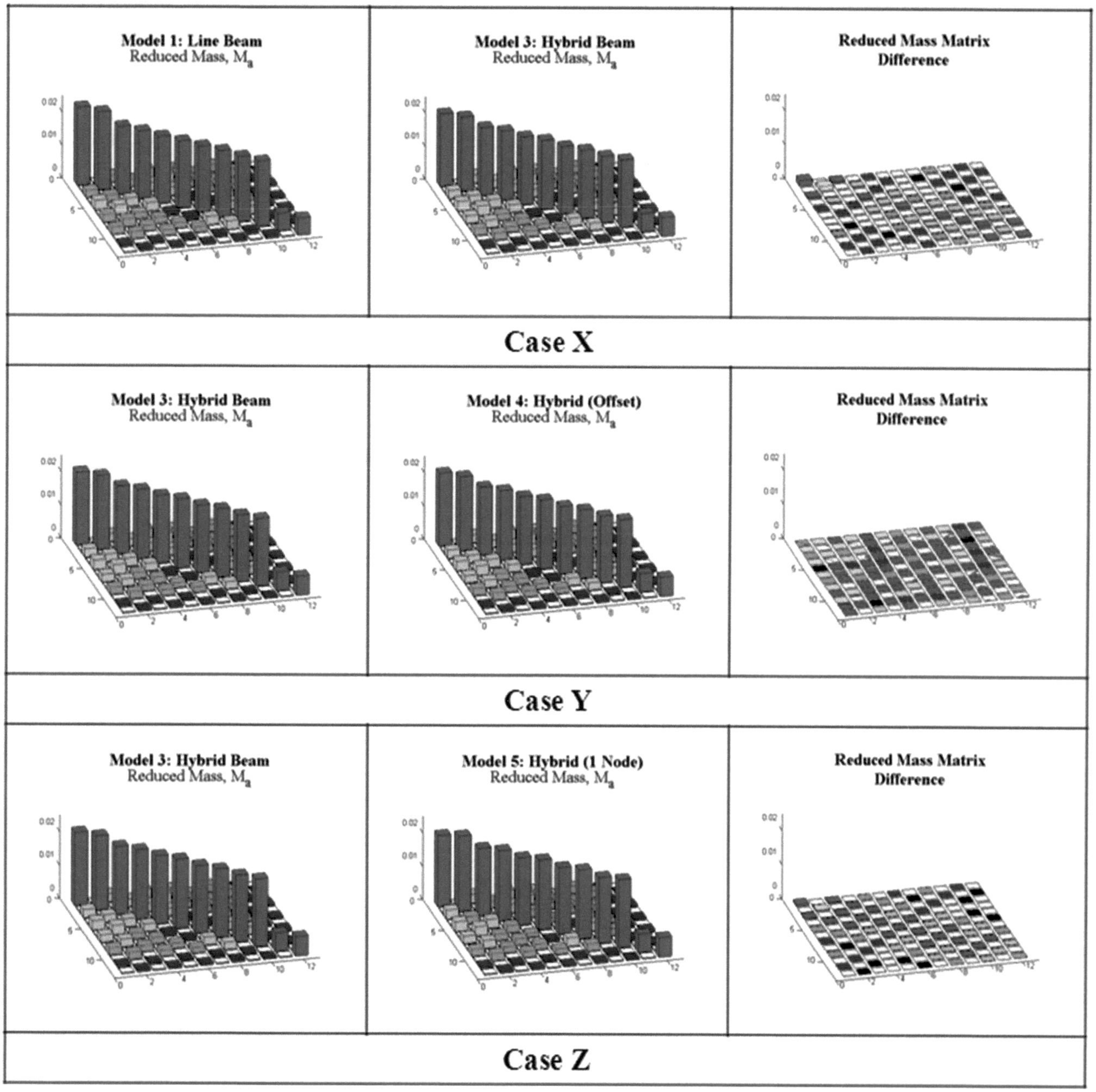

Fig. 1.3 Reduced mass matrix plots and their respective differences for Case X, Y, and Z

Table 1.4 POC and MAC values for Case X, Y, and Z

	POC						MAC		
	Case X		Case Y		Case Z		Case X	Case Y	Case Z
Mode	(a) M_a from Model 1	(b) M_a from Model 3	(a) M_a from Model 3	(b) M_a from Model 4	(a) M_a from Model 3	(b) M_a from Model 5	(c)	(c)	(c)
1	0.9997	1.0003	1.0000	1.0000	1.0000	1.0000	1.0000	1.0000	1.0000
2	0.9992	1.0008	1.0000	1.0000	1.0000	1.0000	1.0000	0.9999	1.0000
3	0.9979	1.0020	1.0000	1.0000	1.0003	0.9997	0.9999	0.9999	1.0000
4	0.9943	1.0057	1.0000	1.0000	0.9999	1.0001	0.9999	0.9996	1.0000
5	0.9959	1.0039	1.0000	1.0000	1.0019	0.9981	0.9995	0.9998	1.0000
7	0.9881	1.0120	1.0000	1.0000	0.9990	1.0010	0.9994	0.9992	1.0000
8	0.9952	1.0037	1.0000	1.0000	1.0066	0.9934	0.9980	0.9996	1.0000
10	1.0037	0.9967	1.0000	1.0000	1.0163	0.9839	0.9951	0.9994	0.9998
11	0.9830	1.0169	1.0000	1.0000	0.9966	1.0035	0.9979	0.9986	1.0000
13	0.9917	1.0097	1.0000	1.0000	1.0325	0.9685	0.9990	0.9992	0.9994
14	0.9891	1.0123	1.0000	1.0000	0.9915	1.0086	0.9951	0.9980	1.0000
18	0.9667	1.0361	1.0000	1.0000	0.9830	1.0173	0.9983	0.9973	1.0000

1.5 Conclusion

This work demonstrates that an exact reduced order model can be developed and reduced to a region where there is no specific model defined such as the neutral axis of a hollow rectangular section. The technique relies on the development of a "phantom" set of elements that can be used to define the neutral axis. This work also shows that the neutral axis of the phantom beams need not be exactly defined on the neutral axis of the hollow structure i.e. the "phantom elements" need not be geometrically located at the neutral axis. Several cases were presented to demonstrate and validate the technique.

The approach is useful for reduction of any large model where the specific neutral axis is unknown. The approach is also useful for reduction of large complicated models that need to be drastically reduced to very low order model for complicated analyses such as flutter analysis. This is useful for large airframe structures, airframe wing like structures and wind turbine blade structures to name a few examples.

Acknowledgements Some of the work presented herein was partially funded by Air Force Research Laboratory Award No. FA8651-10-1-0009 "Development of Dynamic Response Modeling Techniques for Linear Modal Components". Any opinions, findings, and conclusions or recommendations expressed in this material are those of the authors and do not necessarily reflect the views of the particular funding agency. The authors are grateful for the support obtained.

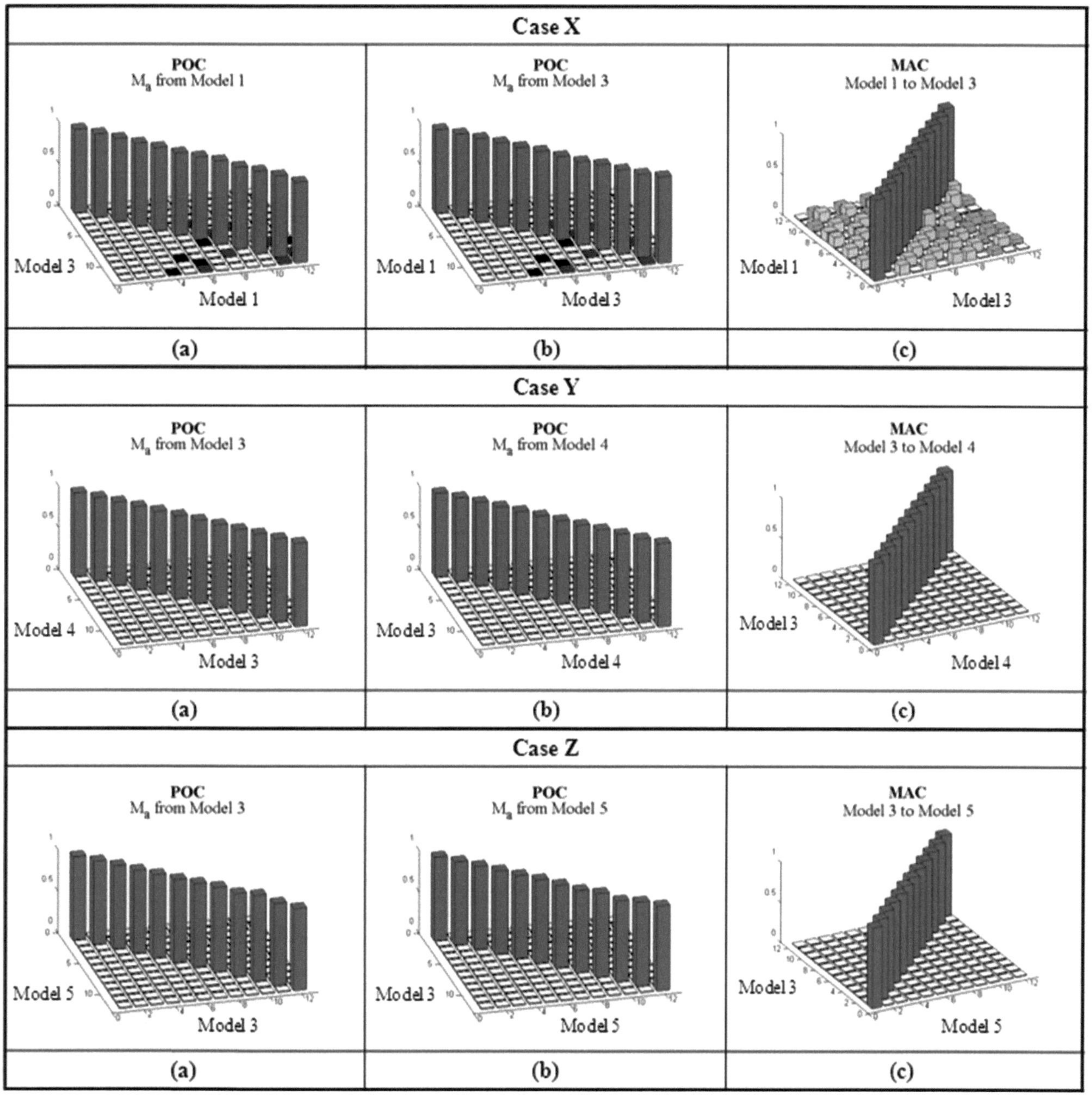

Fig. 1.4 POC and MAC plots for Case X, Y, and Z

References

1. Guyan RJ (1965) Reduction of stiffness and mass matrices. AIAA J 3(2):380
2. O'Callahan JC (1989) A procedure for an Improved Reduced System (IRS) model. In: Proceedings of the seventh international modal analysis conference, Las Vegas, publisher is Society for Experimental Mechanics, February 1989
3. O'Callahan JC, Avitabile P, Riemer R (1989) System equivalent reduction expansion process. In: Proceedings of the seventh international modal analysis conference, Las Vegas, publisher is Society for Experimental Mechanics, February 1989
4. MATLAB Matrix analysis software. The MathWorks, Inc., Natick
5. FEMtools FEMtools 3.0. Dynamic Design Solutions, Leuven

Chapter 2
Prediction of Forced Response Using Expansion of Perturbed Reduced Order Models with Inexact Representation of System Modes

Sergio E. Obando, Peter Avitabile, and Jason Foley

Abstract Variability in measured data is a common problem in the engineering practice. Changes in the mass and stiffness of the same structural component can occur due to minor variability in the tolerances used during the production/manufacturing process. Differences can exist between the real physical structure and its mathematical model representation (FEM) as well as the predicted response and the actual dynamic behavior of the system. For models in which limited data exists or is collected, the quality of the equivalent reduced order model is dependent on the retained modal parameters as well as the level of correlation of the mode shapes. Prediction of system level forced response from the expansion of these reduced order models can be affected by the use of inexact representations of the system modes such as those from Guyan reduced models. Furthermore, the reduction methodology used, the degrees of freedom selected, as well as the number of retained modes can play an important role in the accuracy of the predicted dynamics of the system.

In this work, a truth model (real answer) is created from the perfect analytical representation of a cantilevered beam. A perturbed variation of the analytical representation of the cantilever beam model is also created to correspond to the simulated imperfections of a FEM of the system. The analytical models will be created to investigate the prediction of the full field dynamic response obtained from the expansion of reduced model information (or data at limited number of DOF) and using the inexact mode shapes of the perturbed model (FEM). The perturbed system representation will have the same geometry and properties as the original unmodified beam (perfect analytical model) but imperfections will be introduced by the addition of mass. The models will be created first at full space as a reference and then reduction techniques will be used to determine the necessary information in order to accurately predict the response at all DOF. Aspects involved in model reduction/expansion, DOF selection, and number of retained modes for the analytical cantilever models are investigated for common reduction techniques such as Guyan condensation and SEREP. The use of a perturbed model (not perfectly correlated to the model) for the expansion of measured real time response data will be shown to produce very accurate full field response even though the model does not perfectly correlate to the real truth model.

Keywords Forced linear response • Reduced order modeling • Perturbed models

Nomenclature

Symbols

$\{X_n\}$ Full Set Displacement Vector
$\{X_a\}$ Reduced Set Displacement Vector
$\{X_d\}$ Deleted Set Displacement Vector
$[M_a]$ Reduced Mass Matrix
$[M_n]$ Expanded Mass Matrix
$[K_a]$ Reduced Stiffness Matrix
$[K_n]$ Expanded Stiffness Matrix

S.E. Obando (✉) • P. Avitabile
Structural Dynamics and Acoustic Systems Laboratory, University of Massachusetts Lowell, One University Avenue, Lowell, MA 01854, USA
e-mail: sergio.e.obando@gmail.com

J. Foley
Air Force Research Laboratory, Munitions Directorate Fuzes Branch, Eglin Air Force Base, 306 W. Eglin Blvd., Bldg 432, Eglin AFB, FL 32542-5430, USA

© The Society for Experimental Mechanics, Inc. 2015

R. Allemang (ed.), *Special Topics in Structural Dynamics, Volume 6*, Conference Proceedings of the Society for Experimental Mechanics Series, DOI 10.1007/978-3-319-15048-2_2

$[U_a]$	Reduced Set Shape Matrix
$[U_n]$	Full Set Shape Matrix
$[U_a]^g$	Generalized Inverse
$[T]$	Transformation Matrix
$[T_U]$	SEREP Transformation Matrix
$\{p\}$	Modal Displacement Vector
$[M]$	Physical Mass Matrix
$[C]$	Physical Damping Matrix
$[K]$	Physical Stiffness Matrix
$\{F\}$	Physical Force Vector
$\{\ddot{x}\}$	Physical Acceleration Vector
$\{\dot{X}\}$	Physical Velocity Vector
$\{x\}$	Physical Displacement Vector
$\vec{\ddot{x}}_0$	Initial Acceleration Vector
$\vec{\dot{x}}_0$	Initial Velocity Vector
$\vec{x}_0$	Initial Displacement Vector
$\vec{F}_0$	Initial Force Vector
α	Parameter for Newmark Integration
β	Parameter for Newmark Integration
Δt	Time Step

Acronyms

ADOF	Reduced Degrees of Freedom
NDOF	Full Space Degrees of Freedom
DOF	Degrees of Freedom
ERMT	Equivalent Reduced Model Technique
FEM	Finite Element Model
MAC	Modal Assurance Criterion
SEREP	System Equivalent Reduction Expansion Process
TRAC	Time Response Assurance Criterion
VIKING	Variability Improvement of Key Inaccurate Node Groups

2.1 Introduction

Structural components of typical systems are mass produced with allowable imperfections/tolerances in dimension and physical properties. Within the constraints of the manufacturing design, this variability should not affect the overall nominal values of performance of the component or part. However, often times during qualification testing or whenever components of similar specifications are being replaced/tested, the degree of variation and its effect on the operation of the system are highly important. In particular, if experimental data is being correlated with existing finite element models (FEM) these minor perturbations can result in inaccurate model characterization, poor correlation and poor predicted system response. Thus, there is significant motivation in understanding the effect of imperfections in the system's representation (model perturbations) in the accuracy of reduced order model characteristics, expanded system mode shapes, and predicted full system response.

Expansion of limited sets of data to the full NDOF of a large FEM is often performed to correlate test data with existing analytical models. The expanded mode shapes and the predicted response are affected by the level of correlation of the FEM and the tested structure as well as the location of the preserved ADOF in the reduced model. Recent work has investigated important aspects in the expansion of system modes used in the SEREP methodology for cases where poor correlation exists between test and FEM [1–5]. Mode shapes processed with the Variability Improvement of Key Inaccurate Node Groups (VIKING) method were shown to produce much more accurate results than the shapes expanded with contaminated/perturbed expansion matrices [5]. The most significant aspect of these studies lies in the proper selection of sufficient modal information in the preserved projection vectors resulting in the smoothing of noise and variance of the inaccurate experimental data.

In this paper, the predicted full field forced response of a cantilevered beam structure (from information at a limited number of points) is analyzed in the case where the expansion is performed with inexact system modes from a perturbed model representing an imperfect FEM of the truth model (real answer). The perfect analytical representation of the truth model is formed from the full space FEM of the cantilevered beam. Different reduction methodologies (Guyan condensation [6], SEREP [7], and KM_AMI [8]) are first studied with emphasis in DOF selection and accurate prediction of time response from reduced order models of this reference beam model. Then, a new system model is formed from the same geometry and physical properties of the original (unmodified reference) model but with simulated imperfections due to the addition of a point mass at the tip of the beam. Therefore, the modified/perturbed model will have different frequencies and mode shapes as well time response compare to the perfect analytical model. The forced response at selected DOF will be extracted from the full model response of the truth/unmodified model (much like having measurements of a system at a limited set of points) and the predicted response at full space obtained using both the perturbed and the accurate projection vectors (mode shapes) from a SEREP reduction process of the models. This study will show that when the preserved mode shapes contain sufficient information to span the solution space of the system, the perturbations do not distort the predicted expanded response at full space. Moreover, the selection of degrees of freedom during the reduction process will be explored in cases where the retained DOF are at less than ideal locations. This is of particular importance in real world measurements, as often times, transducers can only be placed at limited locations which may be highly susceptible to noise.

2.2 Theory

The fundamental theory in the study of forced response estimation and expansion/reduction of the type of linear systems analyzed here require a variety of theoretical topics. The summary starts with a description of linear multiple degree of freedom systems and continues with an overview of structural dynamic modification, analytical model reduction and expansion, model updating and forced time response computations. Further information can be found in the respective references.

2.2.1 Equations of Motion for Multiple Degree of Freedom System

The general equation of motion for a multiple degree of freedom system written in matrix form is

$$[M_1]\{\ddot{x}\} + [C_1]\{\dot{x}\} + [K_1]\{x\} = \{F(t)\} \tag{2.1}$$

Assuming proportional damping, the eigensolution is obtained from

$$[[K_1] - \lambda\,[M_1]]\{x\} = \{0\} \tag{2.2}$$

The eigensolution yields the eigenvalues (natural frequencies) and eigenvectors (mode shapes) of the system. The eigenvectors are arranged in column fashion to form the modal matrix $[U_1]$. Often times, only a subset of modes is included in the modal matrix to save on computation time and due to the fact that only certain modes actually contribute to the response. Exclusion of modes results in truncation error which can be serious if key modes are excluded; truncation error will be discussed in further detail in the structural dynamic modification section.

The physical system can be transformed to modal space using the modal matrix as

$$[U_1]^T[M_1][U_1]\{\ddot{p}_1\} + [U_1]^T[C_1][U_1]\{\dot{p}_1\} + [U_1]^T[K_1][U_1]\{p_1\} = [U_1]^T\{F(t)\} \tag{2.3}$$

Scaling to unit modal mass yields

$$\begin{bmatrix} \ddots & & \\ & I_1 & \\ & & \ddots \end{bmatrix}\{\ddot{p}_1\} + \begin{bmatrix} \ddots & & \\ & 2\xi\omega_n & \\ & & \ddots \end{bmatrix}\{\dot{p}_1\} + \begin{bmatrix} \ddots & & \\ & \Omega_1^2 & \\ & & \ddots \end{bmatrix}\{p_1\} = \left[U_1^n\right]^T\{F(t)\} \tag{2.4}$$

where $[I_1]$ is the diagonal identity matrix, $[\Omega_1{}^2]$ is the diagonal natural frequency matrix and $[2\xi\omega_n]$ is the diagonal damping matrix (assuming proportional damping). More detailed information on the equation development is contained in Ref. [9].

2.2.2 *Structural Dynamic Modification*

Structural Dynamic Modification (SDM) is a technique that uses the original mode shapes and natural frequencies of a system to estimate the dynamic characteristics due to changes in the mass and/or stiffness of the system; only mass and stiffness changes are considered in this work. First, the change of mass and stiffness are transformed to modal space as shown

$$\left[\Delta\overline{M}_{12}\right] = [U_1]^T\ [\Delta M_{12}]\ [U_1] \tag{2.5}$$

$$\left[\Delta\overline{K}_{12}\right] = [U_1]^T\ [\Delta K_{12}]\ [U_1] \tag{2.6}$$

The modal mass and modal stiffness changes are added to the original modal space equations to obtain

$$\left[\begin{bmatrix} \ddots & & \\ & \overline{M}_1 & \\ & & \ddots \end{bmatrix} + \left[\Delta\overline{M}_{12}\right]\right]\{\ddot{p}_1\} + \left[\begin{bmatrix} \ddots & & \\ & \overline{K}_1 & \\ & & \ddots \end{bmatrix} + \left[\Delta\overline{K}_{12}\right]\right]\{p_1\} = [0] \tag{2.7}$$

The eigensolution of the modified modal space model is computed and the resulting eigenvalues are the new frequencies of the system. The modified vectors can be obtained from the original vectors with the use of $[U_{12}]$ from the eigensolution of (2.7) and is given as

$$[U_2] = [U_1]\ [U_{12}] \tag{2.8}$$

The new mode shapes are $[U_2]$. The new mode shapes are formed from linear combinations of the original mode shapes $[U_1]$. The $[U_{12}]$ matrix shows how much each of the $[U_1]$ modes contributes to forming the new modes. Figure 2.1 shows the formation of the new mode shapes as seen in Eq. 2.8. See [9] for additional information on SDM.

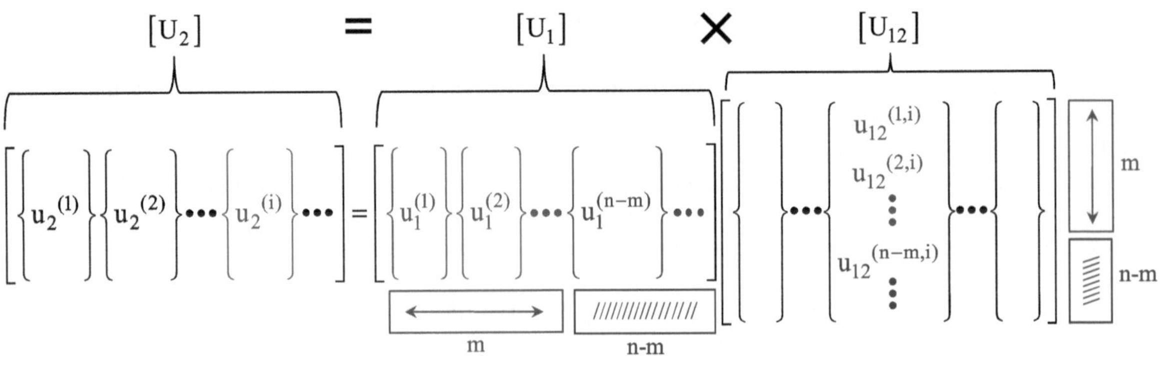

Fig. 2.1 Structural dynamic modification, mode contribution identified using U_{12} [10]

2.2.3 *General Reduction/Expansion Methodology and Model Updating*

Model reduction is a tool used to reduce the number of degrees of freedom (DOF) in order to reduce the required computation time of an analytical model, while attempting to preserve the full DOF dynamic characteristics. The relationship between the full space and reduced space model can be written as

$$\{X_n\} = \begin{Bmatrix} X_a \\ X_d \end{Bmatrix} = [T]\{X_a\} \tag{2.9}$$

where subscript 'n' signifies the full set of DOF (NDOF), 'a' signifies the reduced set of DOF (ADOF) and 'd' is the deleted or embedded DOF (those DOF not used during the reduced computation process). The transformation matrix [T] relates the full set of NDOF to the reduced set of ADOF. The transformation matrix is used to reduce the mass and stiffness matrices as

$$[M_a] = [T]^T [M_n][T] \text{ and } [K_a] = [T]^T [K_n][T] \tag{2.10}$$

The eigensolution of these 'a' set mass and stiffness matrices are the modes of the reduced model. These modes can be expanded back to full space using the transformation matrix

$$[U_n] = [T][U_a] \tag{2.11}$$

If an optimal 'a' set is not selected when using methods such as Guyan Condensation [6] or Improved Reduced System Technique [11], the reduced model may not perfectly preserve the dynamics of the full space model. If System Equivalent Reduction Expansion Process (SEREP) [7] is used, the dynamics of the selected modes will be perfectly preserved regardless of the 'a' set selected as long as the matrix is formed from a linearly independent set of vectors. Furthermore, a model improvement technique, KM_AMI, has been recently developed to update the mass and stiffness matrices obtained through the Guyan methodology and obtained exact modes and frequencies at the reduced 'a' space [8].

2.2.3.1 System Equivalent Reduction Expansion Process (SEREP)

The SEREP modal transformation relies on the partitioning of the modal equations representing the system using selected DOFs and modes to obtain a reduced model that perfectly preserves the eigenvalues and eigenvectors of interest [7]. The SEREP technique utilizes the mode shapes from a full finite element solution to map to the limited set of active DOF. SEREP is not performed to achieve efficiency in the solution but rather is intended to perform an accurate mapping matrix for the transformation. The SEREP transformation matrix is formed using a subset of modes at full space and reduced space as

$$[T_U] = [U_n][U_a]^g \tag{2.12}$$

where $[U_a]^g$ is the generalized inverse and $[T_U]$ is the SEREP transformation matrix. When the SEREP transformation matrix is used for model reduction/expansion as outlined in the previous section, the reduced model perfectly preserves the full space dynamics of the modes in $[U_n]$ [7].

2.2.3.2 KM_AMI Reduction

A more recent technique has been developed that utilizes Guyan Reduction (or any reduced matrix for that matter) along with direct updating of the reduced system matrices with the full space modal vectors as targets for the updating process [11]. This reduction technique also overcomes some of the rank problems associated with SEREP and provides a reduced set of ADOF that retain all the eigenvalues and eigenvectors of the full system matrices. The Guyan reduced mass and stiffness matrices are updated using

$$[M_I] = [M_S] + [V]^T \left[[I] - \left[\overline{M_S}\right] \right] [V] \tag{2.13}$$

and

$$[K_I] = [K_S] + [V]^T \left[\left[\Omega^2_{REF}\right] + \left[\overline{K}_S\right] \right] [V] - [[K_S] [U_{REF}] [V]] - [[K_S] [U_{REF}] [V]]^T \tag{2.14}$$

with

$$[V] = \left[\overline{M}_S\right]^{-1} [U_{REF}]^T [M_S] \tag{2.15}$$

2.2.4 *System Forced Response Analysis*

The computation of the time response developed in this paper is based on the Equivalent Reduced Model Technique (ERMT), a technique developed by Avitabile and Thibault [10, 12]. This technique uses an exact reduced model representation for the calculation of the system response. Newmark integration technique [13] is used to perform the direct integration of the equations of motion for the ERMT solution process due to similarity with the HHT (Hilber-Hughes-Taylor [14]) method commonly used in FEA software. From the known initial conditions for displacement and velocity, the initial acceleration vector is computed using the equation of motion and the applied forces as

$$\ddot{\vec{x}}_0 = [M]^{-1} \left(\vec{F}_0 - [C]\, \dot{\vec{x}}_0 - [K]\, \vec{x}_0 \right) \tag{2.16}$$

Choosing an appropriate Δt, α, and β, the displacement vector is

$$\begin{aligned} \vec{x}_{i+1} = {} & \left[\tfrac{1}{\alpha(\Delta t)^2} [M] + \tfrac{\beta}{\alpha(\Delta t)} [C] + [K] \right]^{-1} \left\{ \vec{F}_{i+1} + [M] \left(\left(\tfrac{1}{\alpha(\Delta t)^2} \right) \vec{x}_i + \left(\tfrac{1}{\alpha(\Delta t)} \right) \dot{\vec{x}}_i + \left(\tfrac{1}{2\alpha} - 1 \right) \ddot{\vec{x}}_i \right) \right. \\ & \left. + [C] \left(\left(\tfrac{\beta}{\alpha(\Delta t)} \right) \vec{x}_i + \left(\tfrac{\beta}{\alpha} - 1 \right) \dot{\vec{x}}_i + \left(\tfrac{\beta}{\alpha} - 2 \right) \tfrac{\Delta t}{2} \ddot{\vec{x}}_i \right) \right\} \end{aligned} \tag{2.17}$$

The values chosen for α and β were ¼ and ½, respectively. This assumes constant acceleration and the integration process is unconditionally stable, where a reasonable solution will always be reached regardless of the time step used. However, the time step should be chosen such that the highest frequency involved in the system response can be characterized properly to avoid numerical damping in the solution.

Following the displacement vector calculation, the acceleration and velocity vectors are computed for the next time step using

$$\dot{\vec{x}}_{i+1} = \dot{\vec{x}}_i + (1-\beta)\,\Delta t\, \ddot{\vec{x}}_i + \beta \Delta t\, \ddot{\vec{x}}_{i+1} \tag{2.18}$$

$$\ddot{\vec{x}}_{i+1} = \frac{1}{\alpha(\Delta t)^2} \left(\vec{x}_{i+1} - \vec{x}_i \right) - \frac{1}{\alpha \Delta t} \dot{\vec{x}}_i - \left(\frac{1}{2\alpha} - 1 \right) \ddot{\vec{x}}_i \tag{2.19}$$

This process is repeated at each time step for the duration of the time response solution desired.

2.2.5 *Time Response Correlation Tools*

In order to quantitatively compare two different time solutions, two correlation tools were employed: The Modal Assurance Criterion (MAC) and the Time Response Assurance Criterion (TRAC).

2.2.5.1 Modal Assurance Criterion (MAC)

The Modal Assurance Criterion (MAC) [15] is widely used as a vector correlation tool. In this work, the MAC was used to correlate all DOF at a single instance in time. The MAC is written as

$$\mathrm{MAC}_{ij} = \frac{\left[\{X1_i\}^T \{X2_j\}\right]^2}{\left[\{X1_i\}^T \{X1_i\}\right]\left[\{X2_j\}^T \{X2_j\}\right]} \tag{2.20}$$

where X1 and X2 are displacement vectors. MAC values close to 1.0 indicate strong similarity between vectors, where values close to 0.0 indicate minimal or no similarity.

2.2.5.2 Time Response Assurance Criterion (TRAC)

The Time Response Assurance Criterion (TRAC) [16] quantifies the similarity between a single DOF across all instances in time. The TRAC is written as

$$\mathrm{TRAC}_{ji} = \frac{\left[\{X1_j(t)\}^T \{X2_i(t)\}\right]^2}{\left[\{X1_j(t)\}^T \{X1_j(t)\}\right]\left[\{X2_i(t)\}^T \{X2_i(t)\}\right]} \tag{2.21}$$

where X1 and X2 are time response vectors for a particular DOF. TRAC values close to 1.0 indicate strong similarity between vectors, where values close to 0.0 indicate minimal or no similarity.

In this work, the MAC is calculated between the shapes of the full space reference solution and estimated solution obtained from the reduced order model at each time step. Similarly the TRAC is used to compare the time response from the reduced order model to the time response from the full space finite element solution at each degree of freedom. A diagram detailing the two comparison techniques is shown in Fig. 2.2.

2.3 Model Description

Analytical models of a cantilevered beam were created to investigate the prediction of the dynamic response using different reduction/expansion methodologies. Planar element beam models of the cantilevered beam were generated using MAT_SAP [17], which is a FEM program developed for MATLAB [18], and forced response calculations were performed in MATLAB using Newmark integration scripts. The beam models were set to have dimensions and characteristics as described in Fig. 2.3. For all models, 1 % of critical damping was used in the time response computation.

The system was subjected to a double sided force pulse at the tip of the beam and this input force was designed as to only excite the modes in a frequency band of approximately 200 Hz as shown in Fig. 2.4.

With all 20 elements of the system (i.e. 40 DOF) the full n-space reference solution to the system was calculated and served as a point of comparison for all subsequent reduced order model calculations. The frequencies of the beam are shown in Fig. 2.3.

The modified/perturbed (imperfect FEM) cantilevered beam was created using a point mass of approximately 2 % of the total mass of the reference (unmodified perfect model representation) beam placed at the tip of the beam. All other properties and input force remained the same. Figure 2.5 depicts the two models (reference/unmodified and modified) and compares the frequencies for the first 6 modes of the system.

2.4 Cases Studied

This work will be divided in two parts. Part A is a study of reduced order modeling in the context of prediction of forced time response. The forced time response of the full space reference model with 40 DOF was first calculated. Reduction techniques (Guyan, SEREP and KM_AMI) were used to reduce the active DOF of the system and preserve a selected number of modes. The time response was computed at the limited set of DOF and expansion used to predict the response at all NDOF. Two groups of ADOF were studied, illustrating instances of poor or limited DOF location and the effect on the predicted (through expansion) response. Part B focuses on the prediction of full NDOF time response from the expansion of data at limited points of the model for cases in which the projection matrices (i.e. mode shape vectors in SEREP methodology) differ from

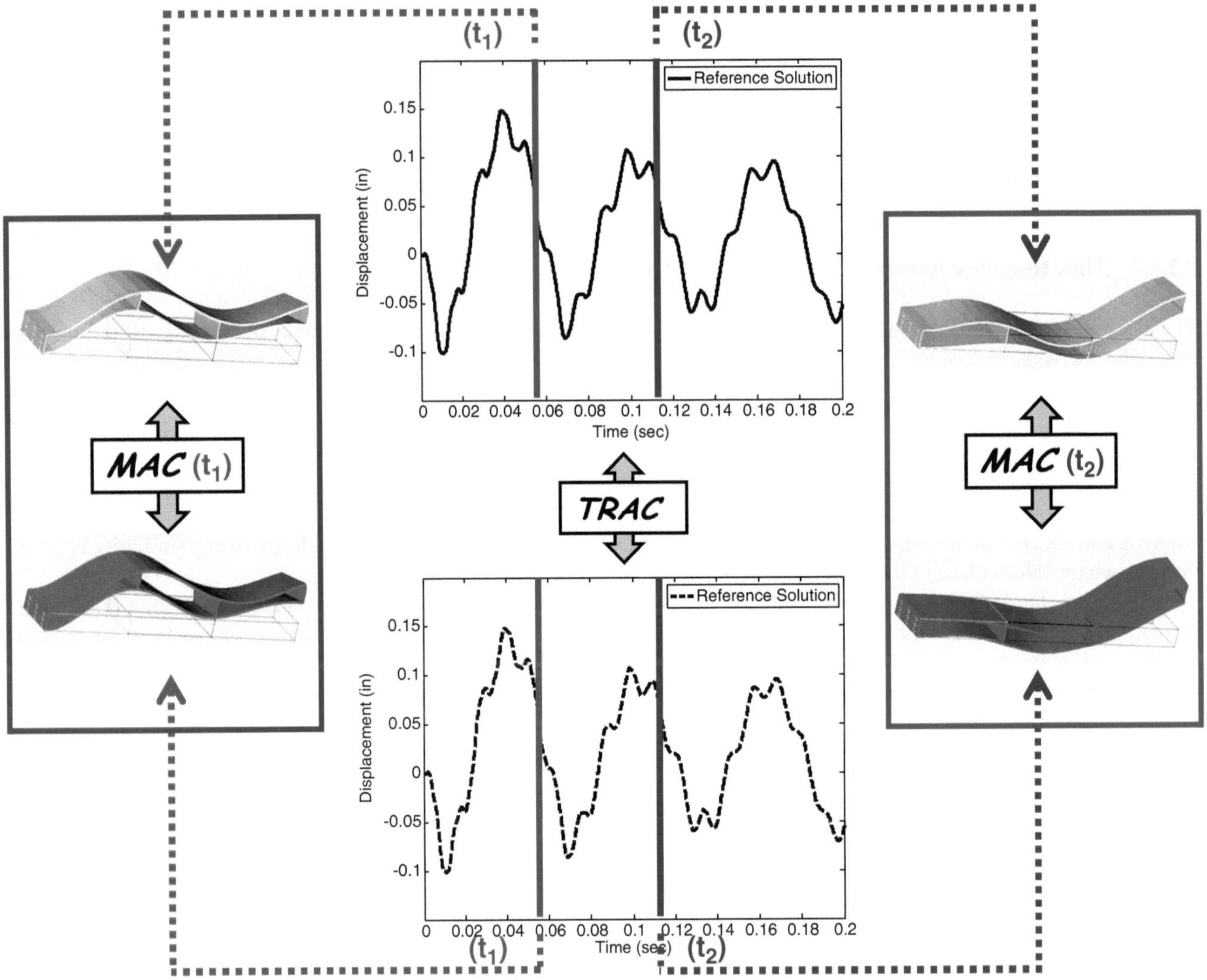

Fig. 2.2 Physical interpretation of MAC and TRAC

Fig. 2.3 Schematic and properties of analytical cantilevered beam model (not to scale)

Beam Dimensions and Physical Properties	
L – Length	40 in
H – Height	0.25 in
W–Width	0.5 in
E –Young's Modulus	10e6 lb/in^2
ρ –Weight density	0.1 lb/in^3
Number of DOF	40

Frequency (Hz)	
Mode 1	4.96
Mode 2	31.09
Mode 3	87.06
Mode 4	170.62
Mode 5	282.08
Mode 6	421.46

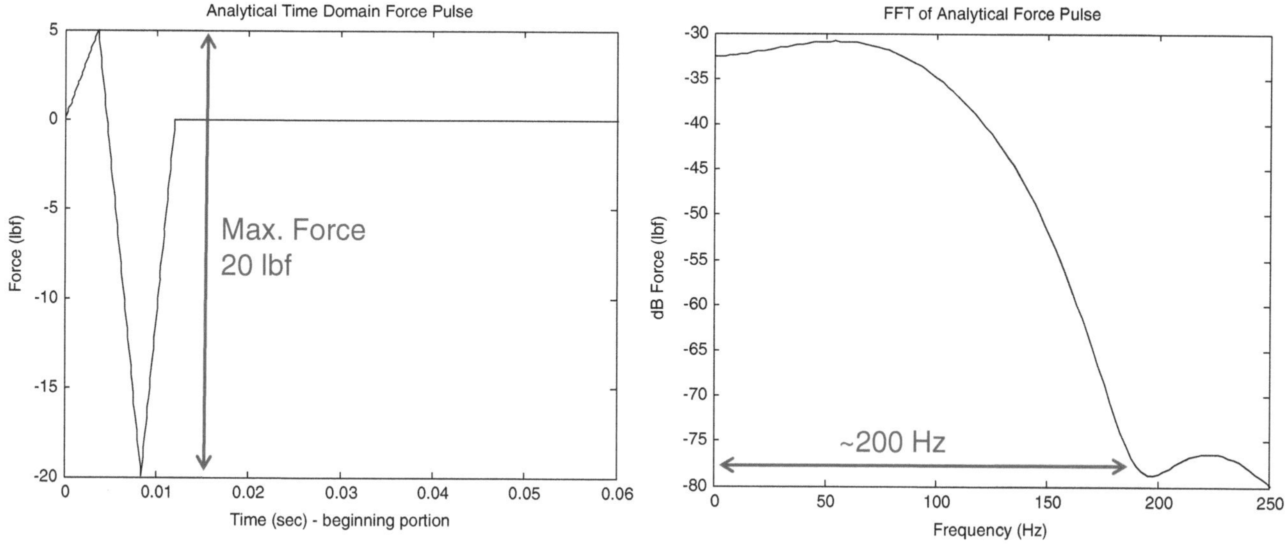

Fig. 2.4 Time (*left*) and frequency (*right*) domain plots of input analytical force pulse

Fig. 2.5 Comparison of modified and unmodified cantilevered beam models

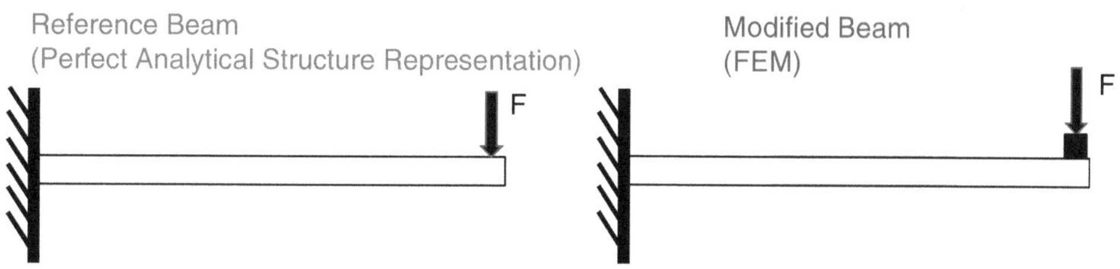

Mode #	Ref./Unmodified Beam Freq. (Hz)	Modified Beam Freq.(Hz)	% Difference
1	4.96	4.77	3.8
2	31.09	29.99	3.6
3	87.06	84.14	3.4
4	170.62	165.20	3.2
5	282.08	273.58	3.0
6	421.46	409.40	2.9

the original model due to perturbations in the model (added tip mass). This portion will only make use of SEREP models because as will be seen in Part A, Guyan reduced models do not accurately preserve the dynamic characteristics of the system. Expansion of reduced order models using both ideal and poor ADOF locations will be analyzed as well as the importance of a properly selected set of mode shape vectors spanning the space of the full space solution.

The cases presented here are summarized as:

Case A-1: Reference Model (Perfect Analytical Representation of Truth Model)

Unmodified cantilevered beam with 40 DOF Total.

Case A-2: Guyan Reduced Order Models

- Case A-2.1: 4 DOF Total; ADOF 33, 35, 37, and 39; Modes 1–4
- Case A-2.2: 6 DOF Total; ADOF 1, 3, 33, 35, 37, and 39; Modes 1–4

Case A-3: SEREP Reduced Order Models

- Case A-3.1: 4 DOF Total; ADOF 33, 35, 37, and 39; Modes 1–4
- Case A-3.2: 6 DOF Total; ADOF 1, 3, 33, 35, 37, and 39; Modes 1–4

Case A-4: KM_AMI Model Improvement from Guyan Reduced Order Models

- Case A-4.1: 6 DOF Total; ADOF 1, 3, 33, 35, 37, and 39; Modes 1–4; Expansion with SEREP [T_U]
- Case A-4.2: 6 DOF Total; ADOF 1, 3, 33, 35, 37, and 39; Modes 1–4; Expansion with Guyan [T]

Case B-1: Modified/Perturbed Reference Model (Imperfect FEM Representation of Truth Model)
Modified cantilevered beam with 40 DOF Total.

Case B-2: Reduced Order Models of Reference Beam (Unmodified)

- Case B-2.1: 5 DOF Total; ADOF 19, 33, 35, 37, and 39; Modes 1–5; Expansion with [T_U] from ***modified*** beam
- Case B-2.2: 5 DOF Total; ADOF 7, 17, 23, 31, and 39; Modes 1–5; Expansion with [T_U] from ***modified*** beam
- Case B-2.3: Considerations for the use of additional DOF and mode shapes

2.4.1 *Case A-1: Reference Model*

The forced response of the cantilever beam of Fig. 2.3 was calculated using Newmark direct integration at full physical space (40 DOF). The cantilever boundary condition was created by constraining 2 DOF at the base of the beam (these 2 DOF are not included in the 40 DOF). The input force primarily excited the first 4 modes of the system in a frequency band of 200 Hz (as shown in Figs. 2.3 and 2.4). The predicted response of this model was used as a reference for correlation with the predicted response of the reduced order models.

2.4.2 *Case A-2: Guyan Reduced Order Models*

Model reduction of the reference cantilever beam (unmodified) was performed using 4 and 6 ADOF. The selected ADOF in the reduction were chosen at locations with poor level of response or representing an uneven distribution across the length of the beam. These ADOF sets exemplify experimental situations where transducer placement is not ideal due limitations in the test structure, mass loading, etc. Because of the poorly selected ADOF locations and the susceptibility of the Guyan method to DOF selection, the reduced model mode shapes are inaccurate and thus the predicted response is expected to be erroneous. A comparison of the Guyan reduced model frequencies and the reference model (Case A-1) is shown in Table 2.1.

2.4.2.1 Case A-2.1

A Guyan reduced model of 4 active transverse DOF was created using nodes 17–20. The aim was to preserved modes 1–4, but as shown in Table 2.1, there is poor correlation between the reference and reduced model. The reduced model was used to calculate the time response at the ADOF and expansion performed to predict the response at all NDOF. Figure 2.6 shows a comparison of the time response at the tip of the beam for the reduced and reference models as well as the MAC and TRAC.

The time response of the predicted expanded Guyan reduced model differs largely from the reference model, in particular at locations that were not included in the reduction process (rotation DOF and locations near the base) as shown by the

Table 2.1 Frequency comparison of reference model and Guyan reduced model with 4 and 6 ADOF

	Frequency (Hz)			% Difference	
Mode #	Reference	Guyan 4 ADOF	Guyan 6 ADOF	4 ADOF vs reference	6 ADOF vs reference
1	4.96	4.97	4.97	0.2	0.2
2	31.09	38.62	35.68	24.2	14.8
3	87.06	448.02	106.51	414.6	22.3
4	170.62	2340.74	502.61	1271.9	194.6
5	282.08	–	1008.30	–	257.5
6	421.46	–	2437.35	–	478.3

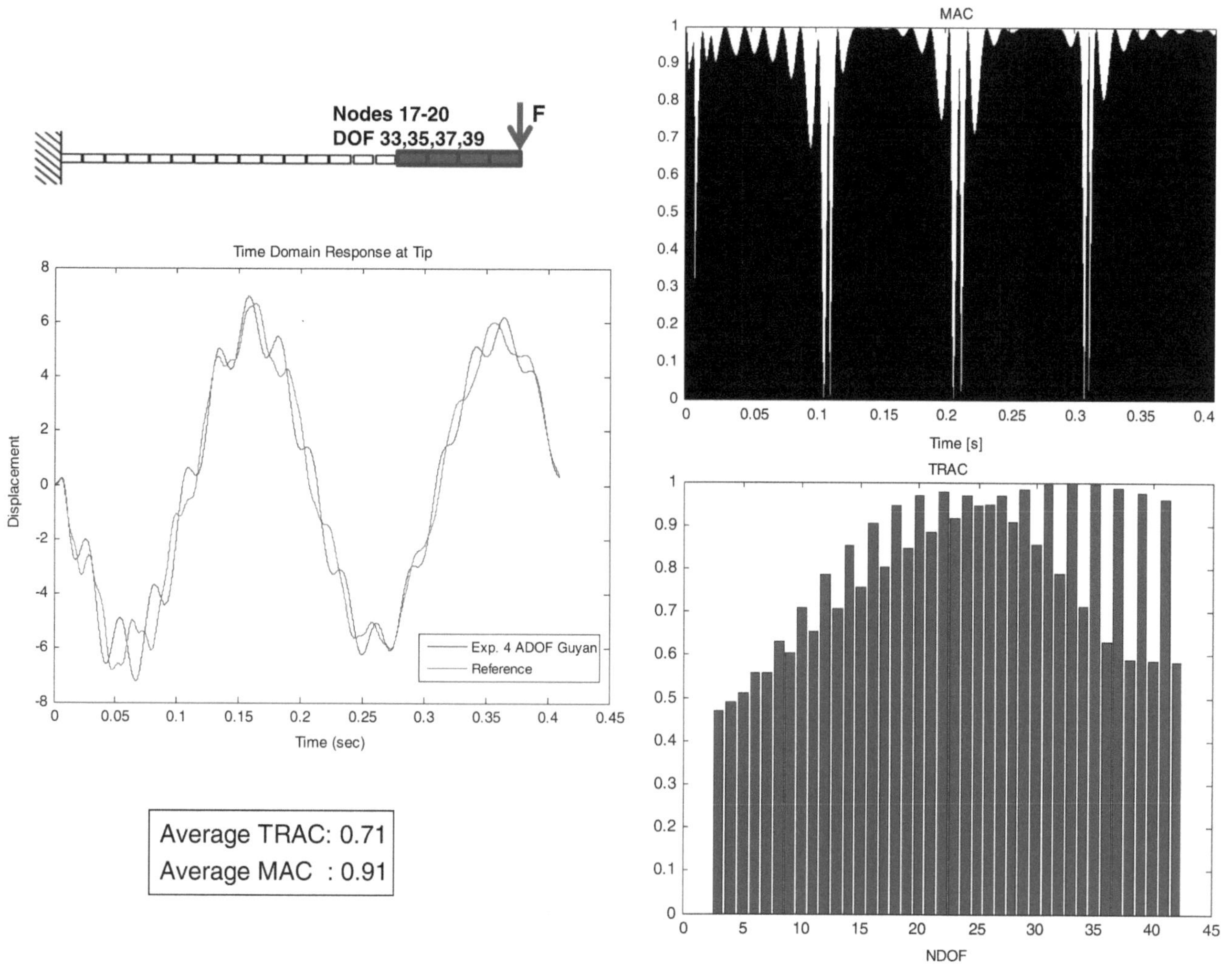

Fig. 2.6 Comparison and correlation of 4 ADOF Guyan reduced and reference cantilever beam models

TRAC of Fig. 2.6 (lower right corner). Clearly, the reduced model must closely preserve the retained modes and a carefully selection of ADOF must be performed. Moreover, a larger set of ADOF is necessary for this reduction methodology.

2.4.2.2 Case A-2.2

A Guyan reduced model of 6 ADOF was created using nodes 1, 2, and 17–20. The aim was again to preserved modes 1–4, but in this instance the number of ADOF is larger than the number of modes of interest. However, note that the two of ADOF locations offer very little response for the preserved modes. As in the previous case, the forced response was computed at the 6 DOF and expanded to predict the response at all 40 DOF. Figure 2.7 shows the comparison of the time response near the base of the beam and the MAC and TRAC correlation with respect to the reference model.

Once again, the predicted response did not compare to the full reference model even for those points preserved in the reduction (e.g., the 2 ADOF near the base of the beam). While there was a slight improvement in the average MAC and TRAC due to the additional DOFs in the reduction, the overall dynamics of the model were not captured through this reduced model.

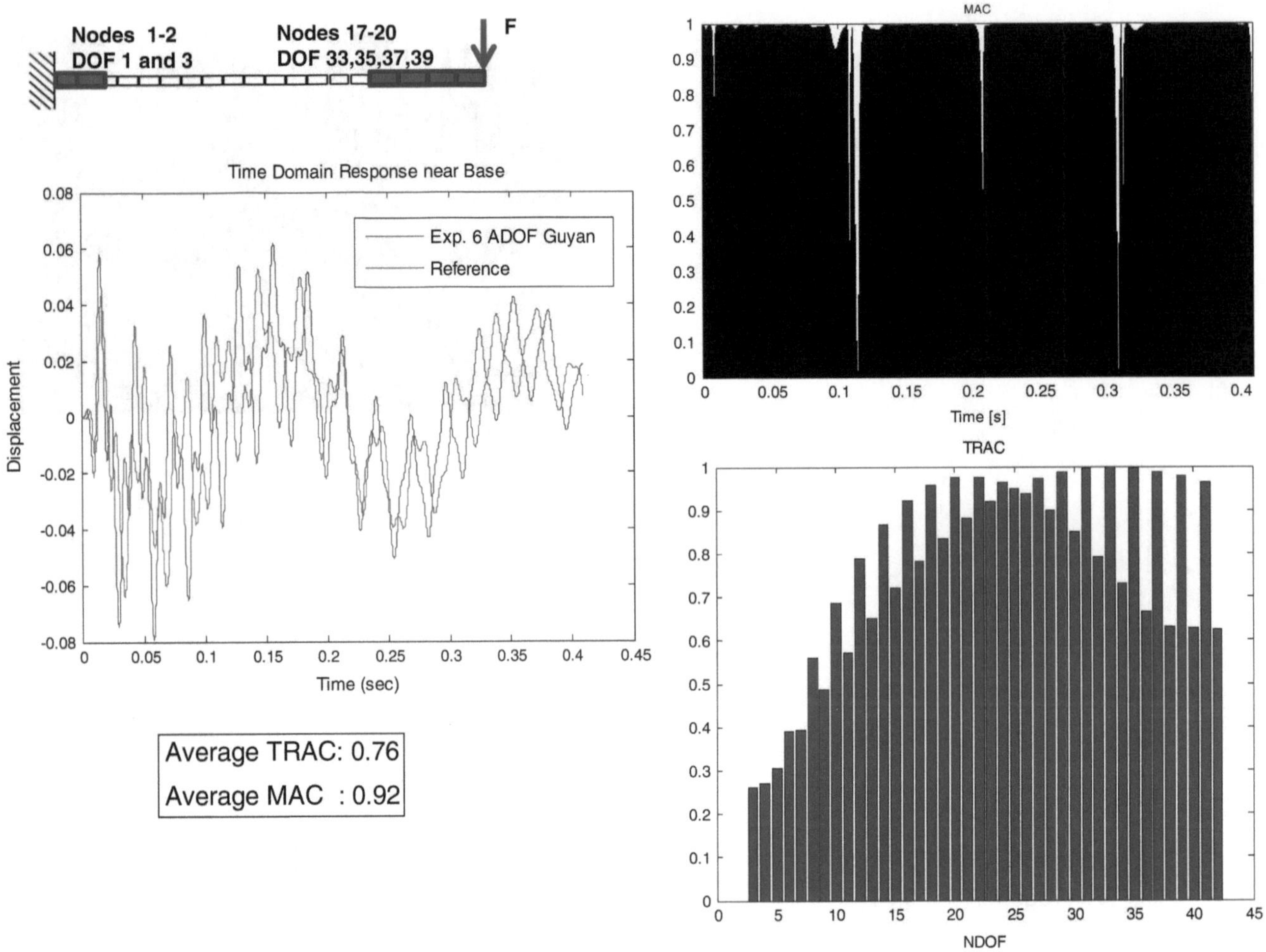

Fig. 2.7 Comparison and correlation of 6 ADOF Guyan reduced and reference cantilever beam models

2.4.3 *Case A-3: SEREP Reduced Order Models*

Model reduction of the reference cantilever beam (unmodified) was performed using 4 and 6 ADOF as in the Guyan reduced models. SEREP reduced models were not affected by the poorly selected ADOF and because the mode shape vectors and frequencies are preserved exactly, good correlation was obtained from the predicted time response.

2.4.3.1 Case A-3.1

In similar manner to that of Case A-2.1, a SEREP reduced model of 4 ADOF was created using nodes 17–20. A comparison of the predicted time response from the expanded reduced order model can be seen in Fig. 2.8.

All DOF were preserved accurately, even those not included in the reduction process. The MAC and TRAC show high correlation for the predicted time response. Thus, SEREP reduction overcomes the limitations encountered in the Guyan reduced models allowing for the accurate prediction of the dynamic characteristics of the system.

2.4.3.2 Case A-3.2

For this case, two additional DOF are included in the reduced model (as in Case A-2.2). The SEREP methodology for this case produces rank deficient matrices and care must be taken in the numerical processing of the models. Nevertheless, accurate results were obtained at all NDOF. Figure 2.9 shows a comparison of the time response near the base of the beam.

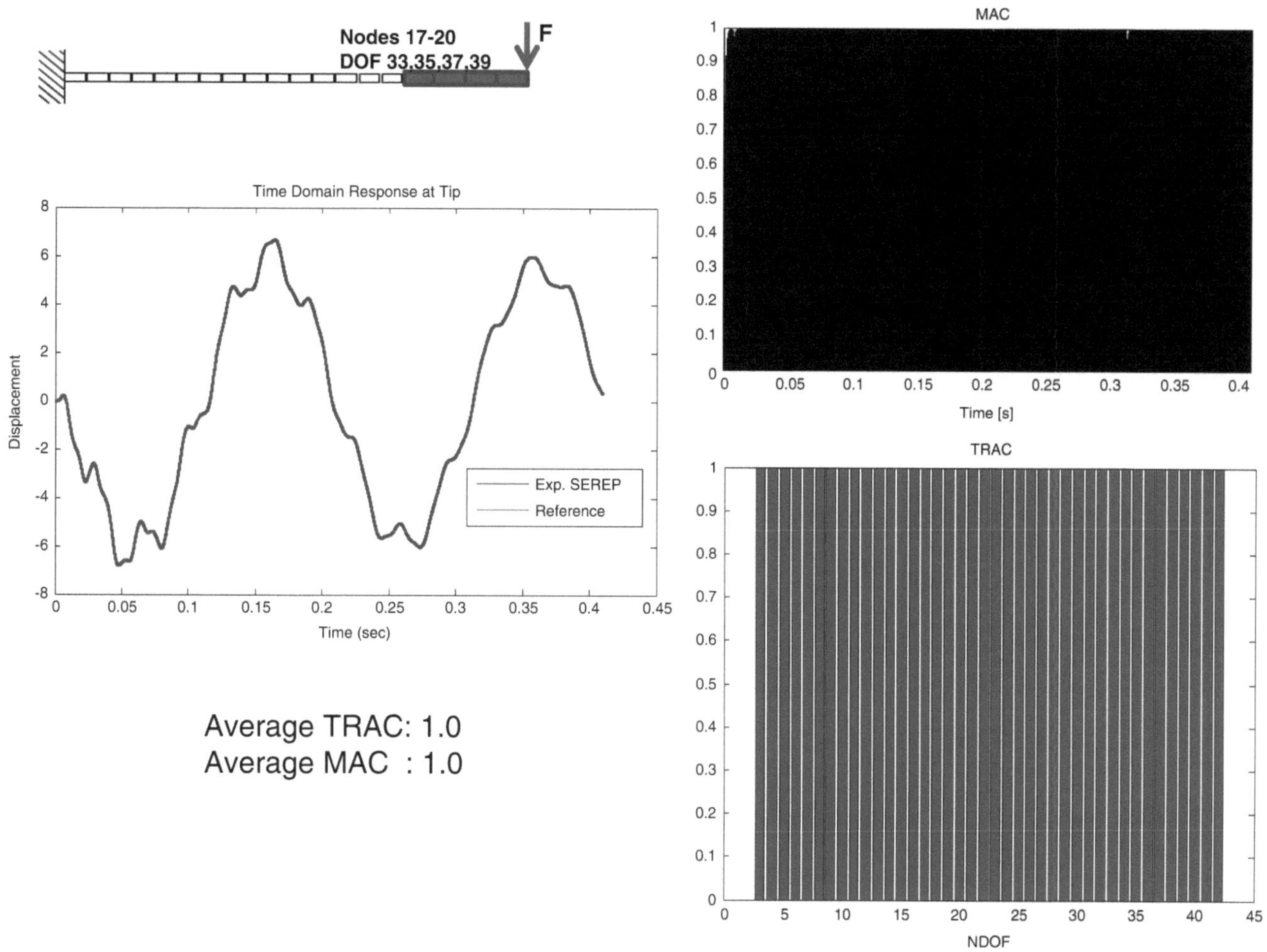

Fig. 2.8 Comparison and correlation of 4 ADOF SEREP reduced and reference cantilever beam models

Comparing Figs. 2.7 and 2.9 shows that the predicted response of the SEREP reduced model preserves accurately all DOF locations including those that have limited response for the modes of interest.

2.4.4 Case A-4: KM_AMI Model Improvement from Guyan Reduced Order Models

The Guyan reduced model of Case A-2 was shown to be significantly affected by the selection of ADOF. The need to use Guyan reduction may arise from limitations of currently available finite element software or from the need to have a fully ranked reduced mass and stiffness matrices; Case A-3.2 produces accurate results but the matrices are rank deficient and can lead to poorly conditioned matrices and numerical inaccuracies. For such situations, this case explores using the exact frequencies and mode shapes of the system to update the Guyan reduced order model using the KM_AMI approach. The target frequencies and mode shapes were directly seeded to the Guyan reduced model of Case A-2. The improved reduced model now accurately preserves the modes of interest. The following cases explore the expansion of the KM_AMI improved model using the transformation matrix from the original Guyan reduction of Case A-2 and using the transformation from the exact SEREP reduction.

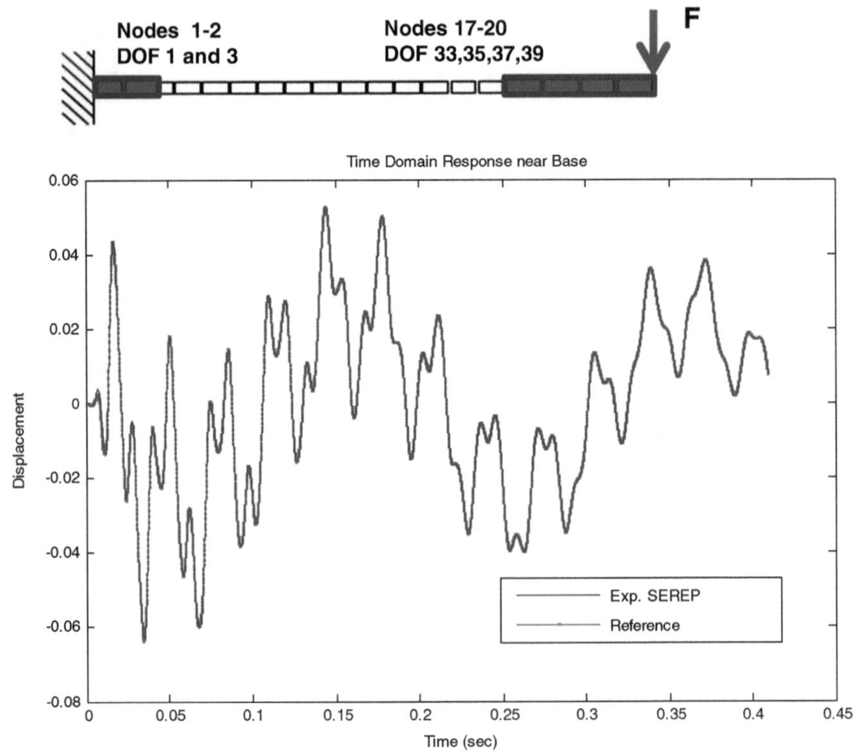

Fig. 2.9 Time response of 6 ADOF SEREP reduced expanded model and reference model

2.4.4.1 Case A-4.1

The time response was calculated at the 6 ADOF and expanded to NDOF using the exact mode shapes and frequencies available from the KM_AMI process ($[T_U]$ transformation matrix). The predicted response produced the exact results encountered in the SEREP reduced models with high correlation across all DOF as seen in Fig. 2.8.

2.4.4.2 Case A-4.2

The time response was calculated at the 6 ADOF and expanded to NDOF using the Guyan [T] transformation matrix used in the reduction of Case A-2. Because the Guyan transformation does not accurately preserved the mode shape vectors of the reference model, the predicted full space response of the KM_AMI model results in decreased correlation values when compared to Case A-4.1. The point of this case is to illustrate the situation in which only the Guyan transformation is available from the original Guyan reduced order model. While the results are not as accurate as those obtained through SEREP, this case shows large improvement when compared to Case A-2.2 which uses the same ADOF but does not employ the KM_AMI model improvement. Figure 2.10 shows the MAC and TRAC correlation for the KM_AMI improved model with Guyan expansion.

The improved reduced models using the KM_AMI approach showed significant gain in accuracy because the frequencies and mode shapes were preserved exactly (as they were directly seeded in the reduced model from the reference solution) compared to the original Guyan reduced models.

Case A has explored the prediction of full space time response from the reduced order model response of a structure. The cases discussed thus far have shown the limitations of the Guyan reduced models as well as the issues encountered with poorly located or limited selection of ADOF. SEREP reduction was shown to bypass these issues and produce accurate results as long as the preserved modes of the system span the full space solution. In other words, the projection vectors used in the expansion must span the space of the problem. Next, the cases of Part B will show how the expansion of the reduced model response (or response at limited DOF) can be performed even for cases where poor correlation exists due to perturbation or inexact projection mode shape vectors.

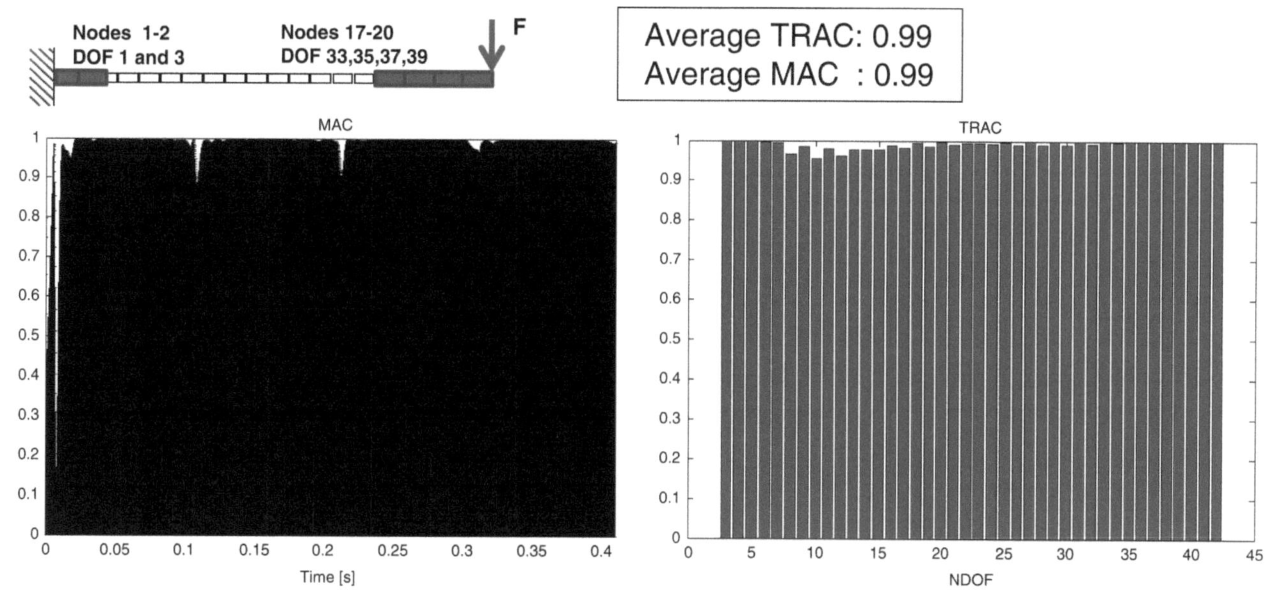

Fig. 2.10 Correlation of KM_AMI improved Guyan reduced model (with Guyan expansion to NDOF) with reference model

2.4.5 *Overview of Case B*

The analytical model (reference unmodified cantilevered beam) used in Part A can be considered the perfect analytical representation of a truth model (real answer) of a structure. The model was created to simulate a system (such as a wind turbine blade as an example) and how its dynamic behavior can be predicted from a reduced model. Part B will explore the case of predicting the full space response (NDOF) of a system using the collected (or in this case simulated) response at limited points of the structure (ADOF) using mode shapes from an inexact/perturbed model of the structure. This is the situation of having collected data on a structure and using the projection vectors of a FEM (that is not a perfect representation of the system) to predict full field displacement. Figure 2.11 shows the case of the dynamic model of a wind turbine blade (as an example) and the analytical model used to represent the structure (reference model) as well as an equivalent imperfect FEM model representation of the structure (due to the addition of mass).

The following cases will show that using the response at ADOF from the reference/unmodified beam model can lead to an accurate prediction of the full space (NDOF) response using the projection mode shape vectors of a perturbed model as long as these mode shape vectors span the space of the full system. This might require the use of additional modes of the modified/perturbed model to fully span the space of the reference model and hence additional DOF may be needed (blue crosses in Fig. 2.11) beyond those used in the original reference model.

2.4.6 *Case B-1: Modified/Perturbed Reference Model (Imperfect FEM Representation of Truth Model)*

The forced response of the modified/perturbed cantilever beam of Fig. 2.5 was calculated using Newmark direct integration at full physical space (40 DOF). The first 2 DOF were constrained to create the cantilever boundary condition. The input force excited the first four modes of the system in a frequency band of 200 Hz (as shown in Fig. 2.5). The predicted response of this model was used as a reference for correlation with the predicted response of the modified and unmodified reduced order models. This model will be referred as modified reference model (inexact FEM) as opposed to the reference model (unmodified beam) of Case A-1. The time response of both modified (with the added mass of Fig. 2.5) and unmodified reference models are shown in Fig. 2.12. A time shift can be seen in the response of the modified system due to the frequency shift introduced by the addition of mass at the tip of the beam.

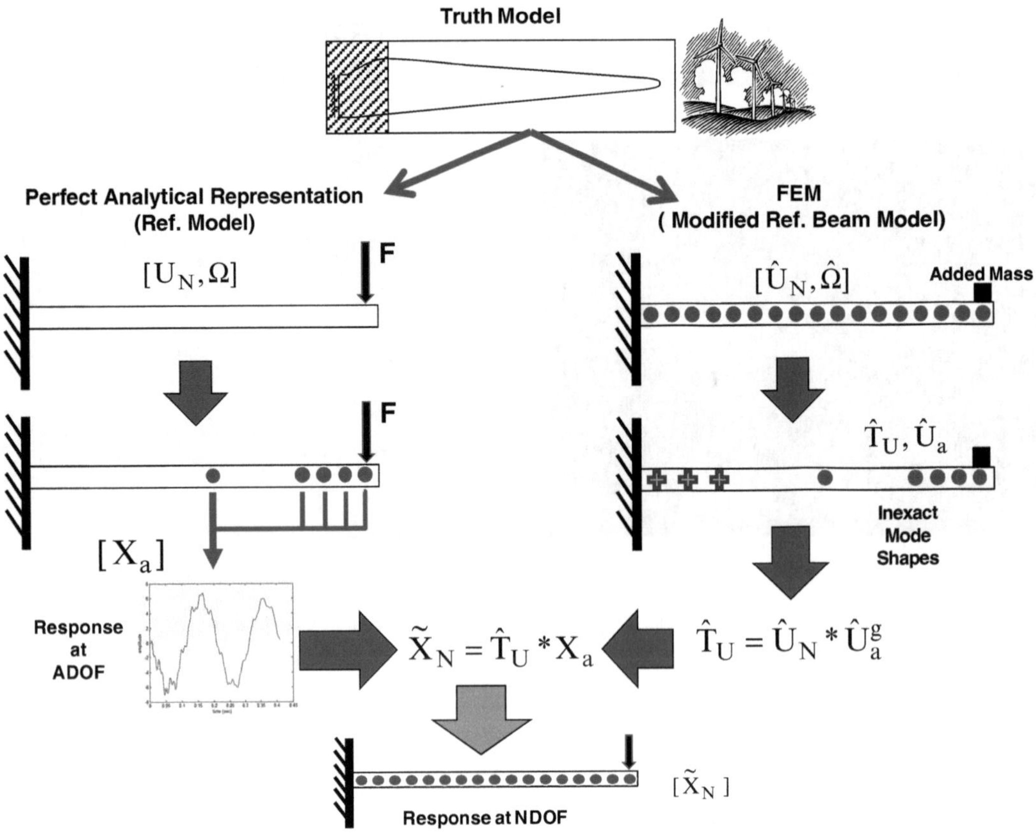

Fig. 2.11 Schematic of reference beam and expansion using transformation matrix from modified beam mode shapes

Fig. 2.12 Comparison of time response at tip of cantilever beam for reference (unmodified) and modified reference models

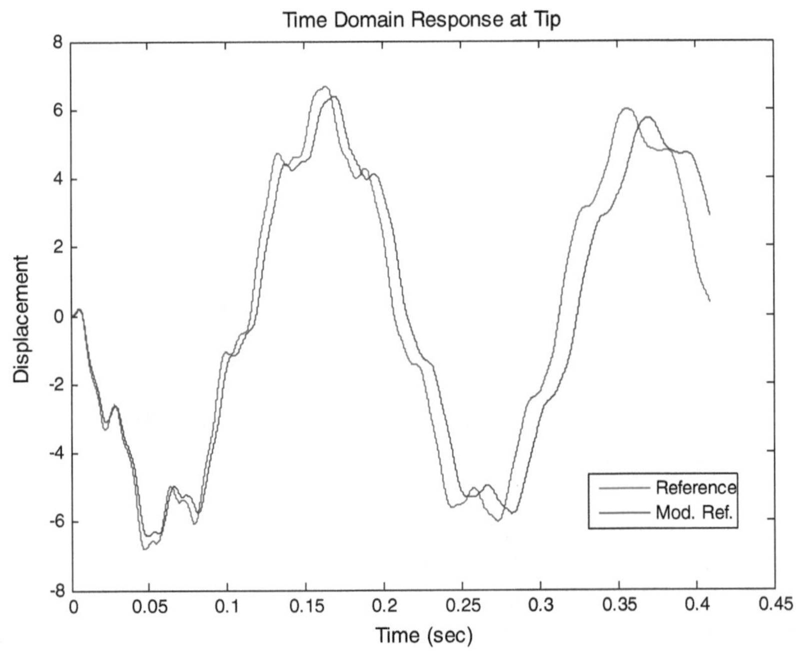

2.4.7 *Case B-2: Reduced Order Models of Reference Beam (Unmodified)*

The reference cantilever beam (unmodified) was used to extract the response at 5 ADOF. In the first case, the ADOF locations are not distributed to uniformly captured the modes of interest (as in Part A) while the second case involves a more appropriate selection of ADOF. Improper selection of DOF leads to poorly conditioned matrices and thus inaccurate characterization of the system. The time response was also accurately calculated for comparison purposes at the reduced space (using SEREP as in Case A-3) but the transformation matrix for the expansion to NDOF originated from the modified cantilever beam mode shapes. The overall reduction and expansion process is illustrated in Fig. 2.11, where the full space response of a structure (such as that of a wind turbine blade) is to be predicted from the measured response at a selected number of DOF (red dots) using the projection vectors (mode shapes) of a modified system.

Figure 2.11 shows that the solution at the selected ADOF, $[X_a]$, can be expanded to NDOF using the transformation matrix $\left[\widehat{T}_U\right]$ obtained from a modified cantilevered beam. The $\left[\widehat{T}_U\right]$ is formed from the modified/perturbed mode shapes of the cantilever beam and therefore corresponds to an inexact representation of the mode shapes of the reference model. The full space solution obtained from this perturbed expansion $\left[\tilde{X}_N\right]$ will be shown to be an approximation of $[X_N]$. Note that additional DOF (blue crosses in Fig. 2.11) and mode shapes may be required to completely span the space of the full space solution of the reference model (unmodified cantilevered beam).

2.4.7.1 Case B-2.1

The forced response at nodes 10 and 17–20 was extracted from the full space solution of the reference cantilever beam (perfect representation of truth model). Expansion of this extracted ADOF space was performed using the transformation matrix $\left[\widehat{T}_U\right]$, obtained from the SEREP reduction of the modified beam (inexact FEM representation of truth model) at the same ADOF. If the unmodified transformation matrix were to be used, as observed in Case A-3, the expansion would produce exact results at all DOF. However, due to the perturbation in the transformation matrix (projection shape vectors) used in the expansion the expanded model does not produce the same equivalent results for this reduced model. Figure 2.13 shows a comparison of the response at the tip of the beam for the expanded and reference models.

The TRAC shown in Fig. 2.13 shows high correlation mainly for locations near DOF preserved in the reduction. The time response at the tip of the beam of the expanded response (in green) accurately overlays with the unmodified reference model response (in red). However, locations near the base of the beam were not approximated well as shown in Fig. 2.14.

The transformation matrix $\left[\widehat{T}_U\right]$ formed from the poorly selected ADOF set was not properly formed to provide a well-conditioned set of matrices. While the perturbations introduced by the addition of mass to the cantilever beam have a strong effect in the accuracy of the expansion, the underlying issue is found in the projection shape vectors not forming

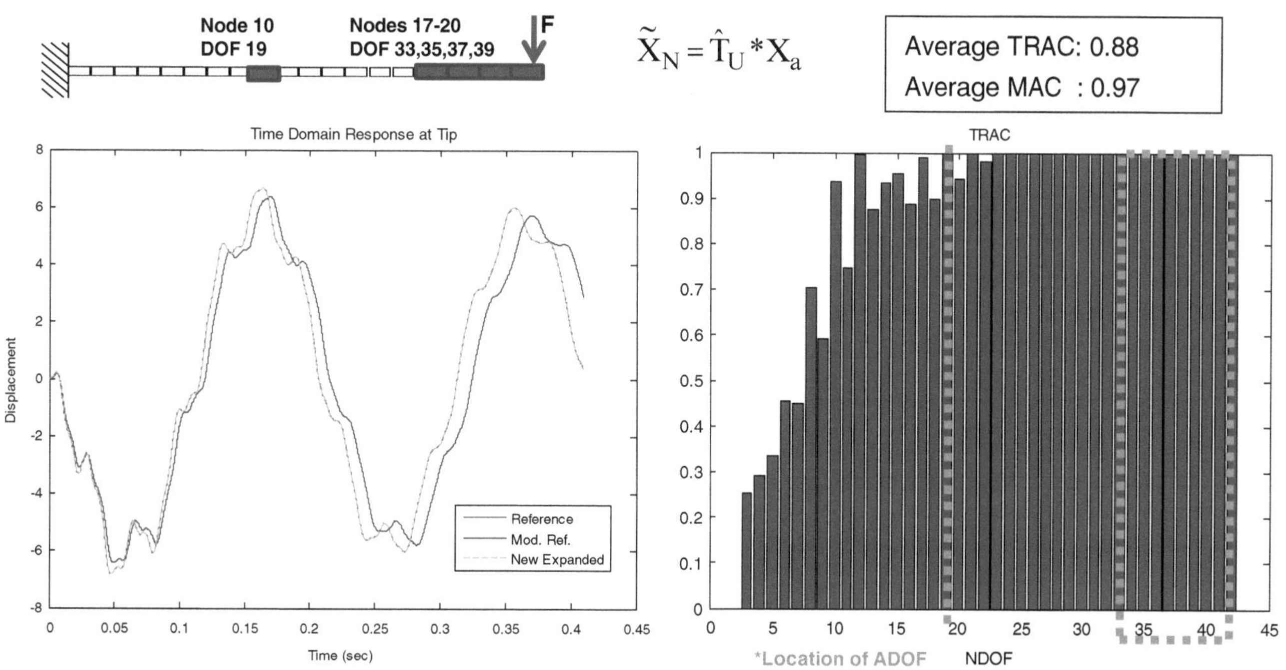

Fig. 2.13 Comparison of expanded ADOF model (in *green*) and unmodified reference model (in *red*). Shown in *blue* is the reference modified model from which the transformation matrix used in the expansion originates

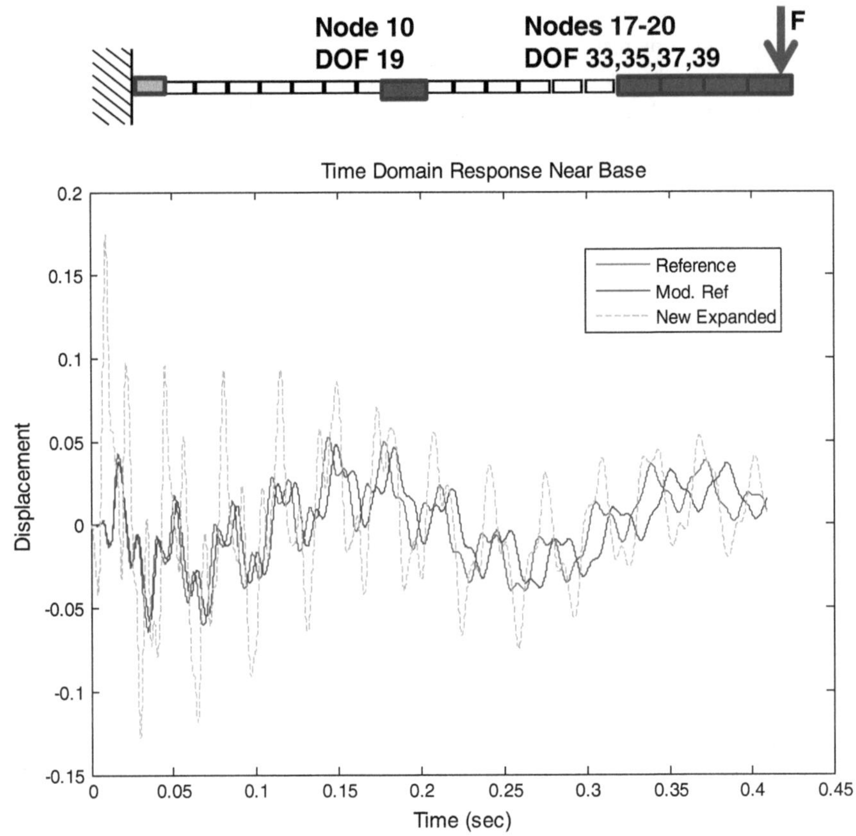

Fig. 2.14 Time response comparison for expanded $\left[\tilde{X}_N\right]$ model response (in *green*) and reference unmodified model (in *red*). Also shown in *blue*, modified reference model forced response

a basis for the solution of the NDOF space. The DOF selected in this case resulted in a large condition number for the transformation matrix indicating that the projection vectors are not linearly independent and the matrix is ill-conditioned leading to large errors during the numerical computations. Thus, the proper number of modes and a redistribution of DOF can be performed to yield a well-conditioned matrix and mitigate errors in the numerical process. Case B-2.2 will show that a properly selected perturbed transformation matrix can yield accurately prediction of the expansion of the unmodified system response (reference model).

2.4.7.2 Case B-2.2

Reference cantilever beam response at nodes 4, 9, 12, 16, and 20 was extracted from the full space solution. The nodes were selected from locations that properly captured the mode shapes of interest and gave a fair distribution of points along the length of the beam. The key point is that the DOF selected lead to well-conditioned and fully ranked transformation matrices and therefore a set of projection mode shape vectors that fully span the space of the system. The forced response was once again expanded to NDOF using the new transformation matrix $\left[\widehat{T}_U\right]$ obtained from the SEREP reduction of the modified beam at the new ADOF space. The DOF locations and the TRAC are shown in Fig. 2.15.

The full NDOF space was accurately predicted as shown in the TRAC of Fig. 2.15. Note that the DOFs near the base of the beam which were previously inaccurate (Fig. 2.14) are now approximated well in the expansion process even though the mode shapes are inexact representations of the unmodified model. Figure 2.16 shows the time response near the base of the beam for the new ADOF set.

As highlighted in this and the previous case (B-2.1), the DOF selection plays an important role in the accuracy of the predicted response. While the carefully selected 5 DOF used in this case produced an accurate full space prediction of the response of the reference model, similar results can be obtained with a less than ideal DOF selection. A uniform distribution of points along the length of the beam works well for the model considered here. However, real world structures with changes in material properties and cross section along the length can lead to large concentrations of DOF at sections of the system where modal information is more abundant. The most important aspect lies in obtaining a transformation matrix that spans

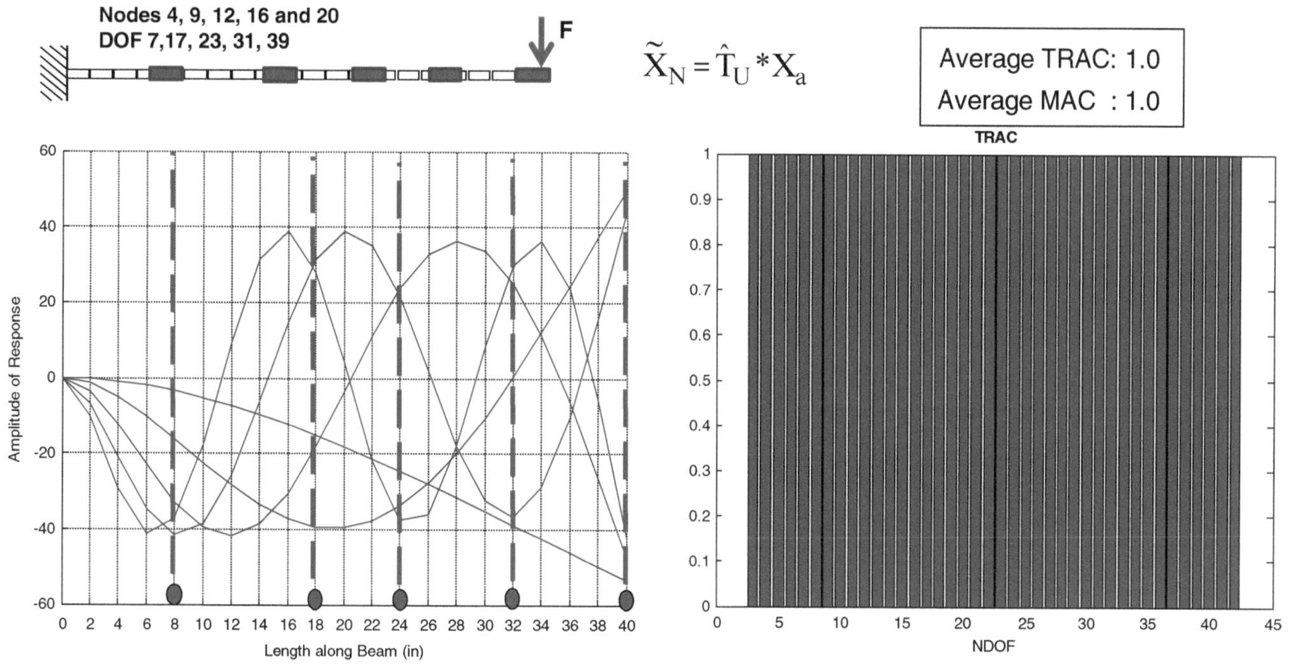

Fig. 2.15 Selection of ADOF for reduced model (*Left*). TRAC correlation of expanded $[\tilde{X}_N]$ reduced order model response and unmodified reference model (*Right*)

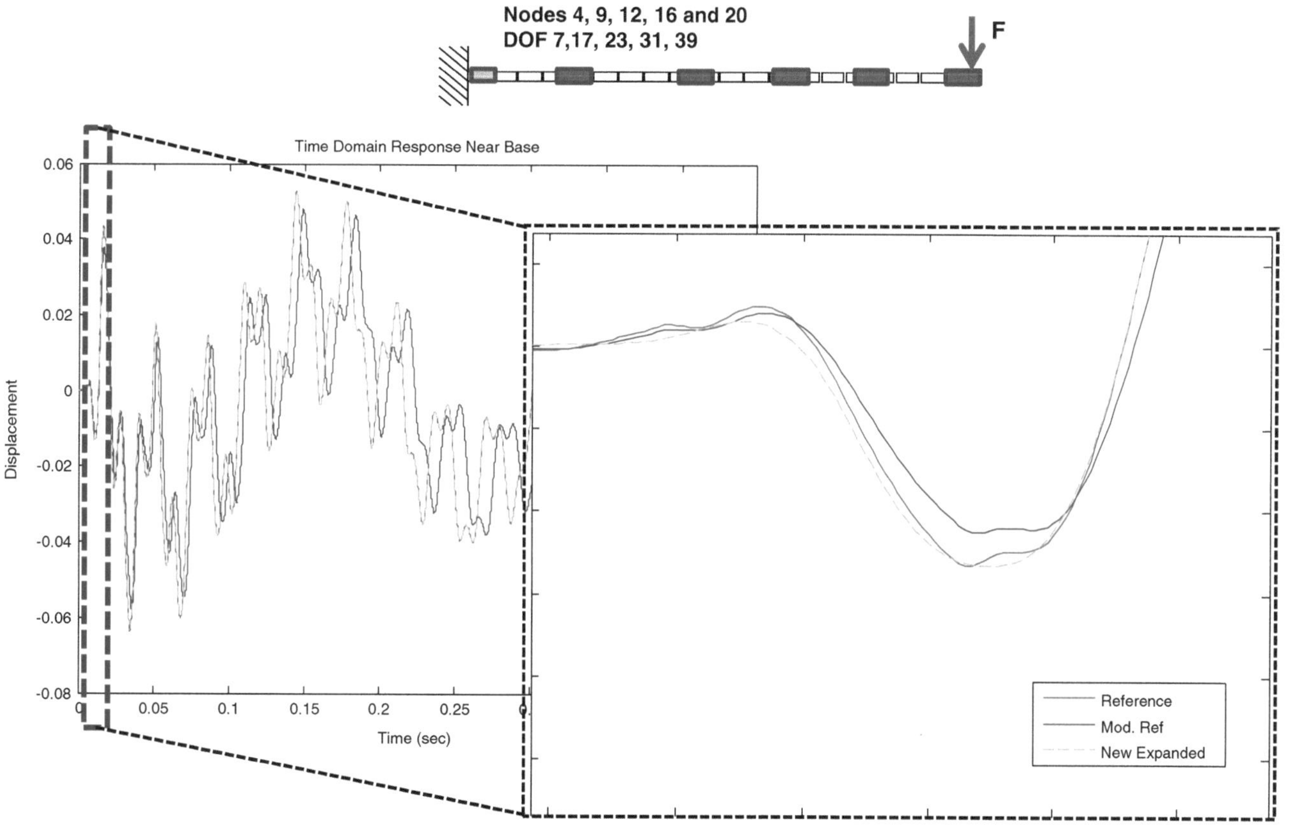

Fig. 2.16 Time response comparison for expanded $[\tilde{X}_N]$ reduced order model response (in *green*) and reference unmodified model (in *red*) using new ADOF set. Also shown in *blue*, modified reference model forced response

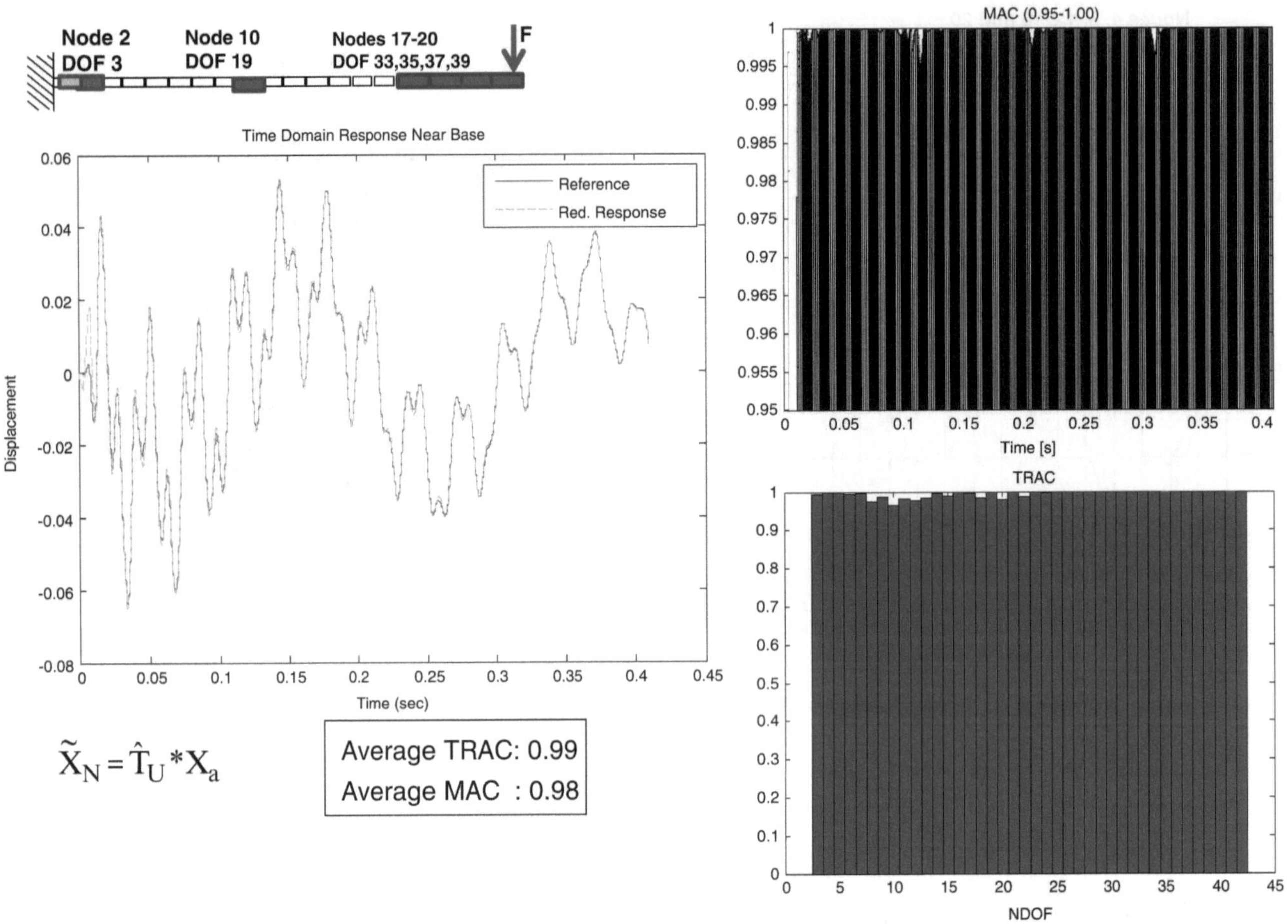

Fig. 2.17 Time response comparison for expanded $[\tilde{X}_N]$ reduced order model response (in *green*) and reference unmodified model (in *red*) using additional mode shapes (6 DOF). MAC (*top right*) and TRAC (*bottom right*) correlation of models

the full space of the response of the system and has a well-conditioned fully ranked transformation matrix. For poorly selected DOF, this may require the use of additional DOF and mode shapes in the expansion process as will be discussed next.

2.4.7.3 Case B-2.3

Additional DOF were added to the "poorly" selected set of points used in case B-2.1 (node 10 and 17–20 as shown in Fig. 2.13). Node 2 (DOF 3) was first added to the previous 5 DOF. The response at the selected ADOF from the unmodified reference model was expanded using the transformation matrix from the modified/perturbed beam model as done in previous cases. Figure 2.17 shows the response at the expanded DOF near the base of the beam (DOF 1) as well as the correlation with the full space reference model.

Comparing Figs. 2.14 and 2.17 shows large improvement in the predicted response at all DOF. The additional mode shape vector and DOF used in the transformation matrix of the modified beam resulted in a better set of projection vectors and hence a more accurate prediction of the response of the original system. However, a larger set of mode shapes may be required if the projection vectors of the additional DOF selected do not span the space of the full space solution. For instance, the set of DOF 7, 19, 23, 27, 33, 35, 37 and 39 and the first 8 mode shapes of the system did not produce an accurate prediction near the base as illustrated in Fig. 2.18.

While additional DOF and modes shapes in the transformation matrix can help smooth the predicted full space response in well correlated models, the expansion using inexact representation of system modes is susceptible to the location of the DOF and more importantly the linear independence of the projection vectors. The mode shapes used for the expansion process must span the space of the full space solution and therefore must be capable of generating any vector in the space as a linear combination of the preserved mode shapes.

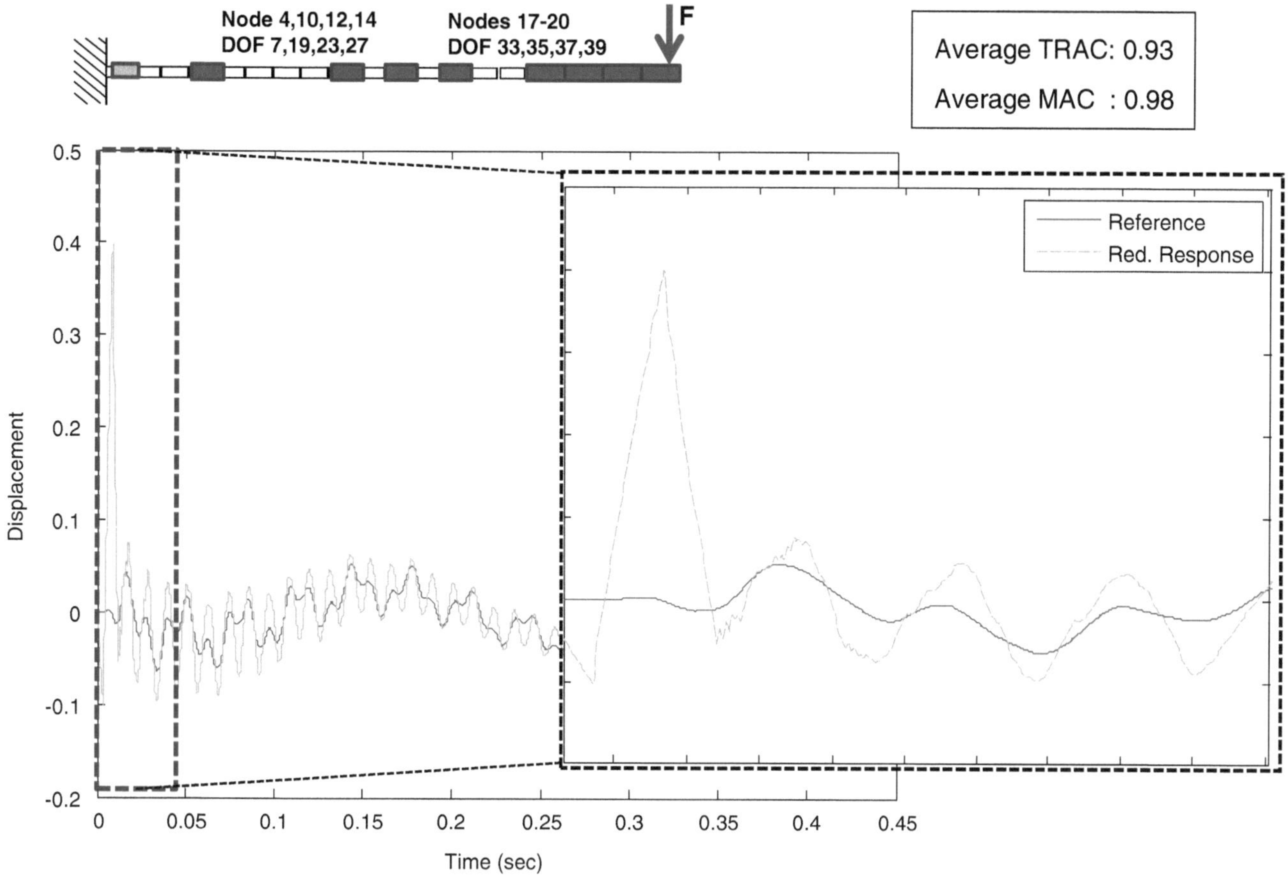

Fig. 2.18 Time response comparison for expanded $[\tilde{X}_N]$ model response (in *green*) and reference unmodified model (in *red*)

2.5 Observations

The formation of the transformation matrix $[T_U]$ from the mode shape vectors of the system at a selected set of DOF must contain sufficient information to accurately project the response at the limited set of points back to full space. In other words, these projection vectors must form a linearly independent set that can approximate any other vector in the space via linear combinations. Poorly selected DOF lead to ill-conditioned transformation matrices, indicating that the matrix is almost singular and among other things that the column vectors of $[T_U]$ do not yield a linearly independent set of mode shape vectors. For the cases B-2.1 (5 poorly selected DOF) and B-2.2 (uniformly selected 5 DOF) the condition number of $[T_U]$ goes from 6,830 to 2 respectively. A large condition number for a matrix is an indication of ill-conditioned with 1 being well-conditioned [19].

2.6 Conclusions

Guyan condensation and SEREP were used to analyze the forced response of a cantilever beam. The reduced models were created with poorly chosen DOF to illustrate commonly encountered issues in structural testing. The Guyan reduction process was shown to be highly affected by the DOF selection and thus produced inaccurate prediction of the dynamics of the system. SEREP and KM_AMI model improvement were successfully used to predict the response at all DOF in cases with poor DOF locations. The key aspect in the accuracy of the expansion was found to be in the proper selection of a basis of modal vectors that span the full space solution of the system. In such cases, the limited or poorly chosen DOF did not affect the fidelity of the reduced model and the subsequent expansion.

Based on this feature of the SEREP expansion, models of reduced modified and unmodified cantilever beams were developed to show that inexact mode shape projection vectors (i.e. transformation matrix [T]) could accurately expand to the respective full DOF space. The reduced space time response of the unmodified beam was accurately mapped back to its full space using the transformation matrix of a model with perturbed mode shapes of the cantilever beam. Thus, this study has proven that when the preserved mode shapes contain sufficient information to span the solution space of the system, the perturbations do not distort the predicted full space time response. Variability in data and minor perturbations in the physical characteristics of a structure, often times encountered in practice, as well as limitations in transducer placement and inexact FEMs of the system can be accounted for while accurately and efficiently predicting full field characteristics of a structure. Very accurate results were obtained when care is used to select appropriate DOF and modes for the expansion process.

Acknowledgements Some of the work presented herein was partially funded by Air Force Research Laboratory Award No. FA8651-10-1-0009 "Development of Dynamic Response Modeling Techniques for Linear Modal Components". Any opinions, findings, and conclusions or recommendations expressed in this material are those of the authors and do not necessarily reflect the views of the particular funding agency. The authors are grateful for the support obtained.

References

1. Butland A (2008) A reduced order, test verified component mode synthesis approach for system modeling applications. Master's thesis, University of Massachusetts Lowell
2. Butland A, Avitabile P (2010) A reduced order, test verified component mode synthesis approach for system modeling applications. Mech Syst Signal Process 24(4):904–921
3. Nicgorski D (2008) Investigation on experimental issues related to frequency response function measurements for frequency based substructuring. Master's thesis, University of Massachusetts Lowell
4. Nicgorski D, Avitabile P (2010) Conditioning of FRF measurements for use with frequency based substructuring. Mech Syst Signal Process 24(2):340–351
5. Thibault L, Butland A, Avitabile P (2012) Variability improvement of key inaccurate node groups – VIKING. In: Proceedings of the thirtieth international modal analysis conference, Jacksonville
6. Guyan RJ (1965) Reduction of stiffness and mass matrices. AIAA J 3(2):380
7. O'Callahan JC, Avitabile P, Riemer R (1989) System equivalent reduction expansion process. In: Proceedings of the seventh international modal analysis conference, Las Vegas
8. Marinone T, Butland A, Avitabile P (2012) A reduced model approximation approach using model updating methodologies. Conference proceedings of the Society for Experimental Mechanics series 2012, Topics in modal analysis II vol 6, pp 625–636
9. Avitabile P (2003) Twenty years of structural dynamic modification – a review. Sound Vib Mag 37(1):14–27
10. Thibault L, Avitabile P, Foley J, Wolfson J (2012) Equivalent reduced model technique development for nonlinear system dynamic response. In: Proceedings of the thirtieth international modal analysis conference, Jacksonville
11. O'Callahan JC (1989) A procedure for an Improved Reduced System (IRS) model. In: Proceedings of the seventh international modal analysis conference, Las Vegas
12. Thibault L (2012) Development of equivalent reduced model technique for linear modal components interconnected with nonlinear connection elements. Master's thesis, University of Massachusetts Lowell
13. Newmark NM (1959) A method of computation for structural dynamics. J Eng Mech Div, Am Soc Civ Eng 85(3):67–94
14. Hilber HM, Hughes TJR, Talor RL (1977) Improved numerical dissipation for time integration algorithms in structural dynamics. Earthq Eng Struct Dyn 5:282–292
15. Allemang RJ, Brown DL (2007) A correlation coefficient for modal vector analysis. In: Proceedings of the first international modal analysis conference, Orlando
16. Van Zandt T (2006) Development of efficient reduced models for multi-body dynamics simulations of helicopter wing missile configurations. Master's thesis, University of Massachusetts Lowell
17. MAT_SAP/MATRIX (1986) A general linear algebra operation program for matrix analysis, Dr. John O'Callahan, University of Massachusetts Lowell
18. MATLAB R2010a (2010) The MathWorks Inc., Natick
19. Cheney E, Kincaid D, (2012) Numerical mathematics and computing. Cengage Learning, p 321

Chapter 3
Estimation of Rotational Frequency Response Functions

T.A.N. Silva and N.M.M. Maia

Abstract As it is widely known, in structural dynamic applications, ranging from structural coupling to model updating, the incompatibility between measured and simulated data is inevitable, due to the problem of coordinate incompleteness. Usually, the experimental data from conventional vibration testing is collected at a few translational degrees of freedom (DOF) due to applied forces, using hammer or shaker exciters, over a limited frequency range. Hence, one can only measure a portion of the receptance matrix, few columns, related to the forced DOFs, and rows, related to the measured DOFs. In contrast, by finite element modeling, one can obtain a full data set, both in terms of DOFs and identified modes. Over the years, several model reduction techniques have been proposed, as well as data expansion ones. However, the latter are significantly fewer and the demand for efficient techniques is still an issue. In this work, one proposes a technique for expanding measured frequency response functions (FRF) over the entire set of DOFs. This technique is based upon a modified Kidder's method and the principle of reciprocity, and it avoids the need for modal identification, as it uses the measured FRFs directly. In order to illustrate the performance of the proposed technique, a set of simulated experimental translational FRFs is taken as reference to estimate rotational FRFs, including those that are due to applied moments.

Keywords FRF estimation • Rotational DOFs • Expansion methods • SEREP • Modified Kidder's method

3.1 Introduction

In structural dynamic applications, it is required the knowledge of the dynamic behavior of the structure under study. Over the years, the use of experimental modal analysis and the finite element (FE) method has become standard, regarding the experimental and theoretical description of the structure to be studied. However the dynamic characteristics of a structure are often derived from a set of measured frequency response functions (FRFs), the spatial model usually contains a more refined information set and thus it allows for the computation of physically inaccessible FRFs, although it needs to be validated. Whenever test and theoretical responses must be matched, the issue of model incompleteness arises, due to the fact that experimental data is usually collected at a few translational degrees of freedom (DOFs) due to applied forces, using hammer or shaker exciters, over a limited frequency range. Hence, one must cope with either the FE model reduction to the set of experimental points, or the expansion of the experimental data over the FE model DOFs [1, 2].

The complete description of the dynamic behavior of a structure, using a response model, is attained only when all the response vs. excitation relations are known. As the responses involve translational and rotational displacements, and the excitations are due to applied forces and moments, it is widely known that a complete response model is inaccessible in practice. The demand for a complete response model has been very active in the structural coupling community, where the significance of assessing FRF at rotational DOFs was addressed in many studies [3]. Within the framework of structural coupling Batista and Maia [4] estimated inaccessible FRFs, namely the ones related to the coupling DOFs, from a set of measured FRFs. In the field of structural dynamic modification and FE model updating, this issue is also discussed [5–7].

T.A.N. Silva (✉)
GI-MOSM: Research Group on Modeling and Optimization of Multifunctional Systems, ISEL – Instituto Superior de Engenharia de Lisboa, IPL, Rua Conselheiro Emídio Navarro, Lisbon 1959-007, Portugal

LAETA, IDMEC, Instituto Superior Tecnico, Universidade de Lisboa, Av. Rovisco Pais 1, Lisbon 1049–001, Portugal
e-mail: tasilva@dem.isel.ipl.pt

N.M.M. Maia
LAETA, IDMEC, Instituto Superior Tecnico, Universidade de Lisboa, Av. Rovisco Pais 1, Lisbon 1049-001, Portugal
e-mail: nmaia@dem.ist.utl.pt

© The Society for Experimental Mechanics, Inc. 2015

R. Allemang (ed.), *Special Topics in Structural Dynamics, Volume 6*, Conference Proceedings
of the Society for Experimental Mechanics Series, DOI 10.1007/978-3-319-15048-2_3

Urgueira et al. [8] applied the generalized transmissibility concept in structures subjected to a certain structural modification in order to estimate unmeasured FRFs. The knowledge of rotational responses proved to be also essential, regarding the identification of the dynamic characteristic of joints, the identification of unknown forces from measured responses or vibration isolation.

However, a significant number of works dealt with either the experimental measurement of rotational displacements or the demand for an excitation due to a pure moment, barely none of these aimed to estimate the entire FRF matrix from few experimental translation/force FRFs. Regarding the subject of model incompleteness, over the years, several model reduction techniques have been proposed and successfully applied [9–11], although experimental data expansion strategies are significantly fewer and the demand for efficient techniques is still an issue. Moreover, almost every reduction/expansion methods were developed in the modal domain. Thus, requiring modal identification and being not suitable to directly use measured FRFs. The need to expand a set of measured FRFs has been addressed by several authors [12]. Regardless of the performance of the proposed expansion strategies, some authors consider very strong assumptions, which are not easy to achieve in practice. This is the case of [13], where the assumptions on the amount of known data can make the procedure unpractical. Hence, one should consider a more general purpose method as the one of Avitabile and O'Callahan [14].

In this work, one proposes a technique for expanding measured FRFs over the entire set of DOFs. This technique is based upon a modified Kidder's method and the principle of reciprocity, and it avoids the need for modal identification, as it directly uses the measured FRFs. In order to illustrate the performance of the proposed technique, a set of simulated experimental translational/force FRFs is taken as reference to estimate FRFs at unmeasured/unexcited DOFs, including those of rotational displacement due to applied moments.

3.2 Theoretical Development

As the experimental data is usually collected from few coordinates and in a limited frequency range, when the comparison of numerical and experimental data is needed, the problem of model incompleteness arises, as discussed.

Assuming a model, with N DOFs, described by the equilibrium equation,

$$\mathbf{M}\ddot{x} + \mathbf{C}\dot{x} + \mathbf{K}x = f \tag{3.1}$$

all the model reduction or data expansion techniques consider that Eq. (3.1) can be recast on its partitioned form as,

$$\begin{bmatrix} \mathbf{M}_{pp} & \mathbf{M}_{ps} \\ \mathbf{M}_{sp} & \mathbf{M}_{ss} \end{bmatrix} \begin{Bmatrix} \ddot{x}_p \\ \ddot{x}_s \end{Bmatrix} + \begin{bmatrix} \mathbf{C}_{pp} & \mathbf{C}_{ps} \\ \mathbf{C}_{sp} & \mathbf{C}_{ss} \end{bmatrix} \begin{Bmatrix} \dot{x}_p \\ \dot{x}_s \end{Bmatrix} + \begin{bmatrix} \mathbf{K}_{pp} & \mathbf{K}_{ps} \\ \mathbf{K}_{sp} & \mathbf{K}_{ss} \end{bmatrix} \begin{Bmatrix} x_p \\ x_s \end{Bmatrix} = \begin{Bmatrix} f_p \\ \mathbf{0} \end{Bmatrix} \tag{3.2}$$

where the subscripts p and s identify the primary and the secondary subsets of model coordinates or DOFs, respectively. Note that forces can only act on the primary DOFs and that this subset is composed by a given number of DOFs to be retained after model reduction ($m \leq N$).

Detailed reviews on model reduction techniques can be found in [9–11]. However, the general framework of all reduction/expansion techniques is to derive a transformation matrix $\mathbf{T}$ capable to describe the complete displacement vector by the displacement vector at primary DOFs subset, as linear combination,

$$\begin{Bmatrix} x_p \\ x_s \end{Bmatrix} = \mathbf{T}x_p \tag{3.3}$$

3.2.1 *Data Expansion Techniques*

It is known that the available model reduction techniques can be inverted and therefore used for expansion purposes [15]. Moreover, the great majority of those techniques are developed in the modal domain, and therefore the experimentally measured FRFs are not directly used, as it is evident from the formulation of the System Equivalent Reduction Expansion Process (SEREP) or the Kidder's method, present in the following subsections.

3.2.1.1 System Equivalent Reduction Expansion Process

SEREP was introduced by O'Callahan et al. [16]. This technique has the ability of being suitable to the used both for model reduction and modal data expansion purposes, with the same level of accuracy. SEREP requires the solution of the generalized eigenproblem in order to obtain the analytical/numerical modal model, defined by the natural frequencies, the mode shapes and the modal damping [2]. The computed analytical eigenvectors must then be partitioned so that,

$$\underset{(n\times 1)}{\boldsymbol{\Psi}^{(r)}} = \left\{ \begin{array}{c} \underset{(m\times 1)}{\boldsymbol{\Psi}_p} \\ \underset{((n-m)\times 1)}{\boldsymbol{\Psi}_s} \end{array} \right\}^{(r)} \tag{3.4}$$

where the superscript r refers to the rth analytical mode shape. Considering the k identified mode shapes, a $m \times k$ matrix $\boldsymbol{\Psi}_p$ is formed. The Moore-Penrose pseudo inverse of $\boldsymbol{\Psi}_p$ is used to expand the mode shapes at the primary DOFs by means of the transformation matrix $\mathbf{T_S}$, given by,

$$\underset{(n\times m)}{\mathbf{T_S}} = \underset{(n\times k)}{\boldsymbol{\Psi}} \underset{(k\times m)}{\left[\boldsymbol{\Psi}_p\right]^+} \tag{3.5}$$

Note that due to ill-conditioning issues, SEREP can only be applied if the number of primary DOFs is greater or equal than the number of identified mode shapes k. However, SEREP has the advantage of preserving the eigenfrequencies and eigenmodes used to form $\mathbf{T_S}$.

In terms of data expansion, SEREP can be used in the modal domain to expand the reference/experimental mode shapes, acquired at the primary DOFs, over the secondary ones, as,

$$\boldsymbol{\Psi}^{(r)} = \left\{ \begin{array}{c} \boldsymbol{\Psi}_p \\ \boldsymbol{\Psi}_s \end{array} \right\}^{(r)} = \mathbf{T_S}\boldsymbol{\Psi}_p^{(r)} \tag{3.6}$$

3.2.1.2 Kidder's Method

Kidder [17] proposed a spatial transformation using the analytical mass and stiffness matrices and the solution of the generalized eigenproblem, as,

$$\left(\begin{bmatrix} \mathbf{K}_{pp} & \mathbf{K}_{ps} \\ \mathbf{K}_{sp} & \mathbf{K}_{ss} \end{bmatrix} - \omega_r^2 \begin{bmatrix} \mathbf{M}_{pp} & \mathbf{M}_{ps} \\ \mathbf{M}_{sp} & \mathbf{M}_{ss} \end{bmatrix} \right) \left\{ \begin{array}{c} \boldsymbol{\Psi}_p \\ \boldsymbol{\Psi}_s \end{array} \right\}^{(r)} = \left\{ \begin{array}{c} \mathbf{0} \\ \mathbf{0} \end{array} \right\} \tag{3.7}$$

where ω_r is the rth natural frequency.

From Eq. (3.7), the expanded mode shapes are given by,

$$\boldsymbol{\Psi}_r = \left\{ \begin{array}{c} \boldsymbol{\Psi}_p \\ \boldsymbol{\Psi}_s \end{array} \right\}^{(r)} = \begin{bmatrix} \mathbf{I} \\ -\left(\mathbf{K}_{ss} - \omega_r^2\mathbf{M}_{ss}\right)^{-1}\left(\mathbf{K}_{sp} - \omega_r^2\mathbf{M}_{sp}\right) \end{bmatrix} \boldsymbol{\Psi}_p^{(r)} = \mathbf{T_K}\boldsymbol{\Psi}_p^{(r)} \tag{3.8}$$

where $\mathbf{T_K}$ is the transformation matrix of the Kidder's method for the rth mode shape.

Note that the Kidder's expansion process is achieved mode by mode, and that the experimental mode shapes can be expanded over the entire set of model coordinates, although the out-of-range modes are neglected.

3.2.1.3 Modified Kidder's Method

A modified version of the Kidder's method, applied in the frequency domain, is here proposed in order to provide an expansion technique suitable to use experimental FRFs as they are collected. The modified Kidder's method is derived from Eq. (3.2), considering a harmonic excitation, which, for an undamped structure, can be recast as,

$$\left(\begin{bmatrix} \mathbf{K}_{pp} & \mathbf{K}_{ps} \\ \mathbf{K}_{sp} & \mathbf{K}_{ss} \end{bmatrix} - \omega^2 \begin{bmatrix} \mathbf{M}_{pp} & \mathbf{M}_{ps} \\ \mathbf{M}_{sp} & \mathbf{M}_{ss} \end{bmatrix}\right) \begin{Bmatrix} \boldsymbol{H}_{pj}(\omega) \\ \boldsymbol{H}_{sj}(\omega) \end{Bmatrix} = \begin{Bmatrix} \boldsymbol{I}_j \\ \mathbf{0} \end{Bmatrix} \tag{3.9}$$

where the primary and secondary coordinates correspond to the measured and unmeasured DOFs, respectively, and I_j is a vector identifying each force location j.

From Eq. (3.9) it can be shown that a FRF vector expanded for the unmeasured DOFs is given by,

$$\boldsymbol{H}_{sj}(\omega) = -\left(\mathbf{K}_{ss} - \omega^2 \mathbf{M}_{ss}\right)^{-1} \left(\mathbf{K}_{sp} - \omega^2 \mathbf{M}_{sp}\right) \boldsymbol{H}_{pj}(\omega) \tag{3.10}$$

Note that the transformation matrix of Eq. (3.8) still applies for the direct expansion of the FRF vector, if one considers it referred to a given spectral line, instead to a mode shape. Moreover, one can formulate an extension of the modified Kidder's method for damped systems and derive the following transformation,

$$\boldsymbol{H}_{sj}(\omega) = -\left(\mathbf{K}_{ss} - \omega^2 \mathbf{M}_{ss} + i\omega \mathbf{C}_{ss}\right)^{-1} \left(\mathbf{K}_{sp} - \omega^2 \mathbf{M}_{sp} + i\omega \mathbf{C}_{sp}\right) \boldsymbol{H}_{pj}(\omega) \tag{3.11}$$

considering the viscously damped system, although the development also applies for the hysteretic damping model.

Regarding the modified Kidder's method, the expansion process is achieved spectral line by spectral line for a given experimental frequency range, for each FRF vector. As in SEREP, the transformation matrix of the proposed method relies on the analytical/numerical spatial model and its result is only the expansion of the measured FRFs due to a set of forced DOFs over the unmeasured DOFs. The process to expand the set of measured FRFs for unexcited coordinates is explained in the following subsection.

3.2.2 *FRF Expansion Techniques*

For the expansion of measured FRFs in order to attain a complete FRF matrix, a two steps expansion strategy must be addressed. Hence, two approaches are described in the following subsections, the one given by Avitabile and O'Callahan [14] and the one proposed here, the modified Kidder's method. Both expansion strategies rely on a careful definition of all the system DOFs, as there are some specificities related to the response and excitation coordinates in the response model. Note that in a spatial model one has only primary and secondary coordinates, while in a response model the primary and secondary response coordinates have their dual in the excitation coordinate set. Therefore, consider $\mathbf{H}(\omega)$, from now on referred to as $\mathbf{H}$, partitioned wrt translational t and rotational θ response DOFs and DOFs where forces f and moments τ can be applied, as

$$\mathbf{H} = \begin{bmatrix} \mathbf{H}_{tf} & \mathbf{H}_{t\tau} \\ \mathbf{H}_{\theta f} & \mathbf{H}_{\theta\tau} \end{bmatrix} \tag{3.12}$$

In practice, usually, one can only measure a subset of t and excite part of f, and often these subsets are not equal. Note that the possibility of measuring rotational DOFs or applying pure moments is not contemplated here. Hence, if one chooses to organize the subsets of t and f type DOFs, by their nature, one can define four types of DOFs, as schematically represented in Fig. 3.1. These types of DOFs are respectively: A for measured/forced DOFs; B for measured/unforced DOFs; E for unmeasured/forced DOFs; and U for unmeasured/unforced DOFs. Regarding this classification, one may rewrite Eq. (3.12) in a more detailed way, as

$$\mathbf{H} = \begin{bmatrix} \mathbf{H}_{AA} & \mathbf{H}_{AB} & \mathbf{H}_{AE} & \mathbf{H}_{AU} & \mathbf{H}_{A\tau} \\ \mathbf{H}_{BA} & \mathbf{H}_{BB} & \mathbf{H}_{BE} & \mathbf{H}_{BU} & \mathbf{H}_{B\tau} \\ \mathbf{H}_{EA} & \mathbf{H}_{EB} & \mathbf{H}_{EE} & \mathbf{H}_{EU} & \mathbf{H}_{E\tau} \\ \mathbf{H}_{UA} & \mathbf{H}_{UB} & \mathbf{H}_{UE} & \mathbf{H}_{UU} & \mathbf{H}_{U\tau} \\ \mathbf{H}_{\theta A} & \mathbf{H}_{\theta B} & \mathbf{H}_{\theta E} & \mathbf{H}_{\theta U} & \mathbf{H}_{\theta\tau} \end{bmatrix} \tag{3.13}$$

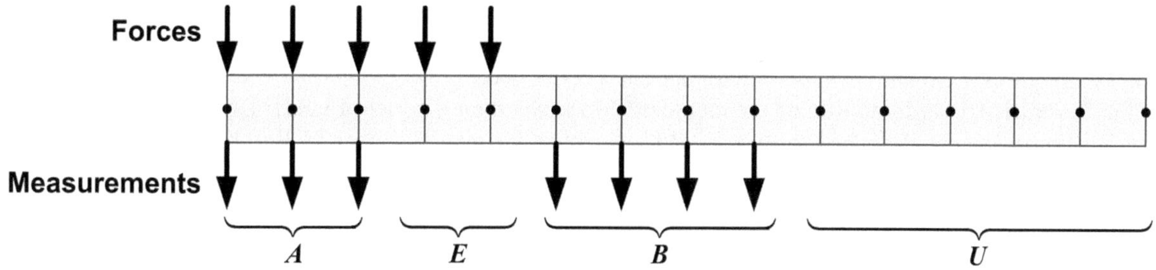

Fig. 3.1 Schematic identification of the different types of DOFs

3.2.2.1 FRF Expansion Using SEREP

The expansion scheme proposed by Avitabile and O'Callahan [14] is based on the modal expansion of the FRF matrix given by,

$$\mathbf{H} = \sum_{r=1}^{k} \left[\frac{\boldsymbol{\phi}_r \boldsymbol{\phi}_r^{\mathrm{T}}}{i\omega - s_r} + \frac{\boldsymbol{\phi}_r^* \boldsymbol{\phi}_r^{\mathrm{H}}}{i\omega - s_r^*} \right] + \sum_{r=k+1}^{n} \left[\frac{\boldsymbol{\phi}_r \boldsymbol{\phi}_r^{\mathrm{T}} + \boldsymbol{\phi}_r^* \boldsymbol{\phi}_r^{\mathrm{H}}}{\omega_r^2} \right] \tag{3.14}$$

with

$$s_r = -\omega_r \xi_r + i\omega_r \sqrt{1 - \xi_r^2} \tag{3.15}$$

where ϕ_r is the rth mass-normalized complex mode shape and ξ_r is the modal damping ratio associated with it. Therefore, SEREP is used to expand the mode shapes over the unmeasured DOFs and the residuals expansion is achieved by a dynamic residual expansion at a given spectral line [18].

Note that this expansion strategy relies on a previous modal parameters identification from the measured FRFs, in order to reconstruct the already measured FRFs. Furthermore, as the identified mode shapes are measured at the primary DOFs, **H** computed using Eq. (3.16) is actually the portion of **H** related to those primary DOFs, and therefore a squared matrix ($\mathbf{H}_{pp}$ in Eq. (3.16)).

Finally, this expansion strategy involves a separated expansion of the reconstructed FRFs, using SEREP, and of the residuals, using one of two schemes based on a dynamic expansion at a given spectral line, in order to estimate the FRFs at the secondary coordinates. A second expansion step is then performed but now for a set of reference coordinates related to the excited DOFs, with the aim to expand the known FRFs over the set of unexcited DOFs.

This expansion process is here illustrated, regarding Eq. (3.13) recast as

$$\mathbf{H} = \begin{bmatrix} \mathbf{H}_{pp} & \mathbf{H}_{pE} & \mathbf{H}_{pU} & \mathbf{H}_{p\tau} \\ \mathbf{H}_{Ep} & \mathbf{H}_{EE} & \mathbf{H}_{EU} & \mathbf{H}_{E\tau} \\ \mathbf{H}_{Up} & \mathbf{H}_{UE} & \mathbf{H}_{UU} & \mathbf{H}_{U\tau} \\ \mathbf{H}_{\theta p} & \mathbf{H}_{\theta E} & \mathbf{H}_{\theta U} & \mathbf{H}_{\theta\tau} \end{bmatrix} \tag{3.16}$$

where p corresponds to the sets of measured DOFs, A and B in Eq. (3.13). Hence, considering a set of measured FRFs taken at the sets of measured and of excited coordinates, after identifying the modal parameters, the reconstructed FRFs correspond to the submatrix $\mathbf{H}_{pp}$, which is expanded to obtain $\mathbf{H}_{Ep}$, $\mathbf{H}_{Up}$ and $\mathbf{H}_{\theta p}$. The second step of expansion process is then initiated with the selection of all FRFs related to unmeasured coordinates from $\mathbf{H}_{\bullet p}$, which are transposed to form the submatrix $\left[\mathbf{H}_{pE} \; \mathbf{H}_{pU} \; \mathbf{H}_{p\tau} \right]$. Finally, this submatrix is expanded using the above described techniques in order to complete the unknown part of **H**.

3.2.2.2 FRF Expansion Using the Modified Kidder's Method

Regarding the expansion of measured FRFs using the modified Kidder's method, it is possible to expand a set of measured FRFs avoiding the need of modal identification. Moreover, in terms of spectral incompleteness, the method is only limited by the acquired spectral lines and not by the number of identified modes, which eliminates residuals computation, and therefore their expansion. However, the computational effort of this expansion strategy is greater than the one of the SEREP based approach, as it requires to expand each FRF vector at each spectral line.

Recalling Eq. (3.13), in order to expand the measured FRFs over the entire set of model DOFs, one considers a similar expansion scheme as the one described in Sect. 3.2.2.1, using now the relations given by Eqs. (3.10) and (3.11). Therefore, it is convenient to assemble the known FRFs on the left upper position of **H** before starting the expansion process.

According to the example that is given in Fig. 3.1, one needs to swap the 2nd and 3rd columns of Eq. (3.13), leading to Eq. (3.17).

$$\mathbf{H} = \begin{bmatrix} \mathbf{H}_{AA} & \mathbf{H}_{AE} & \mathbf{H}_{AB} & \mathbf{H}_{AU} & \mathbf{H}_{A\tau} \\ \mathbf{H}_{BA} & \mathbf{H}_{BE} & \mathbf{H}_{BB} & \mathbf{H}_{BU} & \mathbf{H}_{B\tau} \\ \mathbf{H}_{EA} & \mathbf{H}_{EE} & \mathbf{H}_{EB} & \mathbf{H}_{EU} & \mathbf{H}_{E\tau} \\ \mathbf{H}_{UA} & \mathbf{H}_{UE} & \mathbf{H}_{UB} & \mathbf{H}_{UU} & \mathbf{H}_{U\tau} \\ \mathbf{H}_{\theta A} & \mathbf{H}_{\theta E} & \mathbf{H}_{\theta B} & \mathbf{H}_{\theta U} & \mathbf{H}_{\theta\tau} \end{bmatrix} \tag{3.17}$$

Now one must expand the measured FRFs (in blue) to obtain the FRFs in red, representing the unmeasured coordinates due to the applied excitation set, using the modified Kidder's method.

As it can be noticed, Eq. (3.17) is not symmetric. At this stage, it is advisable to have a symmetric matrix **H**, which is achieved by swapping the 2nd and 3rd rows of Eq. (3.17). Doing this, one obtains Eq. (3.18), where one can observe that symmetry is ensured. In fact, one now have the submatrices in **black**, magenta and green, because the green partition is the transposed of magenta one.

$$\mathbf{H} = \begin{bmatrix} \mathbf{H}_{AA} & \mathbf{H}_{AE} & \mathbf{H}_{AB} & \mathbf{H}_{AU} & \mathbf{H}_{A\tau} \\ \mathbf{H}_{EA} & \mathbf{H}_{EE} & \mathbf{H}_{EB} & \mathbf{H}_{EU} & \mathbf{H}_{E\tau} \\ \mathbf{H}_{BA} & \mathbf{H}_{BE} & \mathbf{H}_{BB} & \mathbf{H}_{BU} & \mathbf{H}_{B\tau} \\ \mathbf{H}_{UA} & \mathbf{H}_{UE} & \mathbf{H}_{UB} & \mathbf{H}_{UU} & \mathbf{H}_{U\tau} \\ \mathbf{H}_{\theta A} & \mathbf{H}_{\theta E} & \mathbf{H}_{\theta B} & \mathbf{H}_{\theta U} & \mathbf{H}_{\theta\tau} \end{bmatrix} \tag{3.18}$$

Finally, one shall expand the green partition of FRFs of Eq. (3.18) to obtain the remaining unknown partition of **H** (in purple);

$$\mathbf{H} = \begin{bmatrix} \mathbf{H}_{AA} & \mathbf{H}_{AE} & \mathbf{H}_{AB} & \mathbf{H}_{AU} & \mathbf{H}_{A\tau} \\ \mathbf{H}_{EA} & \mathbf{H}_{EE} & \mathbf{H}_{EB} & \mathbf{H}_{EU} & \mathbf{H}_{E\tau} \\ \mathbf{H}_{BA} & \mathbf{H}_{BE} & \mathbf{H}_{BB} & \mathbf{H}_{BU} & \mathbf{H}_{B\tau} \\ \mathbf{H}_{UA} & \mathbf{H}_{UE} & \mathbf{H}_{UB} & \mathbf{H}_{UU} & \mathbf{H}_{U\tau} \\ \mathbf{H}_{\theta A} & \mathbf{H}_{\theta E} & \mathbf{H}_{\theta B} & \mathbf{H}_{\theta U} & \mathbf{H}_{\theta\tau} \end{bmatrix} \tag{3.19}$$

3.3 Numerical Simulations

Being the formulation and implementation of the two expansion strategies presented, a series of numerical simulations were prepared and the results are discussed in Sect. 3.4. In order to assess the performance of the proposed strategy, the results given in [14] are here used and referred as benchmark. As in the reference paper, one generates simulated experimental data for a cantilever Bernoulli-Euler beam, considering 14 beam elements with 2 DOFs per node and thus a total of 28 DOFs. Note that the odd DOFs are related to translational DOFs and the even ones to rotational DOFs. Hence, simulated experimental data are obtained for a set of measured translations due to a set of applied forces. In [14] it is considered just one experimental setup or configuration, where the set of measured DOFs is coincident with the one of forced DOFs. This experimental configuration is here identified as Configuration 1 (Fig. 3.2). Note that this experimental configuration is rather inflexible and leads to the idea that one must force every measured coordinate, which is not the case, as shown in Sect. 3.2.2.2. Therefore, Configuration 2 is thought as a general test configuration (Fig. 3.2), as it uses less measurement points than excitation locations and thus it is suitable to be used with experimental data collected with hammer testing. While the results of Configuration 2 only reflect this test possibility, regarding the FRF reciprocity, a configuration where one considers more measurement points than excitation locations is also possible, being this configuration the usual of experimental tests with shaker.

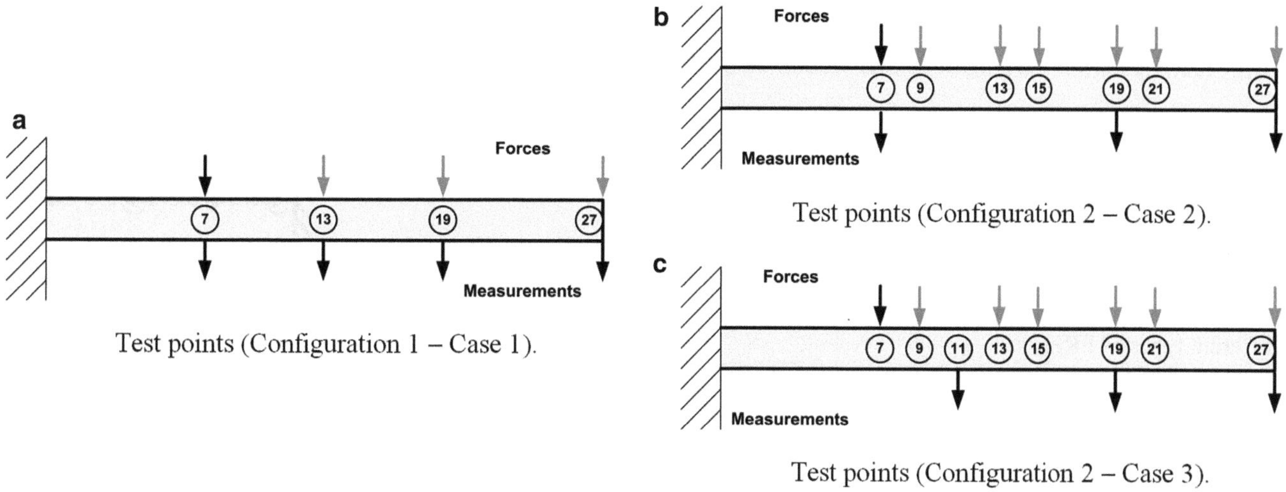

Fig. 3.2 Simulated test points on the cantilever beam (*arrows* in *grey* represent all the possible force locations)

Results were produced for several combinations of number and locations of both measurement and excitation points, although the following set of test cases is considered (Fig. 3.2):

Configuration 1

Case 1: measured DOFs: $\left[7\ 13\ 19\ 27\right]$; forced DOFs: $\left[7\ 13\ 19\ 27\right]$;

$$\therefore A = \left[7\ 13\ 19\ 27\right] \quad \wedge \quad B = E = \varnothing$$

Configuration 2

Case 2: measured DOFs: $\left[7\ 19\ 27\right]$; forced DOFs: $\left[7\ 9\ 13\ 15\ 19\ 21\ 27\right]$;

$$\therefore A = \left[7\ 19\ 27\right] \quad \wedge \quad B = \varnothing \quad \wedge \quad E = \left[7\ 9\ 13\ 15\ 21\right]$$

Case 3: measured DOFs: $\left[11\ \ 19\ 27\right]$; forced DOFs: $\left[7\ 9\ 13\ 15\ 19\ 21\ 27\right]$;

$$\therefore A = \left[19\ 27\right] \quad \wedge \quad B = [11] \quad \wedge \quad E = \left[7\ 9\ 13\ 15\ 21\right]$$

For each case a set of results is presented and discussed in Sect. 3.4. Note that for all the test cases one presents the set of expanded FRFs used in the benchmark work [14], namely:

$H_{23,27}$ (FRF type: *UA* – unmeasured translation/applied force):

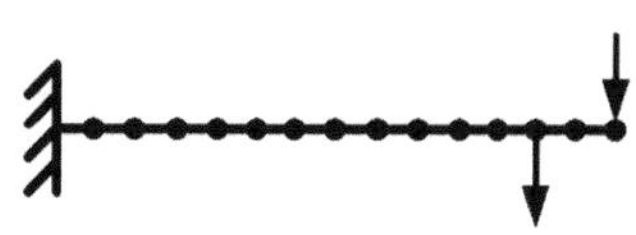

$H_{27,28}$ (FRF type: $A\,\tau$ – measured translation/applied moment):

$H_{28,28}$ (FRF type: $\theta\,\tau$ – unmeasured rotation/applied moment at beam tip):

Additionally, one also shows two more expanded FRFs:

$H_{21,15}$ (FRF type: UU – unmeasured translation/non-applied force):

$H_{2,16}$ (FRF type: $\theta\,\tau$ – unmeasured rotation near clamped end/applied moment at midspan):

The choice of the presented set of expanded FRFs concerns the specificities of each considered FRF, as they are part of different types of DOFs. Hence, one has the possibility to assess the quality of the proposed expansion process, regarding different types of FRFs.

3.4 Results And Discussion

In the current section, one can find results for all the test cases described in Sect. 3.3, including results for the expansion of damped FRFs and for noisy simulated data. The first set of results shows the outcome of the expansion process when applied to all the test cases, neglecting damping and the effect of noise in the simulated data (Figs. 3.3, 3.4, and 3.5). In Fig. 3.6, one can evaluate the robustness to noise of the expansion process, for all the test cases, for FRFs which are related to rotational DOFs, in terms of both measured and excited coordinates ($H_{2,16}$) and just in terms of the excited coordinate ($H_{27,28}$). Figure 3.7 shows results for $H_{2,16}$ ($\theta\,\tau$ type) when a proportionally damped system is considered, using $\mathbf{C} = 5 \times 10^{-5}\mathbf{K}$, for all the test cases. Note that Figs. 3.3, 3.4, 3.5, 3.6, and 3.7 show the exact FRF along with the expanded FRFs for the proposed method and the benchmark one (this with and without residuals compensation).

Globally, from the presented results, one can confirm the expected behavior of the benchmark method, regarding its spectral limitation. Furthermore, one can notice an overestimation of the magnitude of the expanded FRFs by the benchmark method, namely when residuals are considered. However, the inclusion of residuals greatly improves the quality of the expanded FRFs near antiresonances. Good examples of this are given by the results related to $H_{28,28}$ and $H_{27,28}$.

The accuracy of the expansion results obtained by the proposed method is noticeable. As it is evident in all the results, the expanded FRFs related to both translational and rotational DOFs are in very good agreement with the corresponding reference FRFs, especially near resonances. Additionally, one may conclude that the proposed method also provides high quality results in the regions away from the resonances. The results for $H_{2,16}$ should be pointed out as an example of the accuracy of the expanded FRFs over the entire spectral range, even when one addresses the expansion of unmeasured rotations due to applied moments.

However, for specific test configurations, the proposed method can generate spurious solutions due to the expansion process, as it can be seen in Fig. 3.4c–e (Case 2). Note that even for Case 2 the accuracy of $H_{23,27}$ and $H_{27,28}$ is very good, although this is expected, as the DOF number 27 is a measured/forced DOF (A type).

The sensitivity of the expanded FRFs to the considered test configurations can be assessed by the analysis of Figs. 3.3, 3.4, and 3.5. These results may suggest that for a given structure one can observe different levels of accuracy for the expanded results just by placing a sensor or an exciter in a different location, having higher influence in the regions near antiresonances. For the beam structure considered, the more accurate results are attained in Case 1, followed by the results in Case 3, which apparently indicates that it is better to acquire measurements with almost equally spaced sensors. However, it cannot be stated as a general conclusion without further studies.

Figure 3.6 shows results for all the test cases in the presence of noisy data. Note that the reference FRFs are exact and thus noise-free. The expanded results seam very robust to noise, especially away from the antiresonances. Note that one has selected just two of the considered FRFs, namely $H_{27,28}$ and $H_{2,16}$, due to the fact that these FRFs involve rotational DOFs, usually the more difficult to estimate and more susceptible to be inaccurate.

Finally, the results for the expansion of FRFs considering a damped system can be found in Fig. 3.7. As it can be seen the results are in general quite good. Only the results for $H_{2,16}$ are shown, for simplicity, because the results for the rest of the considered FRFs attain similar accuracy.

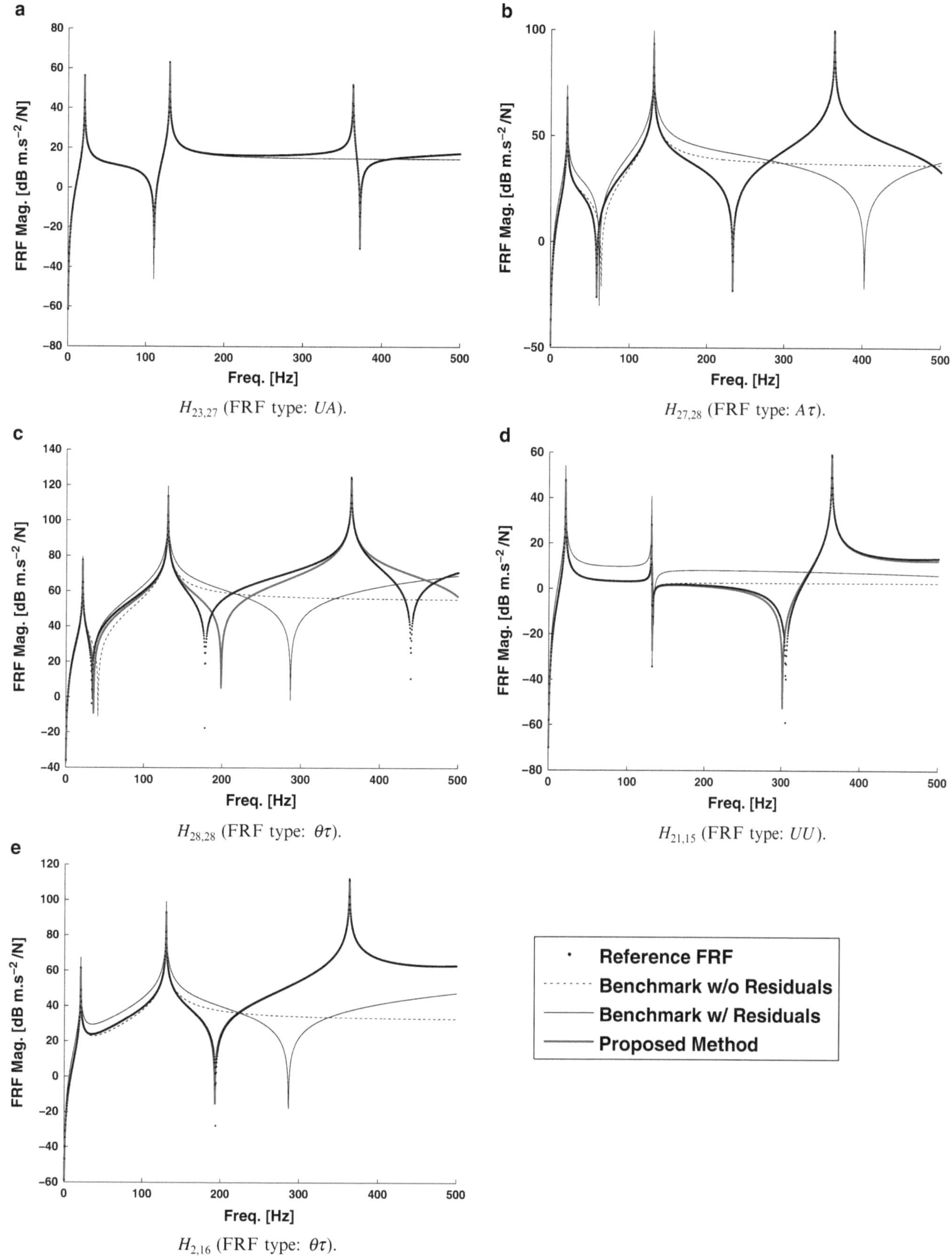

Fig. 3.3 Expanded FRFs for the reference FRFs set of Case 1

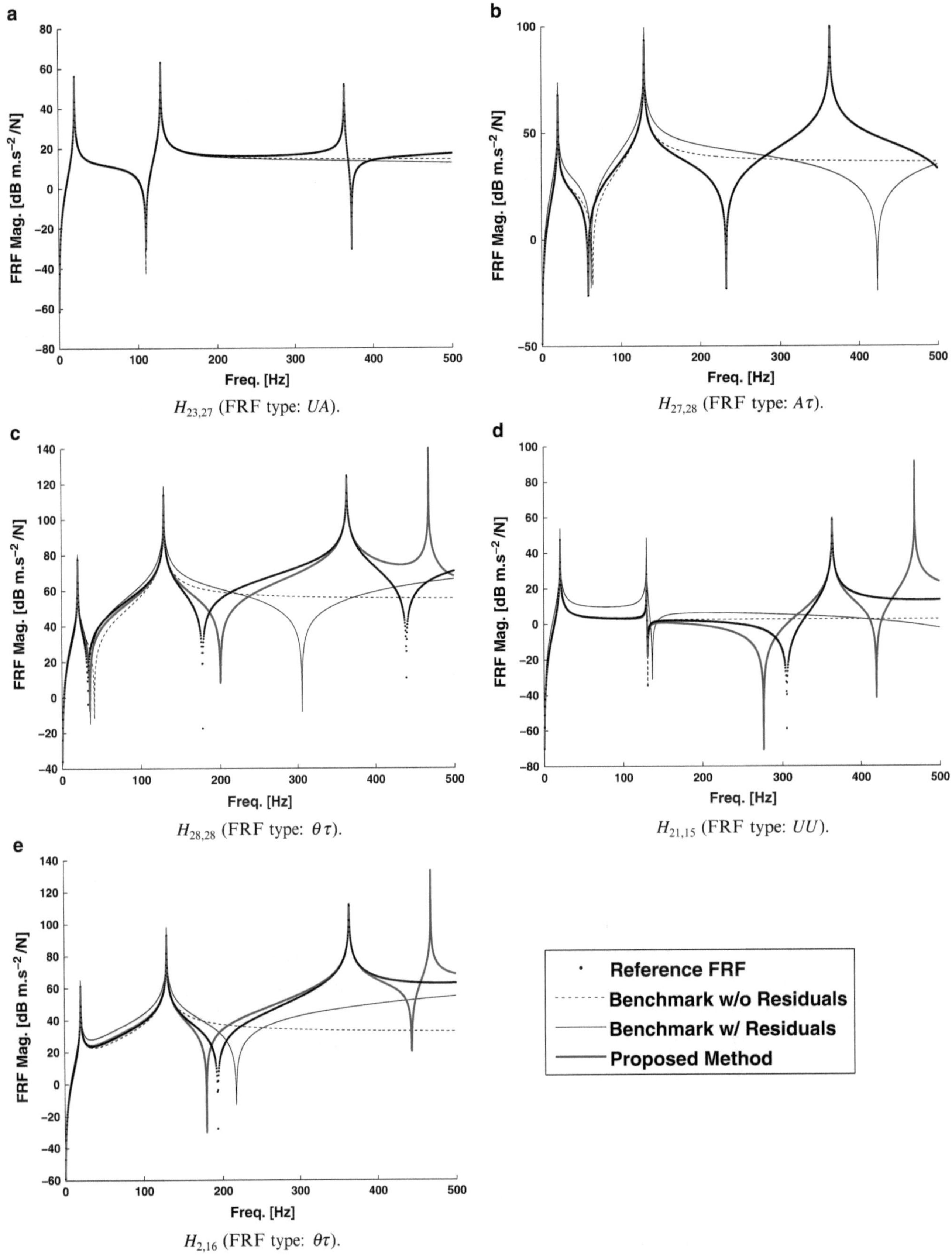

a $H_{23,27}$ (FRF type: UA).

b $H_{27,28}$ (FRF type: $A\tau$).

c $H_{28,28}$ (FRF type: $\theta\tau$).

d $H_{21,15}$ (FRF type: UU).

e $H_{2,16}$ (FRF type: $\theta\tau$).

Fig. 3.4 Expanded FRFs for the reference FRFs set of Case 2

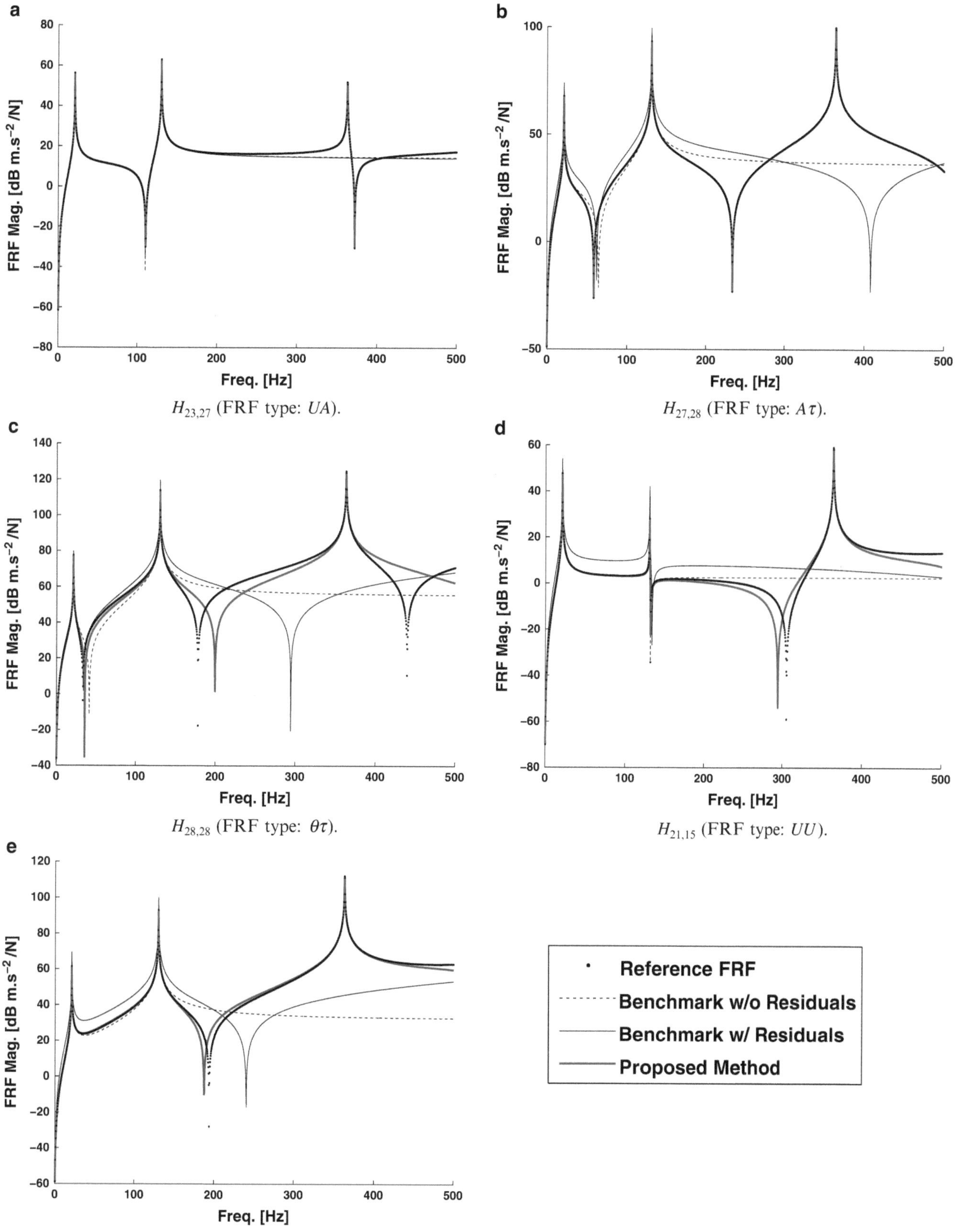

(a) $H_{23,27}$ (FRF type: UA). (b) $H_{27,28}$ (FRF type: $A\tau$). (c) $H_{28,28}$ (FRF type: $\theta\tau$). (d) $H_{21,15}$ (FRF type: UU). (e) $H_{2,16}$ (FRF type: $\theta\tau$).

Fig. 3.5 Expanded FRFs for the reference FRFs set of Case 3

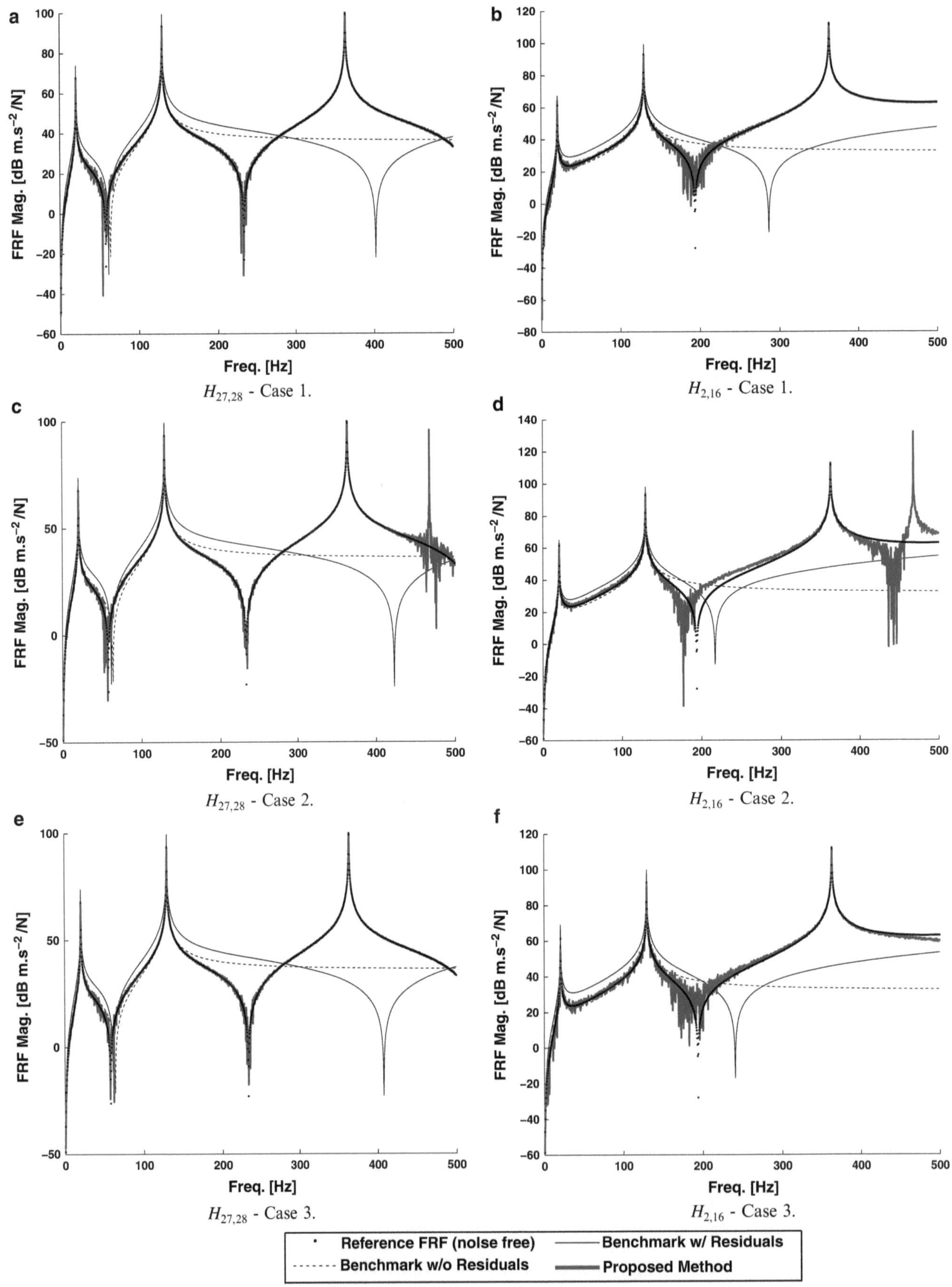

Fig. 3.6 Expanded FRFs for different cases, considering 5% added noise

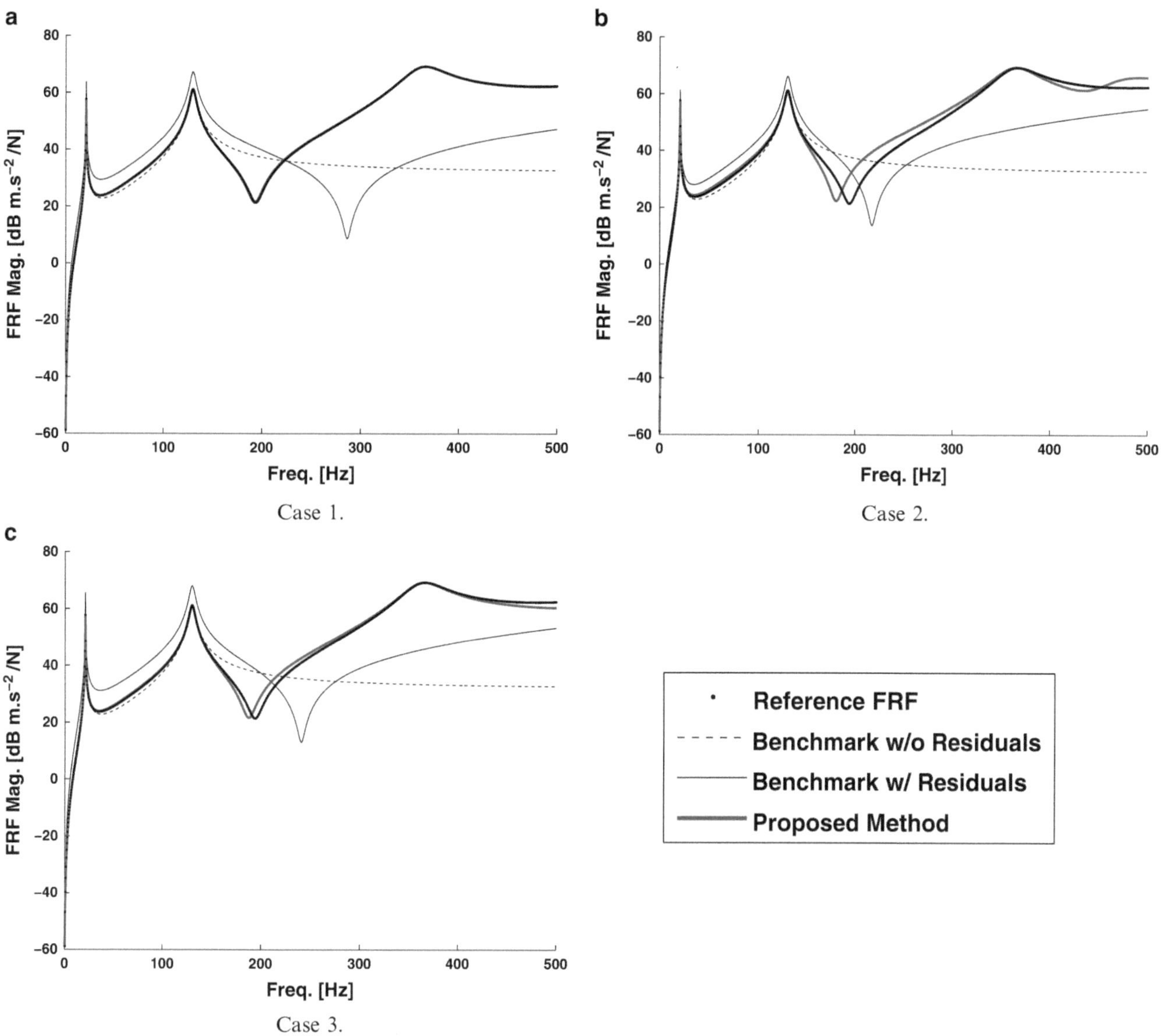

Fig. 3.7 Expanded $H_{2,16}$ for different cases, considering proportional damping ($\mathbf{C} = 5 \times 10^{-5}\mathbf{K}$)

3.5 Conclusions

An expansion process for the response model in the frequency domain is proposed and its performance discussed through a set of numerical test cases. The proposed technique is based upon a modified Kidder's method and it avoids the need for modal identification, as it uses the measured FRFs directly. If compared to the benchmark expansion strategy, the proposed process presents a very good accuracy regarding the expanded FRFs, although it requires a greater computational effort, as it works FRF vector by FRF vector, spectral line by spectral line. The robustness to noise is also addressed and consequently its further application, regarding experimental data collected on a physical structure, is expected to be quite successful. However, from the experimental point of view, to assess the quality of the expanded data is a difficult challenge, as it is difficult to acquire rotational FRFs due to an applied force or pure moment. An issue that also needs to be carefully analyzed is the correctness of the theoretical model, as the transformation matrix relies on it, although if one considers a previous model updating step, it should perform accurately.

Acknowledgments The authors acknowledge the support of the Portuguese Foundation for Science and Technology, FCT, under the projects PEst/OE/EME/LA0022/2011 and PTDC/ATP-AQI/5355/2012, and the PhD grant SFRH/BD/44696/2008.

References

1. Friswell MI, Mottershead JE (1995) Finite element model updating in structural dynamics, solid mechanics and its applications. Kluwer, Dordrecht
2. Maia NMM, Silva JMM (1997) Theoretical and experimental modal analysis, Mechanical engineering series. Research Studies Press, Baldock, Hertfordshire, England
3. Duarte MLM, Ewins DJ (2000) Rotational degrees of freedom for structural coupling analysis via finite-difference technique with residual compensation. Mech Syst Signal Process 14(2):205–227
4. Batista F, Maia NMM (2012) Estimation of unmeasured frequency response functions. In: ICSV19 – 19th international congress on sound and vibration, Vilnius
5. Montalvao D, Ribeiro AMR, Maia NMM, Silva JMM (2004) Estimation of the rotational terms of the dynamic response matrix. Shock Vib 11(3–4):333–350
6. Mottershead JE, Kyprianou A, Ouyang H (2005) Structural modification. Part 1: rotational receptances. J Sound Vib 284(1–2):249–265
7. Silva TAN, Maia NMM, Barbosa JI (2012) A model updating technique based on FRFs for damped structures. In: Sas P (ed) ISMA 2012 international conference on noise and vibration engineering. Katholieke Universiteit Leuven, Leuven
8. Urgueira AP, Almeida RA, Maia NMM (2011) On the use of the transmissibility concept for the evaluation of frequency response functions. Mech Syst Signal Process 25(3):940–951
9. Qu Z-Q (2004) Model order reduction techniques. Springer, London
10. Koutsovasilis P (2009) Model order reduction in structural mechanics: coupling the rigid and elastic multi body dynamics. Ph.D. thesis, Technische Universität Dresden
11. Besselink B, Tabak U, Lutowska A, van de Wouw N, Nijmeijer H, Rixen DJ, Hochstenbach M, Schilders W (2013) A comparison of model reduction techniques from structural dynamics, numerical mathematics and systems and control. J Sound Vib 332(19):4403–4422
12. Bregant L, Sanderson M (2000) Rotational degrees of freedom: an historical overview on techniques and methods. In: Sas P (ed) ISMA 25 international conference on noise and vibration engineering. Katholieke Universiteit Leuven, Leuven
13. Bregant L, Casagrande D (2000) Rotational degrees of freedom data synthesis based on force excitation. In: Sas P (ed) ISMA 25 international conference on noise and vibration engineering. Katholieke Universiteit Leuven, Leuven
14. Avitabile P, O'Callahan J (2003) Frequency response function expansion for unmeasured translation and rotation dofs for impedance modelling applications. Mech Syst Signal Process 17(4):723–745
15. Balmès E (1999) Sensors, degrees of freedom, and generalized modeshape expansion methods. In: IMAC XVII – 17th international modal analysis conference, SEM, Kissimee, Florida, USA
16. O'Callahan J, Avitabile P, Riemer R (1989) System equivalent reduction expansion process (SEREP). In: IMAC VII – 7th international modal analysis conference, SEM, Kissimee, Florida, USA, pp 29–37
17. Kidder RL (1973) Reduction of structural frequency equations. AIAA J 11(6):892
18. Paz M (1984) Dynamic condensation method. Am Inst Aeronaut Astronaut J 22(5):724–727

Chapter 4
Estimation of Spatial Distribution of Disturbances

Yalcin Bulut, Omer F. Usluogullari, and Ahmet Temugan

Abstract The information of spatial distribution of unmeasured disturbances is utilized in controller and observer design. In reality, due to the complexity in the systems, this information is seldom known a priori. Our focus in this study is to estimate the spatial distribution of disturbances from available measurements using a correlations approach that is developed in Kalman filter theory. In this approach one begins by "guessing" a filter gain and then the approach calculates the disturbance covariance matrices from analysis of the resulting innovations. This paper reviews the innovations correlations approach and examines its merit to localize the disturbances.

Keywords Disturbance localization • Process noise • Measurement noise • Kalman filter

4.1 Introduction

The basic idea in estimation theory is to obtain approximations of the true response by using information from a model and from any available measurements. The mathematical structure used to perform estimation is known as an observer. The optimal observer for linear systems subjected to broad band disturbances is the Kalman Filter (KF), [1]. In the classical Kalman filter theory, one of the key assumptions is that a priori knowledge of the spatial distribution of distriburbances and noise covariance matrices are known without uncertainty. In reality, due to the complexity in the systems, this information is seldom known a priori. The objective of this study is to estimate the spatial distribution of distriburbances and the noise covariance matrices using correlations approaches. The two correlations approaches that have received most attention in the noise covariance estimation problem are based on: (1) correlations of the innovation sequence and (2) correlations of the output. In the innovations approach one begins by "guessing" a filter gain and then the approach calculates the noise covariance matrices from analysis of the resulting innovations. The correlations approaches to estimate the covariance matrices of process and measurement noise for Kalman Filtering from the measured data began soon after introduction of the filter. Perhaps the most widely quoted strategies to carry out the estimation of noise covariance matrices are due to Mehra [2] and the subsequent paper by Carew and Bellanger [3]. A noteworthy contribution from this early work is the contributions by Neethling and Young [4], who suggested some computational adjustments that could be used to improve accuracy. Recently, some other contributions to the Mehra's approach on the estimation of noise covariance matrices are presented. Odelson, Rajamani and Rawlings applied the suggestions of Neethling and Young's on Mehra's approach and used the vector operator solution for state error covariance Riccatti equation of suboptimal filter, [5]. Akesson et al. extended their work for mutually correlated process and measurement noise case, [6]. Bulut, Vines-Cavanaugh and Bernal compared the performance of the output and innovations correlations approaches to estimate noise covariance matrices, [7].

The paper is organized as follows: the next section provides a brief summary of the KF particularized to a time invariant linear system with stationary disturbances (which is a condition we have implicitly assumed throughout the previous discussion). The following section reviews the innovations correlations approach for disturbance localization and the paper concludes with a numerical example.

Y. Bulut (✉) • O.F. Usluogullari • A. Temugan
Civil Engineering Department, Turgut Ozal University, Ankara, Turkey
e-mail: ybulut@turgutozal.edu.tr

© The Society for Experimental Mechanics, Inc. 2015

R. Allemang (ed.), *Special Topics in Structural Dynamics, Volume 6*, Conference Proceedings of the Society for Experimental Mechanics Series, DOI 10.1007/978-3-319-15048-2_4

4.2 The Kalman Filter

Consider a time invariant linear system with unmeasured disturbances $w(t)$ and available measurements $y(t)$ that are linearly related to the state vector $x(t)$. The system has the following description in sampled time

$$x_{k+1} = Ax_k + Bw_k \tag{4.1}$$

$$y_k = Cx_k + v_k \tag{4.2}$$

where $A \ \varepsilon \ \mathsf{R}^{nxn}$, $B \ \varepsilon \ \mathsf{R}^{nxr}$ and $C \ \varepsilon \ \mathsf{R}^{mxn}$ are the transition, input to state, and state to output matrices, $y_k \ \varepsilon \ \mathsf{R}^{mx1}$ is the measurement vector and $x_k \ \varepsilon \ \mathsf{R}^{nx1}$ is the state. The sequence $w_k \ \varepsilon \ \mathsf{R}^{rx1}$ is the disturbance known as the process noise and $v_k \ \varepsilon \ \mathsf{R}^{mx1}$ is the measurement noise. In the treatment here, it is assumed that these are mutually correlated Gaussian stationary white noise sequences with zero mean and known covariance matrices, namely

$$E(w_k) = 0 \tag{4.3}$$

$$E(v_k) = 0 \tag{4.4}$$

and

$$E\left(w_k w_j^T\right) = Q\delta_{kj} \tag{4.5}$$

$$E\left(v_k v_j^T\right) = R\delta_{kj} \tag{4.6}$$

$$E\left(w_k v_j^T\right) = S\delta_{kj} \tag{4.7}$$

where δ_{kj} denotes the Kronecker delta function, and $E(\cdot)$ denotes expectation. Q and R are covariance matrices of the process and measurement noise and S is cross-covariance between them. For the system in Eqs. 4.1 and 4.2, the KF estimate of the state can be computed from

$$\widehat{x}_{k+1} = A\widehat{x}_k + K\left(y_k - C\widehat{x}_k\right) \tag{4.8}$$

where $\widehat{x}_k$ is the estimate of x_k and K is the (steady state) KF gain that can be expressed in a number of alternative ways, a popular one is

$$K = \left(APC^T + BS\right)\left(CPC^T + R\right)^{-1} \tag{4.9}$$

where P, the steady state covariance of the state error, is the solution of the Riccati equation

$$P = APA^T - \left(APC^T + BS\right)\left(CPC^T + R\right)^{-1}\left(APC^T + BS\right)^T + BQB^T \tag{4.10}$$

The KF provides an estimate of the state for which trace of is minimal. The difference between measured and estimated output, namely $e_k = y_k - C\widehat{x}_k$ in Eq. 4.8 is known as innovations sequence of the filter which is a white process. The filter is initialized as follows

$$\widehat{x}_0 = E(x_0) \tag{4.11}$$

4.3 Innovations Correlations Approach

We begin with the expression for the covariance function of the innovation process (e_k) for any stable observer with gain K_0. As initially shown by Mehra [2] this function is

$$\mathsf{L}_j = C\overline{P}C^T + R \qquad j = 0 \tag{4.12}$$

$$\mathsf{L}_j = C\overline{A}^j\overline{P}C^T + C\overline{A}^{j-1}B\overline{S} - C\overline{A}^{j-1}K_0R \qquad j > 0 \tag{4.13}$$

where $\overline{P}$ the covariance of the state error in the steady state, is the solution of the Riccati equation

$$\overline{P} = \overline{APA} + K_0RK_0^T + BQB^T - K_0SB^T - BS^TK_0^T \tag{4.14}$$

and

$$\overline{A} = A - K_0C \tag{4.15}$$

Applying *vec* operator to both sides of the auto-correlation function of the innovations in Eqs. 4.12, 4.13 one obtains

$$vec\left(\mathsf{L}_j\right) = (C \otimes C)\, vec\left(\overline{P}\right) + vec(R) \qquad j = 0 \tag{4.16}$$

$$vec\left(\mathsf{L}_j\right) = \left(C\overline{A}^j \otimes C\right) vec\left(\overline{P}\right) + \left(B^T \otimes C\overline{A}^{j-1}\right) vec(S) - \left(I \otimes C\overline{A}^{j-1}K_0\right) vect(R) \qquad j > 0 \tag{4.17}$$

and applying vec operator to error covariance equation in Eq. 4.14, one has

$$vec\left(\overline{P}\right) = [I-(\overline{A}\otimes\overline{A})]^{-1}\left[(K_0 \otimes K_0)\, vec(R) + B \otimes B vec(Q) - (B \otimes K_0)\, vec(S) - (K_0 \otimes B)\, vec\left(S^T\right)\right] \tag{4.18}$$

Substituting Eq. 4.25 into Eqs. 4.23 and 4.24, and adding the terms related to S^T to the terms related to S and canceling S^T, one finds

$$vec\left(\mathsf{L}_j\right) = \left[h_j^Q \; h_j^S \; h_j^R\right]\begin{bmatrix} vec(Q) \\ vec(S) \\ vec(R) \end{bmatrix} \tag{4.19}$$

where

$$h_j^Q = (C \otimes C)\,[I-(\overline{A}\otimes\overline{A})]^{-1}\,(B \otimes B) \qquad j = 0 \tag{4.20}$$

$$h_j^Q = \left(C \otimes C\overline{A}^j\right)[I-(\overline{A}\otimes\overline{A})]^{-1}\,(B \otimes B) \qquad j > 0 \tag{4.21}$$

$$h_j^S = -2I\,(C \otimes C)\,[I-(\overline{A}\otimes\overline{A})]^{-1}\,(B \otimes K_0) \qquad j = 0 \tag{4.22}$$

$$h_j^S = \left(B^T \otimes C\overline{A}^{j-1}\right) - 2I\left[\left(C \otimes C\overline{A}^j\right)\left[I - \left(\overline{A}\otimes\overline{A}\right)\right]^{-1}(B \otimes K_0)\right] \qquad j > 0 \tag{4.23}$$

$$h_j^R = (C \otimes C)\,[I-(\overline{A}\otimes\overline{A})]^{-1}\,(K_0 \otimes K_0) + I \qquad j = 0 \tag{4.24}$$

$$h_j^R = \left(C \otimes C\overline{A}^j\right)[I-(\overline{A}\otimes\overline{A})]^{-1}\,(K_0 \otimes K_0) - \left(I \otimes C\overline{A}^{j-1}K_0\right) \qquad j > 0 \tag{4.25}$$

Listing explicitly the correlation functions in Eq. 4.26 for lags $j = 1, 2, ..p$ and writing in matrix form one has

$$HX = L \tag{4.26}$$

where

$$H = \begin{bmatrix} h_0^Q \; h_0^S \; h_0^R \\ h_1^Q \; h_1^S \; h_1^R \\ h_2^Q \; h_2^S \; h_2^R \\ \vdots \;\; \vdots \;\; \vdots \\ h_p^Q \; h_p^S \; h_p^R \end{bmatrix}, L = \begin{bmatrix} vec\left(\mathsf{L}_0\right) \\ vec\left(\mathsf{L}_1\right) \\ vec\left(\mathsf{L}_2\right) \\ \vdots \\ vec\left(\mathsf{L}_p\right) \end{bmatrix}, X = \begin{bmatrix} vec(Q) \\ vec(S) \\ vec(R) \end{bmatrix} \tag{4.27}$$

Estimates of Q, S and R can be obtained from Eq. 4.26. From its inspection, one finds that H has dimensions $m^2px\left(r^2+m^2+mr\right)$. The sufficient condition for the uniqueness of the solution of Eq. 4.26 is defined as follows in the general case; the number of unknown parameters in Q and S have to be smaller than the product of number of measurements and the state. The error in solving Eq. 4.26 for X is entirely connected to the fact that the L is approximate since it is constructed from sample correlation functions of the innovations which are estimated from finite duration signals, namely

$$\widehat{\mathsf{L}}_j \overset{def}{=} E\left(e_k e_{k-j}^T\right) = \frac{1}{N-j}\sum_{k=1}^{N-j} e_k e_{k-j}^T \tag{4.28}$$

where N is the number of time steps. Substituting $\widehat{L}$ as the estimate of L, the solution of Eq. 4.26 can be presented as in the following.

$$\text{Case \#1 } mn \geq \left(r^2+mr\right) \tag{4.29}$$

In this case H is full rank and there exists a unique minimum norm solution for a weighting matrix I given in the following,

$$\widehat{X} = \left(H^T H\right)^{-1} H^T \widehat{L} \tag{4.30}$$

$$\text{Case\#2 } mn < \left(r^2+mr\right) \tag{4.31}$$

In this case the matrix is rank deficient, and the size of null space of H can be calculated from $t = r^2 - mn$. The solution is written as follows,

$$\widehat{X} = \widehat{X}_0 + null(H)Y \tag{4.32}$$

where $\widehat{X}_0$ is the minimum norm solution given in Eq. 4.30 and $Y \varepsilon R^{tx1}$ is an arbitrary vector. Therefore, we conclude Eq. 4.26 has infinite solution when $mn < \left(r^2+mr\right)$. We note that the innovations correlations approach allows to enforce the positive semi-definiteness when solving for Q, S and R from Eq. 4.26.

4.4 Numerical Experiment: Five-DOF Spring Mass System

In this numerical experiment we use the five-DOF spring mass system depicted in Fig. 4.1 in order to examine the innovations correlations approach for the spatial distribution of distriburbances and noise covariance matrices.

We assume that true stiffness and mass values of the spring-mass system are given in consistent units as $k_i = 100$ and $m_i = 0.05$, respectively. The un-damped frequencies of the system are depicted in Table 4.1.

We obtain results for output sensors at the third masses, which are recording velocity data at 100 Hz sampling.

Case I

The unmeasured disturbances are acting on the masses #2 and #4. The measurement noise is prescribed to have a root-mean-square (RMS) equal to approximately 10 % of the RMS of the response measured (R = 0.030). Unmeasured disturbances and measurement noise are assumed to be mutually uncorrelated, with the covariance matrices,

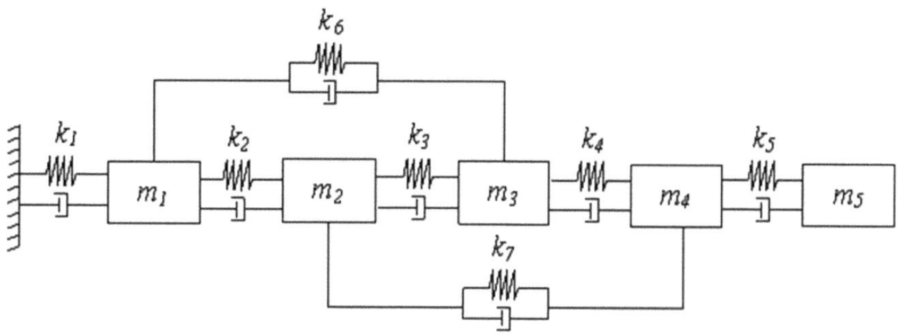

Fig. 4.1 Five-DOF spring mass system, $m_i = 0.05$, $k_i = 100$ (in consistent units). Damping is 2 % in all modes

Table 4.1 The un-damped frequencies of the spring mass system

Frequency no.	Frequency (Hz)
1	0.582
2	1.591
3	2.851
4	3.183
5	3.434

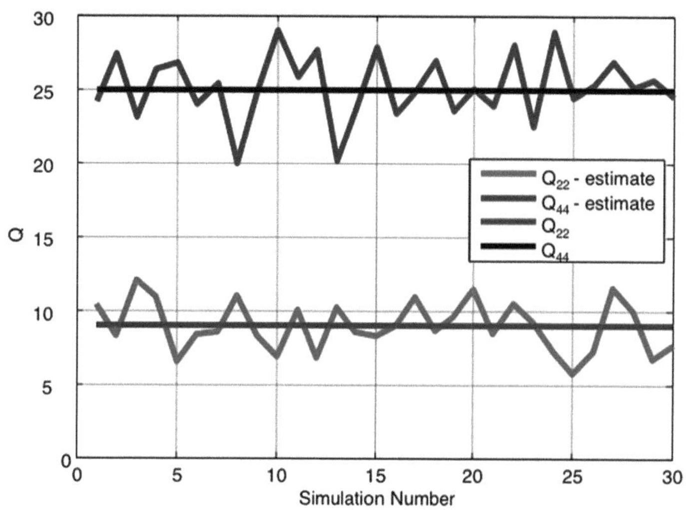

Fig. 4.2 Disturbance covariance estimates ($\hat{Q}$) for 30 simulations

$$Q = \begin{bmatrix} 0 & 0 & 0 & 0 & 0 \\ 0 & 9 & 0 & 0 & 0 \\ 0 & 0 & 0 & 0 & 0 \\ 0 & 0 & 0 & 25 & 0 \\ 0 & 0 & 0 & 0 & 0 \end{bmatrix} \qquad R = 0.030 \qquad S = 0$$

The arbitrary filter gain K_0, that is chosen such that eigenvalues of the matrix $(A - K_0C)$ are assumed to have the same phase as those of A but with a 20 % smaller radius. Eighty lags of correlation functions of innovations process is taken into consideration and the sample innovation correlations functions are calculated using 600 s of data. Thirty simulations are carried out and the disturbance covariance matrices are calculated from innovations correlations approach based on the assumption that the distribution of the unmeasured disturbances, namely input to state matrix (B) is known. The disturbance covariance estimates obtained from the innovations correlations approach are presented in Fig. 4.2.

Case II

In addition to the noise covariance matrices given in Case I, it's assumed that the locations of the disturbances are also unknown (B matrix is used as $I_{5\times5}$). The mean value of the disturbance covariance matrix obtained from 30 simulations is

$$\widehat{Q} = \begin{bmatrix} 0.12 & 0 & 0 & 0 & 0 \\ 0 & 9.46 & 0 & 0 & 0 \\ 0 & 0 & 0.05 & 0 & 0 \\ 0 & 0 & 0 & 24.13 & 0 \\ 0 & 0 & 0 & 0 & 0.38 \end{bmatrix}$$

It's obvious that the large diagonal elements at positions 2 and 4 point to the position of the disturbances.

4.5 Conclusions

This paper attempts to give a concise description of innovations correlation approach for estimation of noise covariance matrices using Kalman filter. The classical innovations covariance technique to estimate the noise covariance matrices from output measurements was reviewed. The method leads to the solution of a linear system of equations based innovations correlation function. The operating assumptions of the method are that the system is linear time invariant and it is subjected to unmeasured Gaussian stationary disturbances and measurement noise, which are mutually correlated. In the numerical example signals with duration on the order of 100 times the period of the slowest mode proved inadequate. When the duration is 300 times the fundamental period the mean of 30 simulations proved in good agreement with the covariance of the disturbances. Numerical results suggest that the method can be effectively used for disturbance localization and covariance estimation.

References

1. Kalman RE (1960) A new approach to linear filtering and prediction problems. Trans ASME J Basic Eng 82:35–45
2. Mehra RK (1970) On the identification of variance and adaptive Kalman filtering. IEEE Trans Autom Control 15:175–184
3. Carew B, Belanger PR (1974) Identification of optimum filter steady-state gain for systems with unknown noise covariances. IEEE Trans Autom Control 18:582–587
4. Neethling C, Young P (1974) Comments on identification of optimum filter steady-state gain for systems with unknown noise covariances. IEEE Trans Autom Control 19:623–625
5. Odelson BJ, Rajamani MR, Rawlings JB (2006) A new autocovariance least-squares method for estimating noise covariances. Automatica 42(2):303–308
6. Åkesson BM, Jørgensen JB, Poulsen NK, Jørgensen SB (2008) A generalized autocovariance least-squares method for Kalman filter tuning. J Process Control 18(7–8):769–779, ISSN 0959–1524
7. Bulut Y, Vines-Cavanaugh D, Bernal D (2010) Process and measurement noise estimation for Kalman filtering. In: IMAC XXVIII, a conference and exposition on structural dynamics, Jacksonville

Chapter 5
Body Wise Time Integration of Multi Body Dynamic Systems

Wolfgang Witteveen

Abstract Several papers have been published in the past on the issue of decomposing a nonlinear system into subsystems for more efficient time integration. In this paper each body of a multi body system is considered as one subsystem. The subsystems (the bodies) are interacting via connection forces. The sources of such connection forces are constraints or directly applied forces. This contribution is restricted to constraint forces only. During a step which is named "body iteration", those forces are considered as constant and the state of the system is computed for each body separately. This can be massively parallelized which can be an efficiency advantage in case of computational costly problems like the ones occurring in parameter estimation. During an "constraint update step" the constraints are evaluated based on the body's current state. If the error is not small enough the interface forces are updated and the inner loop is executed once again until the error of the constraints is negligible. It turns out, that the constraints can be updated separately as well, which can be used again for parallel computing. In the paper, the theory will be outlined and implemented using an N body pendulum. Finally, the advantages and disadvantages of this approach are critically discussed.

Keywords Time integration • Multi body systems • Multi body dynamics • Parallel computing

Nomenclature

N	Number of bodies
M	Number of constraints
a, b	Numbers of bodies which are involved in a certain constraint
t_n	Denotes the time at time step n
h	Step size of time integration ($h = t_n - t_{n-1}$)
α,β,γ	HHT parameters
$\mathbf{q}_i$	State vector of body number i
$\mathbf{C}^j$	Constraint equation j
$\mathbf{P}^j$	Constraint force j
$^{a}\mathbf{P}^{j}_{a,b}$	Constraint force j acting on body a with involved bodies a and b
$\boldsymbol{\lambda}^j$	Lagrange multipliers for constraint j
$\underline{C}^{j}_{\mathbf{q}^j}$	Constraint Jacobian for constraint j
$\underline{E}$	Identity matrix
$\underline{M}$	Mass matrix
$\mathbf{Q}_i$	Forces acting on body i
$\underline{\mathbf{J}}_i$	Jacobian of body number i
$\mathbf{e}_i$	Residuum of equation of motion for body i
$\mathbf{c}^j$	Residuum of constraint equation j
$\mathbf{p}^j$	Residuum of constraint force j

W. Witteveen (✉)
University of Applied Sciences Upper Austria – Wels Campus, Stelzhammerstr. 69, Wels 4600, Austria
e-mail: wolfgang.witteveen@fh-wels.at

© The Society for Experimental Mechanics, Inc. 2015

R. Allemang (ed.), *Special Topics in Structural Dynamics, Volume 6*, Conference Proceedings of the Society for Experimental Mechanics Series, DOI 10.1007/978-3-319-15048-2_5

5.1 Introduction

Computer based multi body dynamics consists of two main procedures. In a first step, the necessary equations are automatically assembled. In the subsequent step numeric time integration of these equations needs to be done. Normally, the time integration is the dominant task in terms of CPU time. This is getting even worse in case of special problems, like parameter identification, when several time integrations of the same system are necessary. Therefore it is obvious to think about parallelization in order to distribute the work onto several, parallel working CPUs. A review paper on that issue has been published by Negrut et al. [1]. Beside the parallelization on code implementation and code compilation level the time integration algorithm itself can be designed with the intention of a latter parallelization. An example for that strategy can be found in [2]. Inspired by the co-simulation algorithm in [3], this contribution is concerned with the question whether each body of a multi body system can be seen as separate subsystem. The equations will be derived using a N-body pendulum. With certain assumptions it turns out, that even the constraints can be computed separately. So to say, the formulation leads to a body and constraint wise time integration which can be parallel executed.

5.2 Theory

The theory in this chapter is outlined using a planar N-pendulum example as depicted in Fig. 5.1. All pendulums are of the same geometry and mass and the symbol N holds the number of pendulums.

The location and orientation of body number i (1 .. i .. N) is given by its generalized coordinate vector $\mathbf{q}_i$. The bodies are connected via M constraints. For the N-pendulum example the number M is equal to N. Each constraint needs to fulfill corresponding constraint equations in the form of

$$\mathbf{C}^j = \mathbf{C}^j_{a,b} = \mathbf{0} \qquad 1 \leq j \leq M \qquad 1 \leq a,b \leq N \tag{5.1}$$

The right superscript j holds the constraint number while the subscripts a and b denote the body numbers which are connected via the constraint. If there is no value for b, the constraint j connects the body number a with ground. The involved coordinate vectors $\mathbf{q}_a$ and $\mathbf{q}_b$ are collected into the vector $\mathbf{q}^j$. This vector is given in (5.2) for body – body constraints and in (5.3) for body ground constraints:

$$\mathbf{q}^j = \left[\mathbf{q}_a \ \mathbf{q}_b\right]^T \tag{5.2}$$

$$\mathbf{q}^j = \mathbf{q}_a \tag{5.3}$$

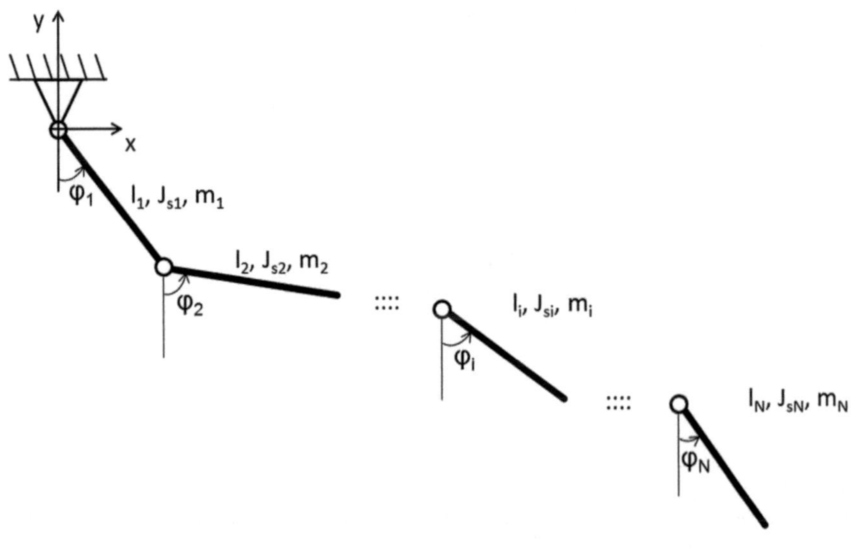

Fig. 5.1 N body pendulum

Note again, that the subscript denotes a body number while the superscript denotes a constraint number. The j-th constraint leads to constraint forces in the form of

$$\mathbf{P}^j = \mathbf{P}^j_{a,b} = \left[\underset{-}{\mathbf{C}}^j_{\mathbf{q}^j}\right]^{\mathrm{T}} \boldsymbol{\lambda}^j = \begin{bmatrix} \underset{-}{\mathbf{C}}^j_{\mathbf{q}_a} \\ \underset{-}{\mathbf{C}}^j_{\mathbf{q}_b} \end{bmatrix}^{\mathrm{T}} \boldsymbol{\lambda}^j = \begin{bmatrix} {}^a\mathbf{P}^j_{a,b} \\ {}^b\mathbf{P}^j_{a,b} \end{bmatrix} \tag{5.4}$$

or, for a body ground constraint

$$\mathbf{P}^j = \mathbf{P}^j_a = \left[\underset{-}{\mathbf{C}}^j_{\mathbf{q}^j}\right]^{\mathrm{T}} \boldsymbol{\lambda}^j = \left[\underset{-}{\mathbf{C}}^j_{\mathbf{q}_a}\right]^{\mathrm{T}} \boldsymbol{\lambda}^j = {}^a\mathbf{P}^j_a \tag{5.5}$$

Furthermore, there are N equations of motions which can be given as

$$\begin{aligned} &\underset{-}{\mathbf{M}}_1(\mathbf{q}_1)\ \ddot{\mathbf{q}}_1 + {}^1\mathbf{P}^1_1 + {}^1\mathbf{P}^2_{1,2} - \mathbf{Q}_1 = \mathbf{0} \\ &:: \qquad :: \qquad :: \qquad :: \\ &\underset{-}{\mathbf{M}}_i(\mathbf{q}_i)\ \ddot{\mathbf{q}}_i + {}^i\mathbf{P}^i_{i-1,i} + {}^i\mathbf{P}^{i+1}_{i,i+1} - \mathbf{Q}_i = \mathbf{0} \\ &:: \qquad :: \qquad :: \qquad :: \\ &\underset{-}{\mathbf{M}}_N(\mathbf{q}_N)\ \ddot{\mathbf{q}}_N + {}^N\mathbf{P}^N_{N-1,N} - \mathbf{Q}_N = \mathbf{0} \end{aligned} \tag{5.6}$$

In this contribution the force vector $\mathbf{Q}_i$ is constant. It contains the gravity and an imposed moment around φ_1. In the non-planar case it would contain velocity inertia forces as well.

The basic idea of the proposed time integration algorithm can be seen in Fig. 5.2.

During the so called "body iteration" block all N equations of motions are separately equilibrated. The constraint forces are considered as constant during this step. This part can be implemented in a parallel matter because each body can be separately executed. In a next step it is checked weather the constraint equations (5.1) and (5.4) ((5.5)) are fulfilled. If yes, time integration for t_n is done and the overall loop can proceed to time step t_{n+1}. If no, an update of the constraint forces and Lagrange multiplier is needed so that the error will become smaller. After the "constraint update", the "body iteration" starts again with updated constraint forces.

5.2.1 Body Iteration

During the "body iteration" all N equations of motions are solved separately using a constant constraint force. The time integration is done via the HHT method, see exemplarily [4]. Following this time integration approach, the state, the velocity and the acceleration of a certain time step n are related to each other and the quantities at time step n-1 via

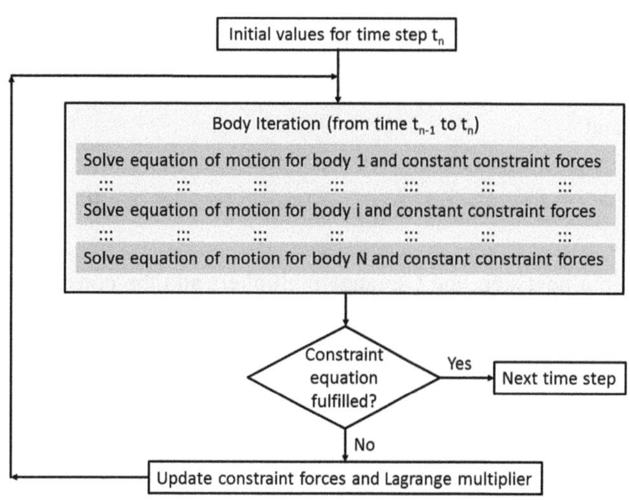

Fig. 5.2 Flow of algorithm

$$\begin{aligned} \mathbf{q}_{\mathrm{n}} &= \mathbf{q}_{\mathrm{n-1}} + \mathrm{h}\dot{\mathbf{q}}_{\mathrm{n-1}} + \tfrac{\mathrm{h}^2}{2}\left[(1\text{-}2\beta)\,\ddot{\mathbf{q}}_{\mathrm{n-1}} + 2\beta\ddot{\mathbf{q}}_{\mathrm{n}}\right] \\ \dot{\mathbf{q}}_{\mathrm{n}} &= \dot{\mathbf{q}}_{\mathrm{n-1}} + \mathrm{h}\left[(1\text{-}\gamma)\,\ddot{\mathbf{q}}_{\mathrm{n-1}} + \gamma\ddot{\mathbf{q}}_{\mathrm{n}}\right] \end{aligned} \tag{5.7}$$

$$\alpha \in \left[-\frac{1}{3}, 0\right]; \quad \gamma = \frac{1-2\alpha}{2}; \quad \beta = \frac{(1-\alpha)^2}{4} \tag{5.8}$$

The derivative of the state and the velocity with respect to the acceleration are needed later and can be given as

$$\frac{\partial \mathbf{q}_{\mathrm{n}}}{\partial \ddot{\mathbf{q}}_{\mathrm{n}}} = \beta \mathrm{h}^2 = k_D \quad \frac{\partial \dot{\mathbf{q}}_{\mathrm{n}}}{\partial \ddot{\mathbf{q}}_{\mathrm{n}}} = \mathrm{h}\gamma = k_V \tag{5.9}$$

One of the key ideas of the HHT method is to evaluate a "damped" equation of motion in the form of

$$\left[\underline{\mathbf{M}}(\mathbf{q}_i)\ \ddot{\mathbf{q}}_i\right]_{\mathrm{n}} + (1+\alpha)\left[{}^{i}\mathbf{P}^{i}_{i-1,i} + {}^{i}\mathbf{P}^{i+1}_{i,i+1} - \mathbf{Q}_i\right]_{\mathrm{n}} - \alpha\left[{}^{i}\mathbf{P}^{i}_{i-1,i} + {}^{i}\mathbf{P}^{i+1}_{i,i+1} - \mathbf{Q}_i\right]_{\mathrm{n-1}} = \mathbf{0} =: \mathbf{e}_i \tag{5.10}$$

Note that the equation looks bit different in case of i = 1 and i = N. Due to numeric the final solution will not lead to a zero sum but to a so called residuum which should be close to zero. The symbol of the residuum for the i-th body is $\mathbf{e}_{\mathrm{i}}$. Under the general assumption of the presence of state depended inertia forces, a Newton iteration can be performed. The relevant equations are:

$$\begin{aligned} \frac{\partial(\mathbf{e}_{\mathrm{i}})}{\partial \ddot{\mathbf{q}}_i}\Delta\ddot{\mathbf{q}}_i &= -\mathbf{e}_{\mathrm{i}} \\ \ddot{\mathbf{q}}_{\mathrm{i,n}} &= \ddot{\mathbf{q}}_{\mathrm{i,n}} + \Delta\ddot{\mathbf{q}}_{\mathrm{i,n}} \end{aligned} \tag{5.11}$$

The Jacobian can be given as

$$\frac{\partial(\mathbf{e}_{\mathrm{i}})}{\partial \ddot{\mathbf{q}}_i} = \frac{1}{1+\alpha}\underline{\mathbf{M}}(\mathbf{q}_i) + \beta\mathrm{h}^2\left(\frac{1}{1+\alpha}\frac{\partial\left[\underline{\mathbf{M}}(\mathbf{q}_{\mathrm{i}})\ \ddot{\mathbf{q}}_i\right]}{\partial \mathbf{q}_{\mathrm{i}}} - \frac{\partial\left[\mathbf{Q}(\mathbf{q}_{\mathrm{i}}, \dot{\mathbf{q}}_i)\right]}{\partial \mathbf{q}_{\mathrm{i}}}\right) - \gamma\mathrm{h}^2\frac{\partial\left[\mathbf{Q}(\mathbf{q}_{\mathrm{i}}, \dot{\mathbf{q}}_i)\right]}{\partial \dot{\mathbf{q}}_{\mathrm{i}}} = \underline{\mathbf{J}}_i \tag{5.12}$$

At the end of the "body iteration" the residuum of all N equations of motion (5.10) are below a certain, user defined limit. Note that all N bodies can be executed by its own. Therefore a massively parallel implementation is thinkable.

5.2.2 Constraint Update

After the "body iteration" (5.1) and (5.4) ((5.5)) are, in general, not fulfilled. The corresponding errors $\mathbf{c}^{\mathrm{j}}$ and $\mathbf{p}^{\mathrm{j}}$ are defined as

$$\mathbf{C}^j = \mathbf{c}^j \tag{5.13}$$

and

$$\mathbf{P}^j - \left[\underline{\mathbf{C}}^j_{\mathbf{q}^j}\right]^{\mathrm{T}}\boldsymbol{\lambda}^j = \mathbf{p}^j \tag{5.14}$$

In a next step, the Lagrange multipliers and the constraint forces have to be updated, so that the error is decreasing. This leads to a update equation in the form of:

$$\begin{bmatrix} \dfrac{\partial \mathbf{p}^j}{\partial \mathbf{P}^j} & \dfrac{\partial \mathbf{p}^j}{\partial \boldsymbol{\lambda}^j} \\ \dfrac{\partial \mathbf{c}^j}{\partial \mathbf{P}^j} & \dfrac{\partial \mathbf{c}^j}{\partial \boldsymbol{\lambda}^j} \end{bmatrix}\begin{bmatrix} \Delta\mathbf{P}^j \\ \Delta\boldsymbol{\lambda}^j \end{bmatrix} = -\begin{bmatrix} \mathbf{p}^j \\ \mathbf{c}^j \end{bmatrix} \tag{5.15}$$

The update rule is

$$\begin{aligned} \mathbf{P}^j &= \mathbf{P}^j + \Delta\mathbf{P}^j \\ \boldsymbol{\lambda}^j &= \boldsymbol{\lambda}^j + \Delta\boldsymbol{\lambda}^j \end{aligned} \tag{5.16}$$

In case of a body – ground constraint the former equations have to be interpreted as

$$\Delta\mathbf{P}^j = \Delta^a\mathbf{P}_a^j \qquad \mathbf{p}^j = {}^a\mathbf{p}_a^j \qquad \frac{\partial\mathbf{p}^j}{\partial\mathbf{P}^j} = \frac{\partial^a\mathbf{p}_a^j}{\partial^a\mathbf{P}_a^j} \tag{5.17}$$

and in case of a body-body constraint as

$$\Delta\mathbf{P}^j = \begin{bmatrix} \Delta^a\mathbf{P}_{a,b}^j \\ \Delta^b\mathbf{P}_{a,b}^j \end{bmatrix} \quad \mathbf{p}^j = \begin{bmatrix} {}^a\mathbf{p}_{a,b}^j \\ {}^b\mathbf{p}_{a,b}^j \end{bmatrix} \quad \frac{\partial\mathbf{p}^j}{\partial\mathbf{P}^j} = \begin{bmatrix} \frac{\partial^a\mathbf{p}_{a,b}^j}{\partial^a\mathbf{P}_{a,b}^j} & \frac{\partial^a\mathbf{p}_{a,b}^j}{\partial^b\mathbf{P}_{a,b}^j} \\ \frac{\partial^b\mathbf{p}_{a,b}^j}{\partial^a\mathbf{P}_{a,b}^j} & \frac{\partial^b\mathbf{p}_{a,b}^j}{\partial^b\mathbf{P}_{a,b}^j} \end{bmatrix} \tag{5.18}$$

Using the chain rule and the identity (5.9) the quantities in the coefficient matrix of (5.15) are

$$\frac{\partial\mathbf{p}^j}{\partial\mathbf{P}^j} = \underline{\mathbf{E}} - \frac{\partial\left(\left[\underline{\mathbf{C}}^j_{\mathbf{q}^j}\right]^{\mathrm{T}}\boldsymbol{\lambda}^j\right)}{\partial\mathbf{q}^j} k_D \frac{\partial\ddot{\mathbf{q}}^j}{\partial\mathbf{P}^j} \tag{5.19}$$

$$\frac{\partial\mathbf{p}^j}{\partial\boldsymbol{\lambda}^j} = -\left[\underline{\mathbf{C}}^j_{\mathbf{q}^j}\right]^{\mathrm{T}} \tag{5.20}$$

$$\frac{\partial\mathbf{c}^j}{\partial\mathbf{P}^j} = \underline{\mathbf{C}}^j_{\mathbf{q}^j}\, k_D \frac{\partial\ddot{\mathbf{q}}^j}{\partial\mathbf{P}^j} \tag{5.21}$$

$$\frac{\partial\mathbf{c}^j}{\partial\boldsymbol{\lambda}^j} = \underline{\mathbf{0}} \tag{5.22}$$

Note that (5.19) and (5.21) would be different in case of a state depended force vector $\mathbf{Q}$. Due to the face that the error is indirect influenced by the bodies state it is not surprising that the sensitivity of the state with respect to the constraint force can be found in the equations. These sensitivities can be obtained by the equations of motion (5.10). Note that the term $\frac{\partial\ddot{\mathbf{q}}^j}{\partial\mathbf{P}^j}$ contains actually $\frac{\partial\ddot{\mathbf{q}}_a}{\partial\mathbf{P}^j}$ and in case of a body-body constraint $\frac{\partial\ddot{\mathbf{q}}_b}{\partial\mathbf{P}^j}$. In case of the N-body pendulum under consideration the equations of motion j-1 and j are needed for the constraint j. The derivation of the latter equations with respect to the j-th constraint force gives

$$\mathbf{0} = \frac{\partial\left(\underline{\mathbf{M}}\left(\mathbf{q}_{j-1}\right)\,\ddot{\mathbf{q}}_{j-1} + {}^{j-1}\mathbf{P}^{j-1}_{j-2,j-1} + {}^{j-1}\mathbf{P}^{j}_{j-1,j} - \mathbf{Q}_{j-1}\right)}{\partial^{j-1}\mathbf{P}^j_{j-1,j}} \approx \underline{\mathbf{E}} + \underline{\mathbf{J}}_{j-1} \frac{\partial\ddot{\mathbf{q}}_{j-1}}{\partial^{j-1}\mathbf{P}^j_{j-1,j}} \tag{5.23}$$

$$\mathbf{0} = \frac{\partial\left(\underline{\mathbf{M}}\left(\mathbf{q}_{j}\right)\,\ddot{\mathbf{q}}_{j} + {}^{j}\mathbf{P}^{j}_{j-1,j} + {}^{j}\mathbf{P}^{j+1}_{j,j+1} - \mathbf{Q}_{j}\right)}{\partial^{j}\mathbf{P}^j_{j-1,j}} \approx \underline{\mathbf{E}} + \underline{\mathbf{J}}_{j} \frac{\partial\ddot{\mathbf{q}}_{j}}{\partial^{j}\mathbf{P}^j_{j-1,j}} \tag{5.24}$$

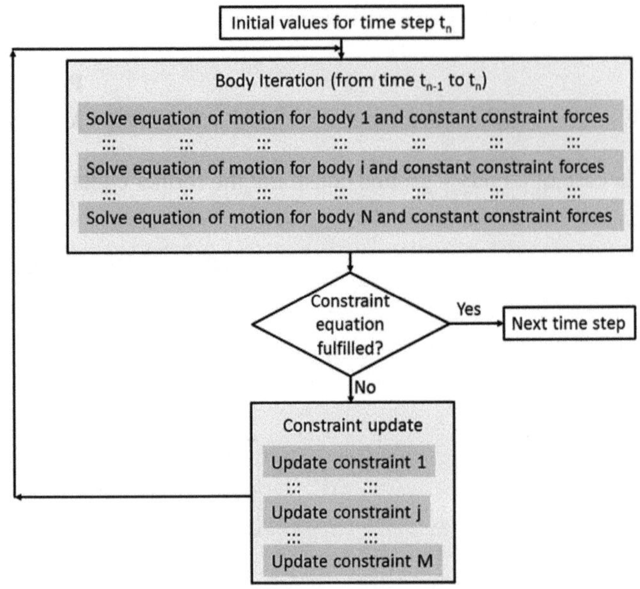

Fig. 5.3 Flow of algorithm with parallel constraint update

Note again, that (5.23) and (5.24) are different when the first and the last constraint are considered. Note furthermore, that

$$\frac{\partial^{j-1}\mathbf{P}^{j-1}_{j-2,j-1}}{\partial^{j-1}\mathbf{P}^{j}_{j-1,j}} \approx \underline{0} \qquad \frac{\partial^{j}\mathbf{P}^{j+1}_{j,j+1}}{\partial^{j}\mathbf{P}^{j}_{j-1,j}} \approx \underline{0} \tag{5.25}$$

is assumed. Obviously these terms are not zero but considering these terms would lead to a coupled system of equations for the desired sensitivities. In other words, assumption (5.25) leads to a formulation where the update of the j-th constraint is only influenced by the bodies involved in the constraint j. The obtained sensitivities can be inserted into (5.19) and (5.21). Finally, (5.15) can be solved and the update (5.16) can be performed.

A direct implication of assumption (5.25) is that all M constraints can be updated separately, see Fig. 5.3. This can be used in a parallel implementation, as it has been suggested for the body iteration.

5.3 Numerical Time Integration

The latter time integration algorithm has been implemented in Scilab [5]. Note that the implementation was quite straight forward, without special attention on efficiency and without a parallel implementation. The intention of this section is to report some basic characteristics of the algorithm which have been observed.

The parameters for the mechanical model are $m = 1$ and $l = 1$. Furthermore a constant torque (=1) on body 1 is acting around φ_1. The time integration interval ranges from 0 to 5 using a fixed step size of 0.01. The HHT parameter α has been set to $-^1/_3$. The error vectors have been considered as small enough in case of a norm lower as 1e-9.

It can be reported, that the integrator is very stable and almost no numerical damping has been observed for the latter set of parameters. Figure 5.4 contains the CPU time with respect to the number of bodies together with the number of required loops (mean value). It can be seen, that the algorithm, in its introduced form, doesn't have a linear runtime complexity with respect to the number of bodies. While the HHT integration for each body ("body iteration") is of linear complexity, the number of outer loop is over linearly increasing. One possible reason could be the assumption given in (5.25). This assumption leads to a separate treatment of the constraints. As a consequence, during one loop a change in the state of body i is transmitted via the connecting constraints to body i-1 and body i + 1. For the further transmission of body i-1 to body i-2 an additional loop is needed, and so on. It needs to be investigated whether the speed up due to parallelization of the constraint update justifies the penalty of the necessary loop in order to eliminate the constraint error.

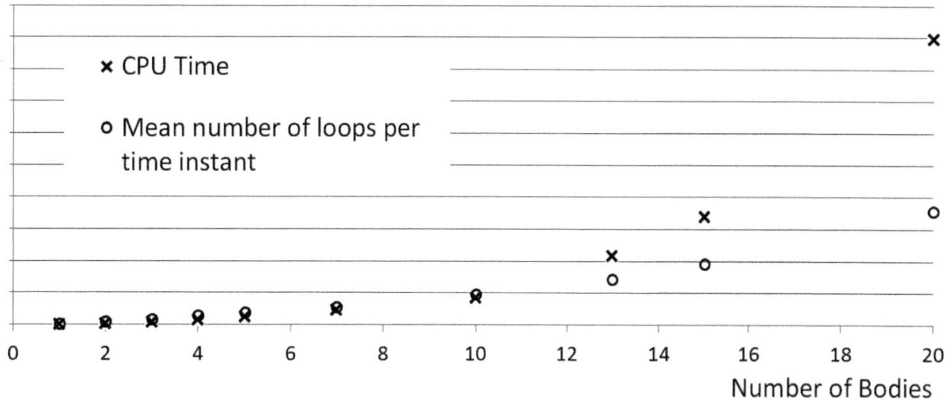

Fig. 5.4 CPU time and mean number of required loops

5.4 Conclusion

In this paper a first framework for a body and constraint wise time integration is presented. The intention is the introduction of a time integration algorithm with high parallelization potential because of its formulation in order to decrease the CPU time. It can be reported that the algorithm works stable and introduces almost no numerical damping in case of a rigid N body pendulum. In its introduced form, an over linear runtime complexity with respect to the number of bodies has been observed. This is probably caused by a decoupled consideration of the constraints and constraint forces.

Acknowledgement We gratefully acknowledge the support from the Austrian funding agency FFG in the Coin-project ProtoFrame (project number 839074).

References

1. Negrut D, Serban R, Mazhar H, Heyn T (2007) Parallel computing in multibody system dynamics: why, when and how. J Comput Nonlinear Dyn 9:12. doi:10.1115/1.4027313
2. van der Valk PLC, Rixon DJ (2014) Towards a parallel time integration for nonlinear systems. In: Allen M, Mayes R, Rixen D (eds) Dynamics of coupled structures. Proceedings of the 32nd IMAC, a conference and exposition on structural dynamics. Springer, vol 1, pp 135–145. ISBN: 978-3-319-04501-6
3. Sicklinger S, Belsky V, Engelmann B, Elmqvist H, Olsson H, Wüchner R, Bletzinger KU (2014) Interface Jacobian based co-simulation. Int J Numer Methods Eng 98:418–444. doi:10.1002/nme.4637
4. Negrut D, Rampalli R, Ottarsson G, Sajdak A (2007) On the implementation of the HHT method in the context of index 3 differential algebraic equations of multi body dynamics. J Comput Nonlinear Dyn 2(1):73–85. doi:10.1115/1.2389231
5. Scilab 5.5.1. www.scilab.com

Chapter 6
Structural Dynamic Modeling: Tales of Sin and Redemption

Robert N. Coppolino

Abstract The great twentieth century mathematician, John von Neumann, once said, "At a great distance from its empirical source, or after much abstract inbreeding, a mathematical subject is in danger of degeneration. Whenever this stage is reached the only remedy seems to me to be the rejuvenating return to the source: the reinjection of more or less empirical ideas." This wisdom is especially applicable to the field of structural dynamics. The present paper takes a look at the historical and empirical bases of key aspects of structural dynamic phenomena including damping of materials and built-up assemblies, behavior of viscoelastic materials, interaction of structures and fluids, and general parametric uncertainties. Migration of misconceptions in engineering practice and, in particular, commercial software products are cited. Illustrative examples of the benefits of recollection of fundamentals in aerospace, marine and civil applications are described.

Keywords Structural Dynamics • Damping Joints Assemblies

6.1 Introduction

The broad discipline of modern structural dynamics modeling is the product of advances in strength of materials, theory of elasticity, and theory of structures [1], and automated computational analysis (primarily the finite element method) [2, 3]. Complementing the above "two legs of the stool" is a third "leg", namely the wealth of empirical data [4], that is often forgotten or ignored. This paper focuses on several areas of structural dynamics modeling, primarily damping and structural joints, that suffer from widespread misconceptions, neglect and errors ("sin") that can best be remedied by recollection of their historical bases ("redemption").

The popular notion of proportional damping is based on Lord Rayleigh's citation, which quoted directly from Theory of Sound [5] (Chapter V, Paragraph 97) states "The first case occurs frequently, *in books at any rate*, when motion of each part of the system is resisted by a retarding force, proportional both to the mass and velocity of the part. The same exceptional reduction is possible when F *(the dissipation force)* is a linear function of T *(kinetic energy)* and V *(strain energy)*". In addition, Lord Kelvin during the late 1800s [1] introduced the notion of internal friction, which he concluded is *not proportional to velocity, as in fluids*. The "Rayleigh" proportional damping model, which generally does not follow experimental data, has been generalized by Caughey and O'Kelly [6] with a complicated, velocity dependent formulation. At the present time, "Rayleigh" proportional damping remains an option for time domain structural dynamic analysis in most commercial finite element codes.

During the 1960s demands in the aerospace industry and memory limitations of digital computers led to introduction of component mode synthesis (CMS) techniques [7–11], which provided means for incorporation of modal damping in structural dynamic analyses that bypassed the conceptual pitfalls of explicit (proportional) damping models. Component mode synthesis tacitly pointed to means for coupling of structural subassemblies at interfaces (joints), which have become routine in modern computed aided engineering (CAE) software tools [12–14]. However, the ease of automation also provides ample opportunity for naïve errors, which ignore local flexibility at structural component interfaces.

The engineering demands introduced during the early years of aeronautics and the subsequent space age caused investigators [15–21] to engage in development of combined fluid–structure modeling techniques, which relied heavily upon empirical data. These activities led to a few isolated situations in which (fluid dynamic perturbation) damping forces were found to be velocity dependent and the overwhelming majority of situations pointing to damping forces that do not fit simplistic "velocity" dependent models. Most notable are the contributions of Kimball and Lovell [22] and Becker and Foppl

R.N. Coppolino (✉)
Measurement Analysis Corporation, 23850 Madison Street, Torrance, CA 90505, USA
e-mail: rcoppolino@macorp.net

© The Society for Experimental Mechanics, Inc. 2015

R. Allemang (ed.), *Special Topics in Structural Dynamics, Volume 6*, Conference Proceedings of the Society for Experimental Mechanics Series, DOI 10.1007/978-3-319-15048-2_6

[23] who independently confirmed Lord Kelvin's observation that structural damping forces are generally not proportional to velocity. In fact the two teams concluded that damping forces appear to be proportional to displacement and in phase with velocity. This model was subsequently expressed in terms of complex variables by Kussner [24] and Kassner [25]. The appropriate, complex formulation of structural damping remains a standard for aeroelasticity [15, 16] and vibroacoustics [19, 20], which are typically expressed in the frequency domain. Viscous (velocity dependent) damping models persist in time domain applications due to mathematical difficulties associated with complex variables outside the frequency domain.

A wealth of empirical data supports the notion of structural damping in metallic structures [26, 27] that is proportional to displacement (strain) and in phase with velocity. Moreover, damping appears to be predominantly concentrated at joints. Important exceptions to the structural damping model are found in viscoelastic materials, shock and vibration isolators and (welded, bolted, riveted and bearing) structural joints. Alternative empirically based models have been developed to describe these phenomena in the time domain [28–30], and recently Genta and Amati [31] introduced an approximate model for structural damping in the time domain that employs a general viscoelastic model. Two key facts result from all empirical models of damping in solid structures, namely (a) a simple, linear, velocity dependent damping model does not appear to be physically appropriate, and (b) localized structural flexibility is a close partner with joint damping. Joint flexibility, as noted by many investigators, represents a strong influence on the modal characteristics of structural systems [32]. Physically consistent formulations of material damping and dynamic stiffness (especially for viscoelastic materials), and localized damping and flexibility at joints are essential for parametric sensitivity analyses, system identification, and structural damage assessments.

6.2 Nomenclature

Symbol	Definition	Symbol	Definition
[B]	Damping matrix	{q}	Modal displacement array
{F}	Force array	{u}	Physical displacement array
[K]	Stiffness matrix	[Φ]	Modal matrix
[M]	Mass matrix	η	Structural damping coefficient
T	Kinetic energy	ω_n	Natural frequency (rad. s)
V	Strain energy	ζ_n	Modal critical damping ratio

6.3 Structural Dynamics Modeling: The Present State-of-the-Art

The systematic development of structural mechanics theory [1], finite element analysis [2], and computer aided engineering tools [3], has resulted in an engineering community characterized by high productivity and (in many cases) a blind faith in automation. The overwhelming majority of structural dynamic models describing behavior of systems are expressed in terms of the following matrix equations:

$$[M]\{\ddot{u}\} + [B]\{\dot{u}\} + [K]\{u\} = \{F_e(t)\} \tag{6.1}$$

For many typical structures, the mass, and stiffness matrices are appropriately assumed to be constant under "normal" operating conditions. Definition of mass and stiffness coefficients, for typical rod, beam, plate and solid components, is quite well-established [1, 2] especially when the system is composed of metallic materials and there are no significant interactions with other media (especially fluids). Formulation of the system damping matrix, however, is generally not based on well-developed theoretical foundations.

Fortunately, there is a wealth of empirical data indicating that the dynamic behavior of many structures may be adequately described in terms of undamped modal vectors (in a relevant frequency band [4]), i.e.,

$$\{u\} = [\Phi]\{q\}, \text{ solutions of } [K][\Phi] = [M][\Phi][\lambda] \tag{6.2}$$

where

$$\left[\Phi^t\right][M][\Phi] = [I], \left[\Phi^t\right][K][\Phi] = \text{diag}\left[\omega_n^2\right], \left[\Phi^t\right][B][\Phi] = \text{diag}\left[2\zeta_n\omega_n\right] \tag{6.3}$$

It should be noted that an explicit damping matrix, that produces the (possibly empirical) damping coefficients, may be constructed by the following operation,

$$[B] = [M\Phi]\,[2\zeta_n\omega_n]\,\left[\Phi^t M\right] \tag{6.4}$$

Note that the constructed damping matrix is fully populated and it does not have a theoretical basis as is the case for the mass and stiffness matrices. Moreover, the above triple product is not related to an explicit "viscous" theoretical damping matrix.

6.3.1 Proportional Damping Formulations

Most engineering organizations employ empirically based values for modal damping (ζ_n), and the modal approach for description of "viscous" damping, when practical, which circumvents difficulties associated with the lack of a theoretical damping matrix.

A variety of artificially constructed mathematical forms for the damping matrix have been defined over the past century. One form that has managed to find its way into most finite element codes, attributed to Rayleigh [5] and known as proportional damping is

$$[B] = \alpha\,[M] + \beta\,[K] \tag{6.5}$$

Application of the modal transformation on this matrix form results in the following distribution of modal damping, which does not resemble typical empirical data records.

$$\zeta_n = \frac{\alpha}{2\omega_n} + \frac{\beta\omega_n}{2} \tag{6.6}$$

Caughey and O'Kelly [6] introduced the extension of Rayleigh's proportional damping formula,

$$[B] = [M]\sum_{j=0}^{N-1}\alpha_j\left[M^{-1}K\right]^j \tag{6.7}$$

which results in the less restrictive distribution of modal damping,

$$\zeta_n = \left(\frac{\alpha_0 + \alpha_1\omega_n^2 + \alpha_2\omega_n^4 + \ldots}{2\omega_n}\right) \tag{6.8}$$

While the Caughey and O'Kelly generalization of proportional damping, in the limit permits any frequency dependent distribution of modal damping, it has two distinct shortcomings, namely (1) the damping matrix is fully populated, introducing computational inefficiencies for large-order dynamic systems, and (2) it does not permit radical differences in the damping of closely-spaced modes that sometimes occur in actual physical systems. At this point, Rayleigh's and Caughey-O'Kelly's proportional damping constructs clearly bring us to Von Neumann's point of a "great distance from its empirical source", and "much abstract inbreeding", resulting in this "mathematical subject … in danger of degeneration". Our only remedy for this situation is to return to empirical sources.

6.3.2 Assembly of Structural Dynamic Models from Discrete Components

During the 1960s structural dynamic models for aerospace systems taxed the capacity of digital computers and a series of component mode synthesis (CMS) techniques were developed to address the challenge. The variety of CMS techniques includes (1) Hurty-Craig-Bampton [7, 8], (2) MacNeal-Rubin [9, 10], and (3) Benfield-Hruda [11]. These techniques define component structural dynamic models, characterized by discrete physical boundary degrees of freedom that facilitate direct connection to adjoining system components. Moreover, CMS techniques provide an easy means for incorporation of empirical modal damping (expressed as generalized viscous damping coefficients).

Computer resources and numerical analysis techniques have now surpassed the limitations of the 1960s, and the utilization of CMS techniques is no longer necessary. However, the integration of system design and analysis through Computer Aided Engineering (CAE) employs strategies for automated assembly of highly detailed component models to form system dynamic models [12–14].

Both CMS and state-of-the-art CAE techniques provide a subtle, yet serious opportunity for falling into Von Neumann's point of "degeneration". Specifically, the ease with which an engineer can automatically stitch components together often simultaneously eliminates opportunities for proper reconciliation of models with empirical data. In particular, the root cause of deviation of a structural dynamic model from reality is often found in component interface flexibilities, which are "shorted out" by elementary stitching of component models.

6.3.3 Interaction of Structures with Fluid Media

The advent of the aircraft age in the early twentieth century and the subsequent space age prompted the engineering community to address aeroelastic instability (flutter) [15, 16], launch vehicle structure-propulsion system instability (Pogo) [17], and structure-control system instability [18]. In addition, vibroacoustic coupling phenomena in aircraft and spacecraft systems [19, 20] represent a serious threat for fatigue failures. During the same time period, safety issues associated with ocean installations (especially oil and gas production platforms) required an understanding of wave induced structural loads [21]. All of the above cited issues have required faithful, empirically based representations of (a) structural dynamic damping and (b) models describing perturbed fluid dynamic phenomena (i.e., fluid mass, damping and stiffness effects). Legitimate characterizations of velocity dependent damping were defined on the basis of flow perturbations in ducts (especially for Pogo dynamic models [17]) and on immersed marine structures [21]. *In these exceptional situations, the damping matrix is roughly proportional to mass and velocity.*

While detailed discussion of fluid–structure interaction is beyond the scope of the present paper, it is of importance to note the fact that progress in the above-cited applications was heavily influenced by close interaction of experimental and theoretical endeavors. This is attributed to the novelty of these challenges (in contrast with a perceived confidence in more "conventional" challenges in the field of "dry" structural dynamics).

6.4 Damping in Structural Assemblies

During the late 1920s Kimball and Lovell [22] and Becker and Foppl [23], independently determined by experiment, that damping in typical structures is simultaneously proportional to displacement (strain) and in phase with velocity, Shortly thereafter, Kussner [24] and Kassner [25] introduced the concept of complex structural damping, which appropriately describes the observations of Kimball, Lovell, Becker and Foppl. In short, the mathematical description of damping in typical structures shifted from a theoretical (viscous) formulation,

$$[M]\{\ddot{u}\} + [B]\{\dot{u}\} + [K]\{u\} = \{F_e(t)\} \tag{6.9}$$

to a hysteretic formulation (known today as structural damping),

$$[M]\{\ddot{u}\} + (1 + i\eta)[K]\{u\} = \{F_e(t)\} \tag{6.10}$$

Note that structural damping and viscous modal damping coefficients are related to one another as

$$\eta = 2\zeta_n \tag{6.11}$$

The contributions of Kimball, Lovell, Becker and Foppl represent a great contrast from the situation which resulted from Rayleigh's citation about proportional damping [5], which quoted directly from *Theory of Sound* (Chapter V, Paragraph 97) states "The first case occurs frequently, *in books at any rate*, when motion of each part of the system is resisted by a retarding force, proportional both to the mass and velocity of the part, The same exceptional reduction is possible when F *(the dissipation force)* is a linear function of T *(kinetic energy)* and V *(strain energy)*". It is ironic that Rayleigh appropriately pointed out the unverified status of proportional damping, yet much of the subsequent technical community succumbed to von Neumann's state of "abstract inbreeding" and "degeneration".

Table 6.1 Typical values of modal damping

		Frequency band (Hz)			Damping, ζ (%)	
Source	Article	Min	Max	Mean	Mean	STD
AGARD conference no. 277 (Wada & DesForges) 1979 [27]	Voyager	10.60	52.90	28.41	2.08	0.74
	Intelsat IV (In orbit)	1.85	195.00	82.19	0.86	0.39
	Hughes "1"	9.01	47.91	27.62	0.83	0.40
	Hughes "2"	5.89	40.15	22.27	1.03	0.49
	Hughes "3"	7.95	46.88	29.20	2.28	1.01
	SCATHA	14.10	42.60	28.30	1.18	0.58
	SEASAT	3.77	16.99	10.93	0.39	0.21
	Ranger III	28.40	94.50	58.16	0.75	0.20
	SKYLAB	4.10	17.02	10.04	1.52	0.82
	Space telescope truss	16.60	106.00	53.50	0.57	0.16
Measurement analysis corporation (Coppolino) 2014	Ford crown vic frame	12.72	49.79	30.75	0.36	0.34
	Boeing ISS-EAS	30.99	97.37	65.79	0.78	0.28
	Boeing ISS-P5	16.94	62.84	42.88	1.07	0.54
	Boeing PLA	32.04	88.19	53.73	0.37	0.09
	Macrolink card cage	269.28	596.96	407.65	1.21	0.86
	(Aerospace) Aries-1X FTV	0.18	4.78	2.55	0.79	0.69
Total of all data sets		0.18	596.96	52.98	1.10	0.78

6.4.1 *Evidence of Displacement Proportional Structural Damping*

The thorough treatment of structural damping, found in the text by Cremer, Heckl, and Ungar [26], provides a wealth of empirical data along with a technical viewpoint that complements the prevailing, automated finite element mindset. Three crucial features inherent in many structural systems are clearly noted in that text, namely:

1. "Solid" structures generally exhibit damping forces that are independent of frequency (displacement dependent structural damping) for a wide range of building materials.
2. Structural damping is often extremely low ($\eta \sim 10^{-4}$) for individual, unattached structural members, such as bars, beams, plates and shells. This is typical for steel, aluminum and other "hard" metals; damping may be two orders of magnitude greater for lead, concrete and brick.
3. Structural damping in assemblies is often on the order of $\eta \sim 0.01$, which is attributed to losses in (welded, bolted, riveted, and bearing type) joints.

Typical values of modal damping for aerospace and other structures, compiled by Wada and Des Forges [27] and by this author, are summarized below in Table 6.1 and Fig. 6.1, respectively.

The probability distribution for all of the above modal damping parameters, shown below in Fig. 6.2, indicates that over 90 % of damping values lies in below $\zeta_n = 2$ %.

The above cited data suggests that (1) the mean modal damping (ζ_n) of aerospace and similar structural systems is on the order of 1 % with (2) lack of any systematic frequency trend over a four decade frequency band. Moreover, (3) the "Rayleigh" proportional damping curve fit is clearly not representative of the collection of modal test data. Finally, (4) it is suggested that mode-to-mode variations in damping for each of the test articles are attributed to exercise of joints in each particular mode; note the test fixture mode for the Boeing ISS-P5, which is associated with localized deformation of two heavy steel plates. The above four observations are consistent with empirical damping data trends cited by Cremer, Heckl, and Ungar [26].

6.4.2 *Viscoelastic Material Behavior and Structural Joint Models*

While the material model based on elasticity theory [1] and structural damping [22–25] describes the dynamics of many structural assemblies, alternative models (e.g., schematic "circuit" models illustrated in Fig. 6.3) are required for viscoelastic materials, shock and vibration isolators and (welded, bolted, riveted and bearing) structural joints.

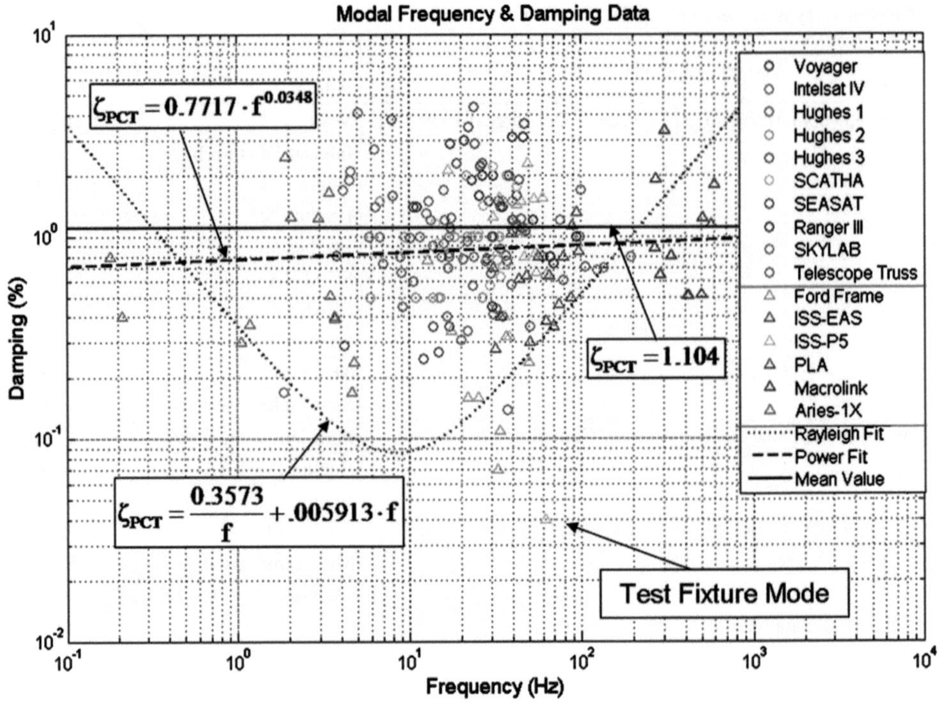

Fig. 6.1 Typical values of modal damping

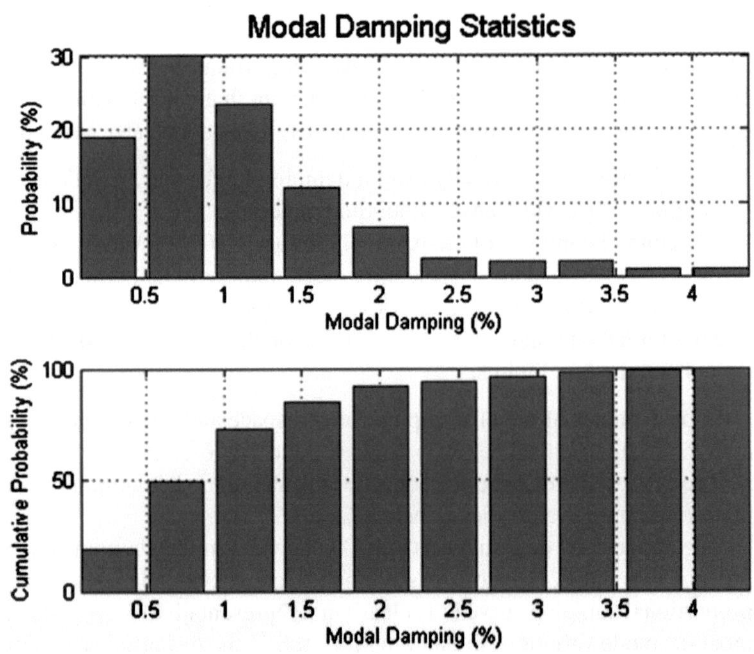

Fig. 6.2 Modal damping statistics

The Kelvin-Voigt model [28] represents the most commonly assumed (and physically erroneous) behavior of structural materials (the correct behavior more closely follows Kassner's form [25], (6.10)). The standard linear solid model (SLS), introduced by Zener [29], describes the fundamental (observable) behavior of viscoelastic materials, as does the more general "Maxwell-Weichert" model [28]. Closely related to the Maxwell-Weichert model is the Iwan model, which describes fundamental behavior of structural joints exhibiting slip-friction characteristics [30].

It is rather interesting to cite some historical notes related to the above four schematic "circuit" models:

1. According to Timoshenko's biographical sketch on Lord Kelvin [1], "He observes that structural materials are not perfectly elastic and, in investigating this imperfection, he introduces the notion of internal friction, which he studies by examining the damped vibrations of elastic systems. From his experiments, he concludes that this friction is not

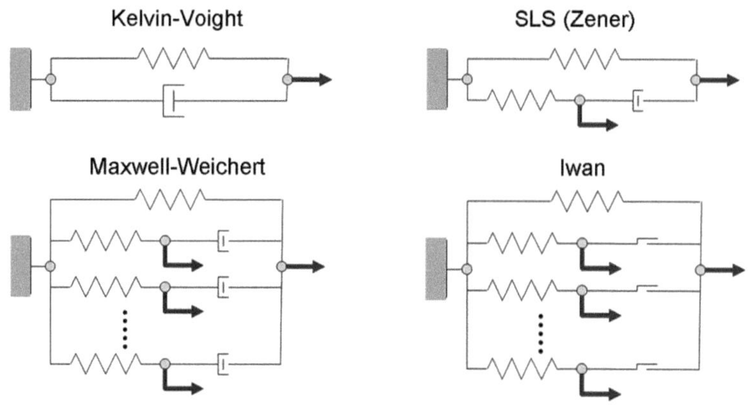

Fig. 6.3 Schematic "circuit" models for viscoelastic and structural joint behavior

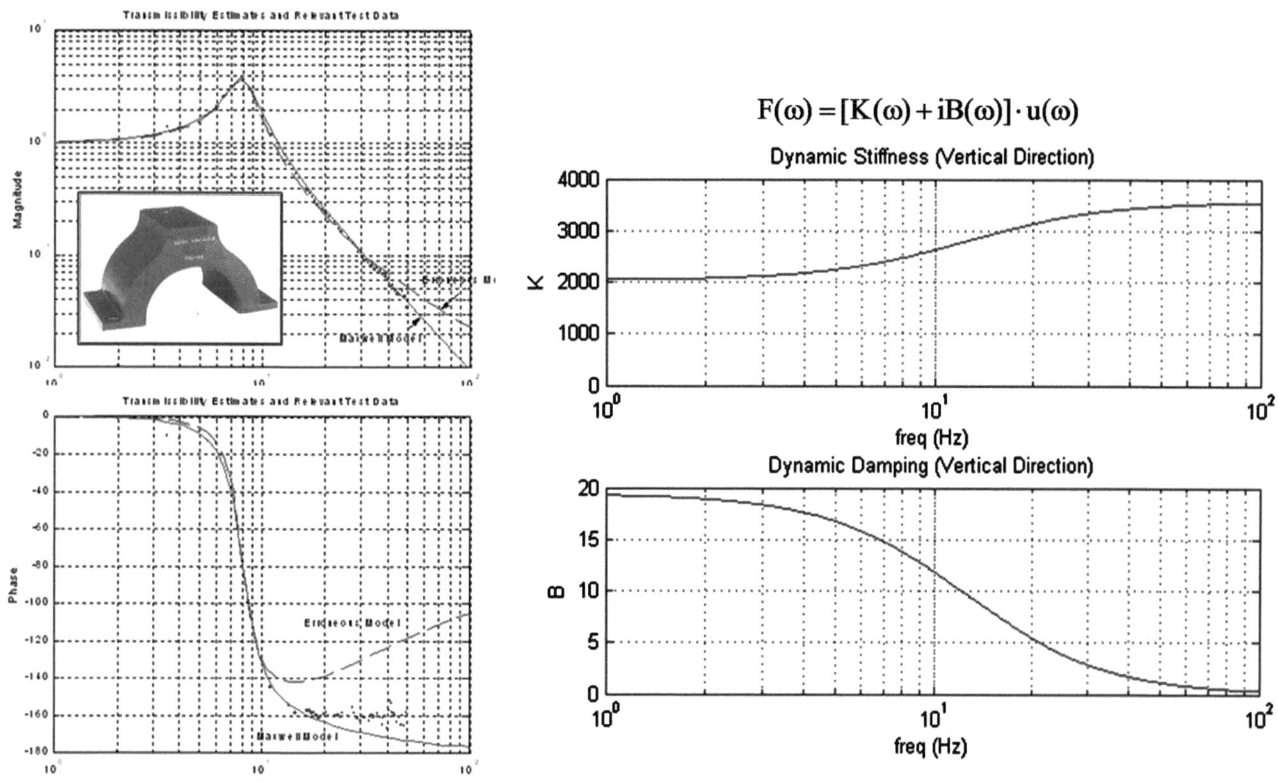

Fig. 6.4 Measured behavior of a typical viscoelastic shock and vibration isolator (colored version of the figure can be obtained from Dr. Coppolino -rcoppolino@macorp.net)

proportional to velocity. as in fluids". This observation predates the contributions of Kimball and Lovell [22] and Becker and Foppl [23] by about 50 years. More importantly, it appears that the so-called Kelvin-Voigt model is named in honor of the two scientists (rather than being a direct product of their works).

2. The SLS (Zener) [29] and Iwan [30] damping models are clearly attributed to the published works of the two respective authors.
3. The Maxwell-Weichert model [28] is obviously named in honor of the two scientists, as this generalization of SLS (Zener) was introduced well after their lifetimes.

An illustration of viscoelastic behavior of a typical shock and vibration isolator, used to protect electronic subassemblies, is summarized below in Fig. 6.4.

The graphics on the left side of Fig. 6.4 illustrate (a) magnitude and phase of frequency response estimated from broadband random data (blue curves), (b) curve fits for an erroneous Kelvin-Voigt model (red curves), and (c) curve fits for an appropriate SLS (Zener) model (green curves). The linear material model for the isolator, described in the frequency domain and illustrated on the right side of Fig. 6.4, indicates that the effective stiffness and damping for the component are frequency dependent.

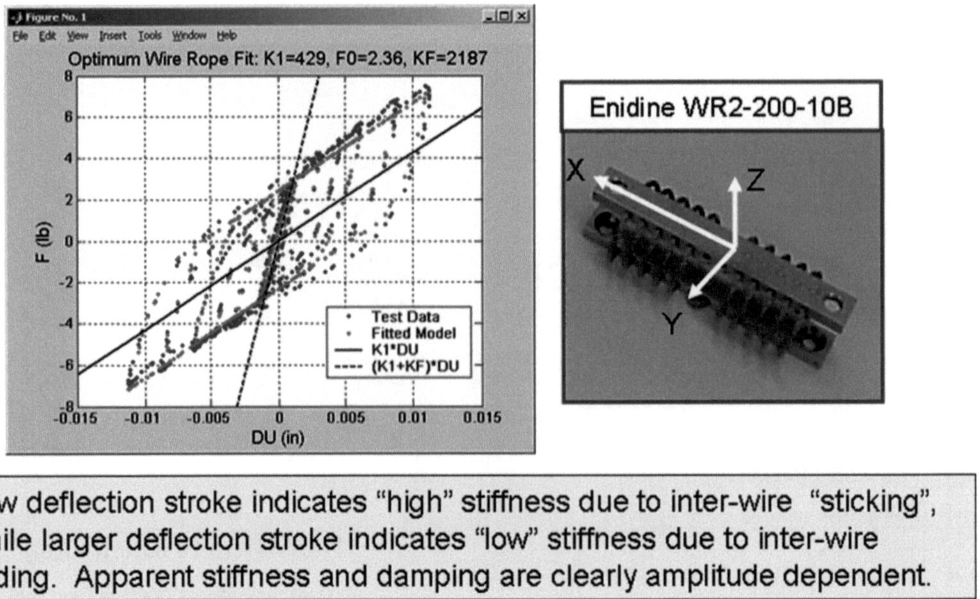

Fig. 6.5 Measured (Z-axis) behavior of a typical wire-rope shock and vibration isolator

An illustration of slip-friction behavior of a typical wire-rope shock and vibration isolator (which is representative of many structural joints) is provided below in Fig. 6.5.

The wire-rope isolator is a revealing example of non-linear characteristics of many structural joints. Not only is the friction damping nonlinear, but the apparent stiffness may be very high for low-level loading and substantially lower for high-level loading. This type of behavior is an appropriate lead in to the subject of structural component interface constraints.

6.4.3 A Fresh Look at Structural Damping

The primary aversion to employment of structural damping in structural dynamic modeling is due to the inconvenience of complex matrix equations for time domain analysis. This contrasts application of the complex formulation in aircraft flutter [16] and vibroacoustic [19, 20] analyses, which are generally performed in the frequency domain. Recently, Genta and Amati [31] published a state-space, time domain formulation of structural damping, which exploits the Maxwell-Weichert model. The Genta-Amati formulation for structural damping provides a welcome approach to structural dynamic modeling in situations for which a modal formulation is not convenient. It should be reiterated, however, that the relationship between modal viscous and structural damping coefficients (6.11), is quite appropriate when the system can be described in terms of modal coordinates.

6.5 Interface Flexibility in Structural Assemblies-A Retrospective

Mathematical solutions for structural components, subjected to a wide variety of boundary conditions are a staple in the historical development of structural mechanics theory [1]. The mathematical solutions are complemented by a wealth of empirical data indicating variability of joint stiffness as well as damping (especially when joints have slip-friction behavior). Significant deviations from assumed ideal joint behavior are also present in structures, which are composed of components that are welded to one another.

A quite revealing illustration of non-ideal boundary conditions is noted in results of a series of modal tests conducted on thin cylindrical shells (see Fig. 6.6) at NASA Langley Research Center in the mid 1970s [32].

Initial NASTRAN mathematical models of the test article were defined with fixed end boundary conditions for all test conditions [(1) empty, unpressurized, (2) empty pressurized, (3) half-filled with water unpressurized and pressurized]. Natural frequencies of shell breathing modes for the initial models were significantly higher than all corresponding test data.

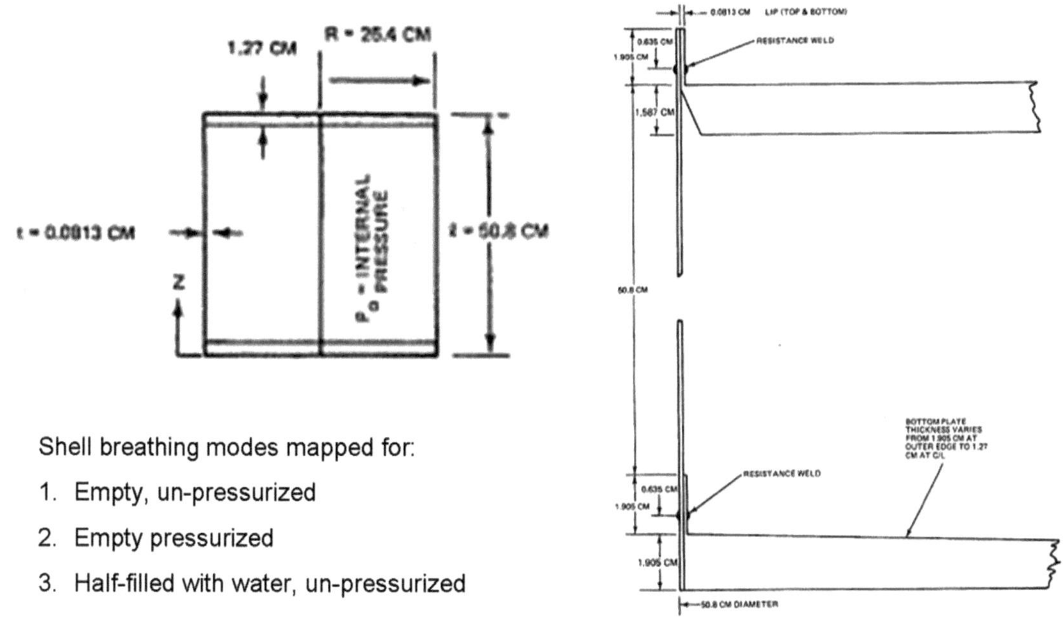

Fig. 6.6 NASA Langley Research Center cylindrical shell test article

After changing the NASTRAN model end boundary conditions to pinned (shear diaphragm), which was not intuitively obvious to the young engineer (Coppolino), all of the analytical natural frequencies closely followed modal test data, as illustrated in Fig. 6.7.

This lesson, experienced by many young engineers, is a clear example of non-ideal boundary conditions that exist in real structures. It is most unfortunate that this point is so easily missed by many practicing engineers due to the high degree of automation in day-to-day utilization of today's CAE tools [3].

6.6 Conclusions

The broad discipline of modern structural dynamic modeling is the product of advances in strength of materials, theory of elasticity, and theory of structures, and automated computational analysis. This paper focuses on several areas of structural dynamics modeling, primarily damping and structural joints, that suffer from widespread misconceptions, neglect and errors. These shortcomings are remedied by recollection of historical foundations of structural dynamic modeling.

A mathematically convenient model for damping forces in structures, namely proportional damping, is attributed to Lord Rayleigh, who cited a hypothesis that "*occurs frequently, in books*". The proportional damping notion, moreover assumes that damping forces are proportional to velocity, which Lord Kelvin during the late 1800s concluded is "not proportional to velocity, as in fluids" on the basis of experiments. By forgetting the actual statements, by Baron Rayleigh and Lord Kelvin, inappropriate models for damping in solid structures persist in the twenty-first century.

The engineering demands introduced during the early years of aeronautics and the subsequent space age caused investigators to engage in development of combined fluid–structure modeling techniques, which relied heavily upon empirical data. These activities led to a few isolated situations in which (fluid dynamic perturbation) damping forces were found to be velocity dependent and an overwhelming majority of situations pointing to damping forces that do not fit simplistic "velocity" dependent models. Most notable are the contributions of Kimball and Lovell, and Becker and Foppl who independently confirmed Lord Kelvin's observation that structural damping forces are generally not proportional to velocity. This model was subsequently expressed in terms of complex variables by Kussner and Kassner. Unfortunately, mathematical difficulties with the complex model in time domain applications limited use of the appropriate "structural dynamic damping" model to aeroelastic stability and vibroacoustic applications, which are typically treated in the frequency domain.

A wealth of empirical data supports the notion of structural damping in metallic structures that is proportional to displacement (strain) and in phase with velocity. Moreover, damping appears to be predominantly concentrated at joints. Important exceptions to the structural damping model are found in viscoelastic materials, shock and vibration isolators and

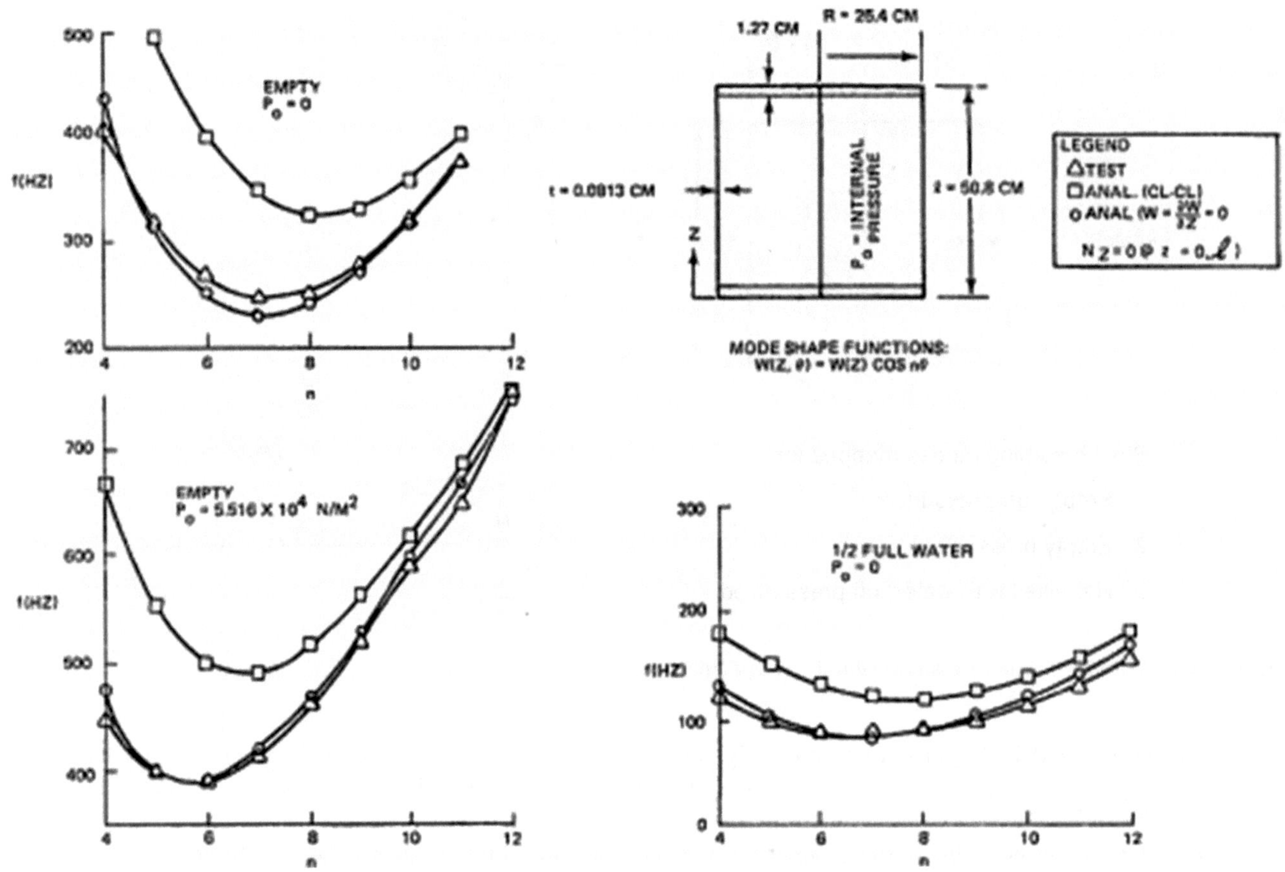

Fig. 6.7 Comparison of predicted and measured shell breathing mode frequencies

(welded, bolted, riveted and bearing) structural joints. Two key facts result from all empirical models of damping in solid structures, namely (a) a simple, linear, velocity dependent damping model does not appear to be physically appropriate, and (b) localized structural flexibility is a close partner with joint damping. Joint flexibility, as noted by many investigators, represents a strong influence on the modal characteristics of structural systems.

Opportunities for erroneous structural dynamic modeling and analysis offered by simplistic models of damping and joint stiffness and automated CAE tools are remedied (or redeemed) by following John von Neumann's advice, namely, "At a great distance from its empirical source, or after much abstract inbreeding, a mathematical subject is in danger of degeneration. Whenever this stage is reached the only remedy seems to me to be the rejuvenating return to the source: the reinjection of more or less empirical ideas". A similar point of council, attributed to Augustine of Hippo (c 400 AD) suggests "that we should be willing to change our mind ... as new information comes up". In the context of structural dynamic models, the apparently "new" information to some listeners is actually somewhat "old". Nevertheless, the empirical sources offer us a clear path from "sin" to "redemption".

References

1. Timoshenko S (1983) History of strength of materials. Dover Publications, New York
2. Zienkiewicz O, Taylor R, Zhou J (2005) The finite element method: its basis and fundamentals, 6th edn. Elsevier, Boston
3. Joshi A (2004) CAE data management using traditional PDM systems. In: 24th computers and information in engineering conference, ASME
4. Piersol A, Paez T (2010) Harris' shock and vibration handbook, 6th edn. McGraw-Hill, New York
5. Rayleigh JWS (1945) The theory of sound, 1st edn. Dover Publications, New York
6. Caughey T, O'Kelly M (1965) Classical normal modes in damped linear dynamic systems. J Appl Mech 32:583–588
7. Hurty W (1965) Dynamic analysis of structural systems using component modes. AIAA J 3(4):678–685
8. Craig R, Bampton M (1968) Coupling of substructures for dynamic analysis. AIAA J 6(7):1313–1319
9. MacNeal R (1971) A hybrid method of component mode synthesis. Comput Struct 1:581–601

10. Rubin S (1975) Improved component mode representation for structural dynamic analysis. AIAA J 13(8):995–1006
11. Benfield WA, Hruda RF (1971) Vibration analysis of structures by component mode substitution. AIAA J 9(7):1255–1261
12. (2012) Solidworks essentials, Dassault Systemes
13. (2009) Abaqus CAE user's manual, Dassault Systemes
14. Carrington J, Bowdon J (2008) NX6 assembly modeling update, Siemens PLM Software
15. Bisplinghoff R, Ashley H (1962) Principles of aeroelasticity. Wiley and Sons, New York
16. Rodden W (2011) Theoretical and computational aeroelasticity. Crest Publishing, Burbank
17. Rubin S (1970) Prevention of coupled structure propulsion system instability. NASA SP-2055, Washington, DC
18. Abrahamson HN (1966) The dynamic behavior of liquids in moving containers. NASA SP-106, Washington, DC
19. Lyon R (1975) Statistical energy analysis of dynamical systems: theory and applications. The MIT Press, Cambridge, MA
20. Beranek L (1971) Noise and vibration control. McGraw Hill, New York
21. Morison J et al (1950) The force exerted by surface waves on piles. Pet Trans 189:149–154
22. Kimball A, Lovell D (1927) Internal friction in solids. Phys Rev 30:705
23. Becker E, Foppl O (1928) Dauerversuche zur Bestimmung der Festigkeitseigenschaften, Beziehungen zwischen Baustoffdampfung und Verformungeschwindigkeit. Forschungsh Ver Deutsch Ing., No. 304
24. Kussner H (1935) Augenblicklicher Entwicklungsstand der Frage des Flugelflatterns. Luftfahrtforsch 12(6):193–209
25. Kassner R (1936) Die Berucksichtigung der inneren Dampfung beim ebenen Problem der Flugelschwingung. Luftfahrtforsch 13(11):388–393
26. Cremer L, Heckl M, Ungar E (1973) Structure borne sound, Springer-Verlag
27. Wada B, DesForges D (1979) Damping effects in aerospace structures. AGARD-CP-277
28. Gutierrez-Lemini D (2014) Engineering viscoelasticity. Springer, New York
29. Zener C (1948) Elasticity and anelasticity of metals. University of Chicago Press, Chicago
30. Iwan W (1967) On a class of models for the yielding behavior of continuous composite systems. J Appl Mech 89:612–617
31. Genta G, Amati N (2008) On the equivalent viscous damping for systems with hysteresis. Meccanica di Solidi
32. Coppolino R (1976) A numerically efficient finite element hydroelastic analysis, vol 1, Theory and results. NASA CR-2662, Washington, DC

Chapter 7
Muscle Property Identification During Joint Motion Using the NL-LTP Method

Michael W. Sracic

Abstract This work develops a simple low order model of the lower leg that incorporates the effects of a nonlinear biceps femoris muscle actuator and explores the feasibility of identifying a nonparametric model for the knee joint's rotational behavior as well as the joint activation function that drives the motion. The NL-LTP algorithm is applied to this system and some promising results are obtained. In particular, the nonlinear equations of motion and the joint activation moment are estimated from the "measured responses". The accuracy of these nonlinear parameter estimates and the implications for future experimental studies of biomechanical systems are discussed.

Keywords Nonlinear identification • Restoring force • Musculo-tendon model • Muscle activation • Hill-type muscle

7.1 Introduction

Biomechanical systems are challenging to model for many reasons, for example the properties of muscles and tendons are nonlinear, influenced by pathology, age, and other factors and even in ideal situations their mechanical properties are difficult to accurately measure. Likewise the skeletal structure, geometry, mass distribution, etc . . . of each subject is typically not known precisely yet it may affect the overall dynamic response. Several nonlinear system identification routines have been proposed for biomechanical systems, which could allow one to estimate these parameters in vivo. The author recently presented the NL-LTP algorithm [1, 2], a semi-nonparametric approach based on approximating the nonlinear system as linear time periodic about a limit cycle, and found encouraging results for some simple systems with masses and nonlinear springs. Over the past few decades, several efforts have been made to identify models for biomechanical systems from measurements [3–6], where the "inverse-dynamics" approach in [3] is frequently utilized (i.e., joint kinematics are measured and used to estimate joint moments). This approach requires one to assume accurate forms for system equation coefficients that scale the joint kinematics, which can be difficult.

While full system identification is not yet feasible, system identification techniques have proved their utility. For example, Dingwell et al. and Hurmuzlu et al. used system dynamics techniques to study the stability of the gait cycle [6–8]. Dingwell and Cusumano identified the maximum Floquet exponent and the maximum Lyapunov exponent from gait measurements and correlated them with gait stability, revealing a difference between the stability of healthy and diabetic neuropathic subjects [9]. Hurmuzlu et al. identified the maximum Floquet exponents for gait dynamics of polio survivors [10]. This work seeks to identify a model for a biomechanical system entirely from measurements, working towards identification of a complete dynamic model for the system.

The following section presents a simple model used to describe the motion of a simple biomechanical system, with special attention given to the manner in which the muscles are modeled, as they contribute to the nonlinearity of the system. The Results section presents the result of applying the NL-LTP identification method to simulated measurements from the system, revealing how one might obtain model of the net joint parameters for a system of muscles acting on a joint. Finally, a discussion and conclusions are presented.

M.W. Sracic (✉)
Milwaukee School of Engineering, 1025 N Broadway, Milwaukee, WI 53202, USA
e-mail: sracic@msoe.edu

© The Society for Experimental Mechanics, Inc. 2015

R. Allemang (ed.), *Special Topics in Structural Dynamics, Volume 6*, Conference Proceedings of the Society for Experimental Mechanics Series, DOI 10.1007/978-3-319-15048-2_7

7.2 Model and Simulation

Many dynamic simulations of human gait use musculo-skeletal models, where the equations of motion govern a skeletal system which is driven by muscle actuators. Models for these muscle actuators mimic the complicated physiological process of skeletal muscle contraction, which was described by Huxley in 1957 using his cross bridge theory [11]. This phenomenological theory is well accepted in the field, but other researchers have found it more practical to use empirically derived relationships for the forces produced during skeletal muscle contraction. For example, many researchers have produced simple "Hill-type" models of skeletal muscle (e.g., [12]) that can be used to accurately describe the forces produced by skeletal muscle. These models (referred to as "Hill-type" herein this work) have been used extensively in simulation [13–15] and have been shown to characterize the actual response of muscle in controlled experiments [16]. The model includes the effect of a contractile element (CE), passive elastic elements in parallel and series with CE, and a nonzero pennation angle (i.e. a muscle that has contractile fibers that are at an oblique angle to the connecting tendon).

The general model for a Hill-type muscle actuator is shown in Fig. 7.1a [15], where F_T describes the tendon force, L_T describes the series elastic tendon length, L_M describes the muscle actuator length, L_{MT} describes the total muscle-tendon length, and α_p describes the muscle fiber pennation angle. Many relationships have been used to describe the force that can be produced by the Hill-type muscle model, but generally the muscle force is assumed to depend on a muscle fiber length relationship and muscle fiber velocity relationship (i.e., muscle fibers that are shortening or lengthening). Since the model assumes that parallel and series elastic muscle-tendon components exist, the total force F_T produced by the muscle-tendon unit also depends on the passive force in these two components. These relationships are shown qualitatively in Fig. 7.1b from [15], but they have been described in many works as highly nonlinear with coupled activation, muscle-tendon property, and muscle-tendon contraction dynamics. See for example [17].

Hill-type muscle dynamics can be difficult to simulate in stable periodic limit cycles, therefore the following simplified version of the Hill-type muscle dynamics are used in this initial work. The model is assumed to contain a contractile element (CE) in parallel with a passive elastic (PE) component, and the series elastic component is omitted. The dynamics of the muscle force are described next with the simulation model.

The biomechanical model considered here is of a simple pendulum that is driven by a muscle actuator; a model such as this is appropriate for modeling a biceps femoris (hamstring) curl exercise. The effect of gravity is included in the model and the rotational degree of freedom, θ, is allowed to be large, so the pendulum system is nonlinear even without the addition of the muscle. The muscle is attached at a position $d_p = 0.05$ m and produces a force vector that is directed along the line of action of the muscle. The muscle is assumed to be straight, but its line of action with respect to the pendulum changes depending on the knee angle.

The muscle actuator is modeled as a simplified Hill-type muscle model and includes the contractile element and a parallel elastic element. The contractile tissues are represented with a function F_{MT} that depends on an activation function u, a scalar force-length relationship $f_l(l_{MT})$ where l_{MT} is the length of the muscle-tendon unit, a scalar force-velocity relationship $f_v\left(\dot{l}_{MT}\right)$ where $\dot{l}_{MT} = dl_{MT}/dt$ is the shortening or lengthening of the contractile component, and a passive spring force relationship to represent the parallel elastic component of the muscle where $k_{PE} = 5$ N/m is the passive spring constant of the muscle and $l_{MT0} = 0.45$ m is a reference muscle-tendon length based on the optimal fiber length and unstretched tendon length [15]. Using this model, the muscle force is expressed as follows,

$$F_{MT}(u, f_l, f_v) = u(t) f_v\left(\dot{l}_{MT}\right) f_l\left(l_{MT}\right) - k_{PE}\left(l_{MT} - l_{MT0}\right) \tag{7.1}$$

The simplified relationships for $f_l(l_{MT})$ and $f_v\left(\dot{l}_{MT}\right)$ are given in the following two equations. First, the force-length relationship is a piecewise linear function based on work from [16] that approximates the active force curve in the left most portion of Fig. 7.1b, which is used in many studies (e.g., [13, 15]).

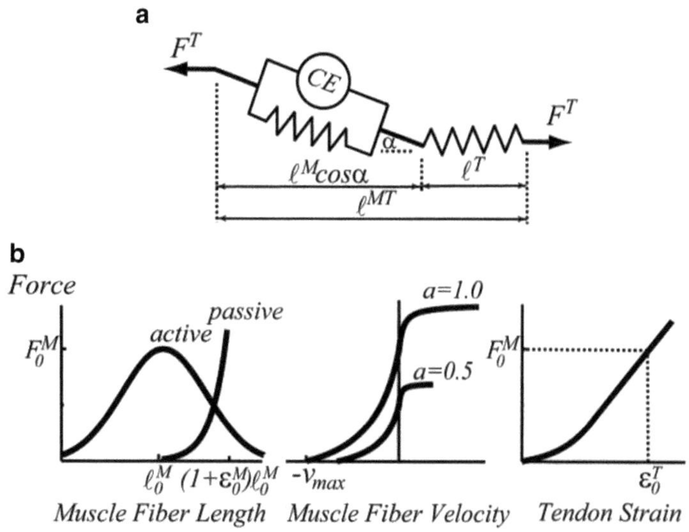

Fig. 7.1 (**a**) Hill-type muscle model schematic and (**b**) muscle force relationships from [15]

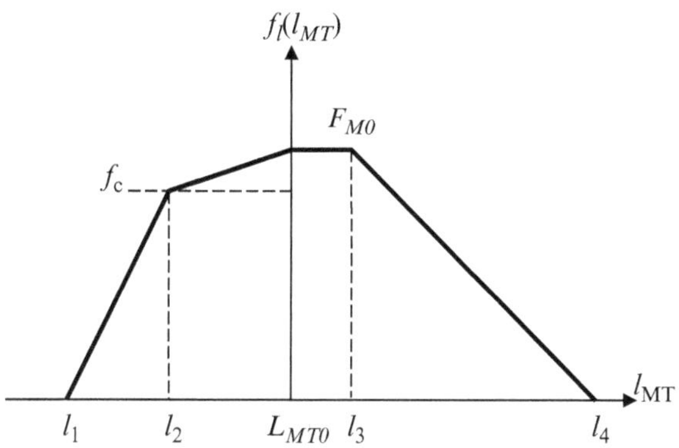

$$f_l(l_{MT}) = \begin{cases} \dfrac{f_c}{l_2 - l_1}(l_{MT} - l_1) & , \quad l_1 \le l_{MT} \le l_2 \\ \dfrac{F_{M0} - f_c}{L_{MT0} - l_2}(l_{MT} - l_2) + f_c & , \quad l_2 \le l_{MT} \le L_{MT0} \\ F_{M0} & , \quad L_{MT0} \le l_{MT} \le l_3 \\ \dfrac{-F_{m0}}{l_4 - l_3}(l_{MT} - l_3) + F_{M0} & , \quad l_3 \le l_{MT} \le l_4 \end{cases} \tag{7.2}$$

In Eq. (7.2), the parameter $F_{M0} = 1334$ N is the maximum isometric muscle fiber force [16] for the biceps femoris long head from [15]. The remaining parameters are chosen to bound the muscle-tendon length curve domain within the domain of the muscle-tendon deflection range. The values are as follows: $l_1 = L_{MT0} - 0.1$ m, $l_2 = L_{MT0} - 0.009$ m, $l_3 = L_{MT0} + 0.005$ m, $l_4 = L_{MT0} + 0.1$ m, and $f_c = 0.75F_{M0}$. The force-velocity relationship in the second portion of Fig. 7.1b is approximated as a simple linear function,

$$f_v\left(\dot{l}_{MT}\right) = \frac{1}{\dot{l}_{MT,\max}}\dot{l}_{MT} + 1 \tag{7.3}$$

where the maximum muscle fiber velocity $\dot{l}_{MT,\max} = 15$ m/s is used from [15]. In this form, the force-velocity relationship acts like a multiplier that at most doubles the muscle-tendon unit force with the muscle is at maximum lengthening and decreases the force when the unit is shortening (i.e., the force is zero at maximum shortening). Even with these simple approximations to the Hill-type muscle relationships, the muscle-tendon force is still quite complicated and nonlinear. The activation function for Eq. (7.1) is chosen as a repeating half-sine function, $u(t) = \sin(2\pi t)$, that only produces non-zero activation when $\sin(2\pi t) > 0$. This choice was made out of convenience in order to produce periodic motion of the system and is not assumed to be representative of true muscle activation dynamics (that topic will be studied in future efforts).

$$u(t) = \begin{cases} \sin(2\pi t) & , \quad n \le t \le (n + 1/2) \quad , \quad n = 0, 1, 2, \ldots \\ 0 & , \quad (n + 1/2) \le t \le (n + 1) \quad , \quad n = 0, 1, 2, \ldots \end{cases} \tag{7.4}$$

Using these descriptions for the muscle-tendon actuator force and the model from Fig. 7.2, the equations of motion for the system then become those seen in Eq. (7.5)

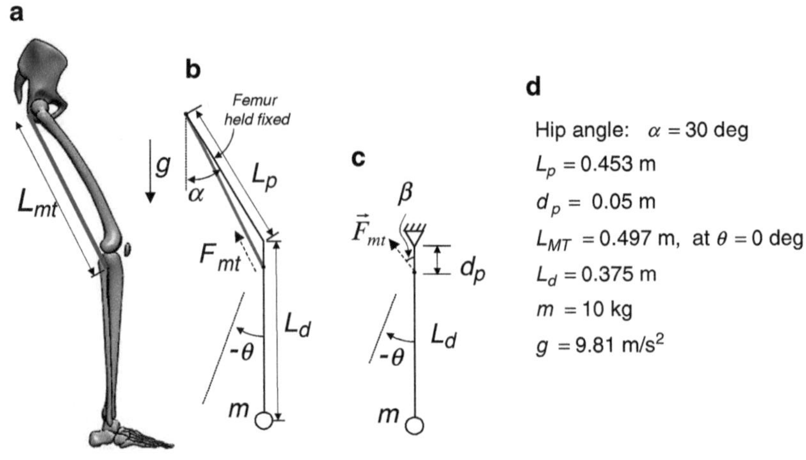

Fig. 7.2 (**a**) Representative musculo-skeleton schematic of model; (**b**) schematic of model geometry including proximal leg (femur) length L_P, distal leg (tibia/fibula) length L_d, hip angle α, and lumped distal leg mass m; (**c**) simple pendulum model for bicep femoris curl including the insertion point at distance d_p from the knee joint and the muscle force vector which is directed along the variable line of action of the muscle; (**d**) musculo-skeletal parameters used in the simulation model which are based on muscle and skeletal parameters from Delp et al. [13]

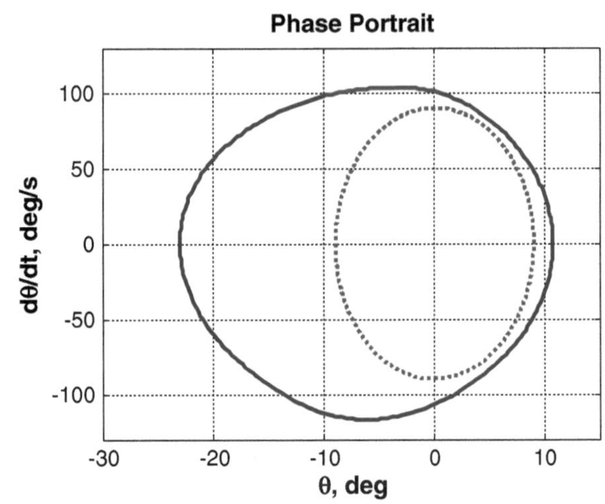

Fig. 7.3 Phase portrait of pendulum system in a periodic orbit (note: the *dotted line* shows a sinusoidal orbit for reference). (color version of the figure available in www.springer.com)

$$ml_d{}^2\ddot{\theta} + c\dot{\theta} + \frac{g}{l_d}\sin\theta = -F_{MT}(t)\sin(\beta)\,d_p \tag{7.5}$$

In the previous equation, $c = 1.2$ kg/[m-s] is an artificial torsional damping parameter that is added to capture the dissipation in the joint due to soft muscle-tendon tissues (musculo-skeletal dynamics here are assumed to be representative of an over-damped structural dynamic system), and β (see Fig. 7.2) is the time varying angle between the muscle-tendon unit and the pendulum (tibia/fibula). This angle can be described by the following function, which is defined by the Law of Cosines.

$$\beta = \arccos\left[\left(l_p^2 - L_{MT}^2 - d_p^2\right)/\left(2L_{MT}d_p\right)\right] \tag{7.6}$$

The following initial conditions can be applied to the model in order to obtain a stable limit cycle: $\left[\theta,\ \dot{\theta}\right]^{\mathrm{T}} = [-0.04491\ \mathrm{rad},\ \ 1.8057\ \mathrm{rad/s}]^{\mathrm{T}}$. This was found using the shooting method described in [18]. The phase portrait of the periodic response is shown in Fig. 7.3. The red-dotted line shows a pure sinusoidal for reference, revealing that the actual response of the system is not a simple harmonic.

In order to apply the NL-LTP method [1, 2], a small disturbance must be applied to the system in order to perturb the response from the periodic orbit. This was accomplished using a short duration impulsive half-sinusoid function with magnitude 1,000 N-m and duration 0.01 s (see for example [1]) applied to the pendulum at time zero with the same limit cycle initial conditions. The early response (i.e. for one cycle) of both the periodic limit cycle (dashed blue curve) and the

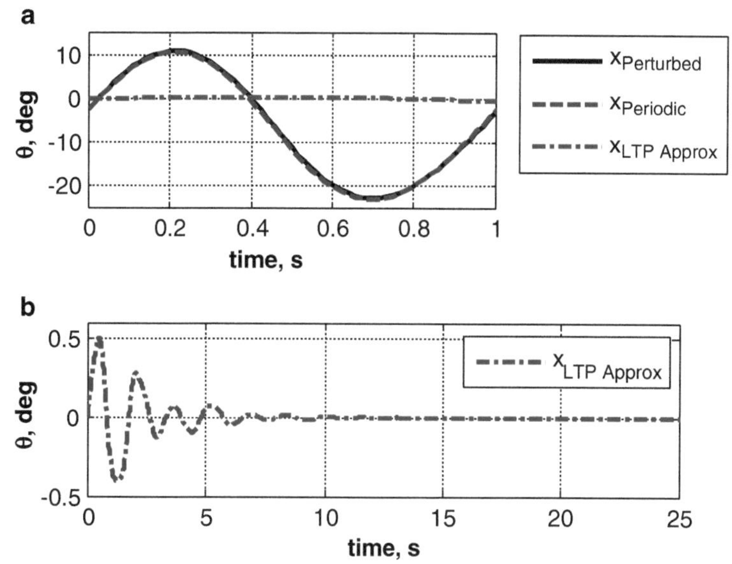

Fig. 7.4 (**a**) Plot of the periodic limit cycle (*dashed blue curve*), the perturbed response (*solid black curve*), and the perturbed-periodic signal (LTP) (*dash-dot green curve*); (**b**) Plot of the first 25 s of the perturbed-periodic (LTP) signal. (color version of the figure available in www.springer.com)

perturbed signal (solid black curve) are shown in Fig. 7.4a. Using the NL-LTP method, the approximate linear time periodic response is the difference between these two signals. This difference signal is also plotted in Fig. 7.4b (dash-dot green curve). Although in this simulation both the periodic and perturbed signals were simulated separately, in practice the perturbed signal is usually measured, and then the late time response of this signal (i.e. when the perturbations have dissipated) can be used to estimate the periodic limit cycle [1, 2].

Figure 7.4 shows that the perturbed-periodic signal in (*b*) has about a peak-to-peak magnitude of approximately 3 % of the periodic peak-to-peak signal magnitude, so the perturbed-periodic response should be well approximated as linear time periodic [2]. In Fig. 7.4b, the LTP signal has dissipated to very low levels within 25 s due to the torsional damping that was added. Using these time histories, the NL-LTP method was applied and the LTP system was identified. Briefly stated, the eigenvalue for the LTP system was estimated to be $\lambda = -0.4284 + 2.323i$, which after accounting for aliasing corresponds to a frequency of 0.634 Hz. The mode vector associated with the identified eigenvalue was expanded in a Fourier series (the Fourier series coefficients are plotted and tabulated in the Appendix) and used to estimate the LTP system. The remaining results from the NL-LTP identification are summarized in the next section. For more information regarding the NL-LTP method, the reader is directed to the following references where the method was developed in detail [1, 2].

7.3 Results

The NL-LTP method was used to identify the time-varying state matrix A(t) that describes the dynamics of the small deviations of the LTP system about the full nonlinear system periodic response. For the pendulum leg model, A(t) is a two-by-two matrix and the A(2,1) and A(2,2) terms contain the time varying coefficients for the state components θ and $\dot{\theta}$, respectively. These coefficients are analogous to the time varying stiffness and damping, respectively, of the system. These terms are plotted versus time in Fig. 7.5 using open blue circles for the terms that were estimated with the NL-LTP method. The equations of motion were used to calculate analytical system state matrix terms, and these are also plotted in the figure (solid blue lines).

In Fig. 7.5a, the A(2,1) coefficient varies by approximately 90 % of its starting magnitude (−26.1 to −2.1 [N/(kg-m)]). The estimated A(2,1) term agrees reasonably well with the analytical values, especially for the first half-cycle where the largest difference is about 10 % and occurs near 0.25 s when the analytical coefficient is approximately −2.1 [N/(kg-m)]. In the second half-cycle, from 0.5 to 1 s, there are some spurious fluctuations about the analytical curve, but the estimate curve agrees in an average sense. The analytical A(2,2) term is an order of magnitude smaller than the A(2,1) coefficient, and the term varies subtly with time in the first 5 s of the cycle (the coefficients varies by about 1 % from approximately −0.854 to −0.864 [(N-s)/(kg-m)]). The estimated coefficient fluctuates spuriously about the analytical coefficient curve, seeming to capture the coefficient value in an average sense.

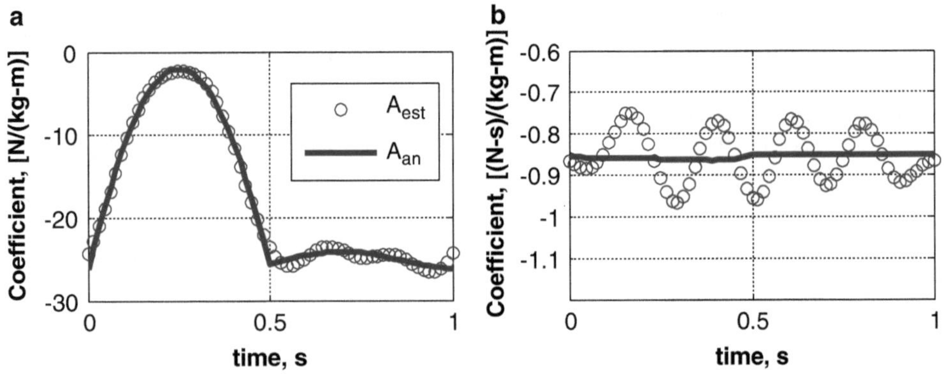

Fig. 7.5 LTP system state matrix terms plotted versus time (*open circles* estimated values, *solid blue line* analytical values) (**a**) A(2,1), time varying coefficient for state component θ; (**b**)) A(2,2), time varying coefficient for state component $\dot{\theta}$. (color version of the figure available in www.springer.com)

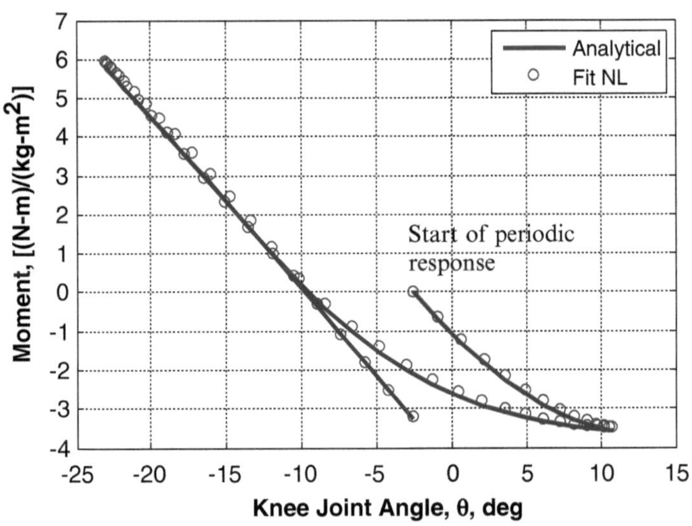

Fig. 7.6 Estimated nonlinear restoring moment due to joint rotation θ (*open circles* NL-LTP estimated values, *solid blue line* analytical values). (color version of the figure available in www.springer.com)

Next, the state matrix terms were integrated with respect to the measured periodic state components (i.e., with respect to θ for A(2,1) and $\dot{\theta}$ for A(2,2)). (Note that in practice, one usually measures one state (e.g., one might measure angular position, θ, with a goniometer), which can be used to estimate the other two states by numerically differentiating or integrating the measured signal. This process is quite straightforward for a periodic response, since one can represent the signal with a Fourier series. This process is discussed further in [1]). By integrating the A(t) matrix terms, one obtains an approximation for the nonlinear restoring moments due to the motion of each degree of freedom, individually. (In reality, the restoring angular acceleration is calculated, but this differs from the restoring moment by the moment of inertia constant, so the term "restoring moment" will be used hereafter). For example, the nonlinear function of restoring moments, $\widehat{f}_1(\theta)$, due to the angular rotation, θ, is plotted in Fig. 7.6. Once again, an analytical curve was also calculated. In this instance, the analytically calculated state matrix coefficients were integrated to obtain the corresponding analytical restoring moments due to knee joint rotation, θ.

The curves shown in Fig. 7.6 appear to be highly nonlinear as well as discontinuous. The estimated values shown with blue circles generally agree well with the analytical values. The A(2,2) term was also integrated to determine the function of restoring moments, $\widehat{f}_2\left(\dot{\theta}\right)$, due to joint rotational velocity, $\dot{\theta}$. This function is not plotted in this case because it is predominately linear. However, both restoring moment functions are needed because their sum provides a partial estimate for the nonlinear equations of motion of the system as discussed in [1]. For example, the following equation describes the full reconstructed equations of motion.

$$\ddot{\theta} = f\left(\theta, \dot{\theta}, u\right) = \widehat{f}_1(\theta) + \widehat{f}_2\left(\dot{\theta}\right) + \widehat{f}_3(u) \tag{7.7}$$

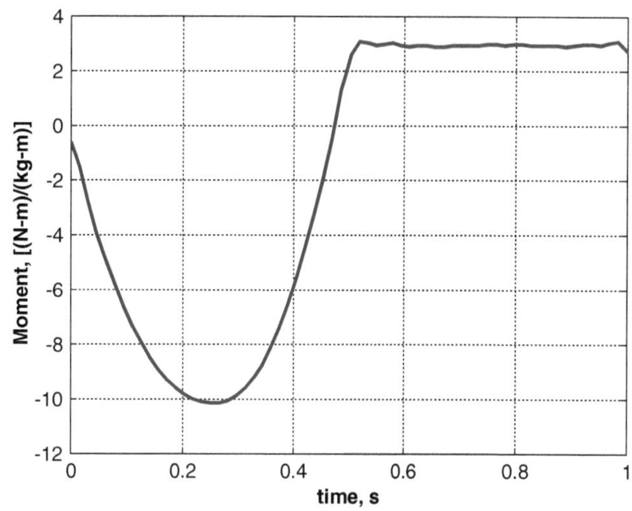

Fig. 7.7 Estimate for the active joint moment (due to muscle activation)

The only term in the previous equation that has not been estimated using the NL-LTP method is $\widehat{f}_3(u)$, which is the restoring moment due to the muscle activation function which drives the periodic response of the system. Therefore, Eq. (7.7) can be rearranged to estimate the activation function (or joint activation moment).

$$\widehat{f}_3(u) = \ddot{\theta} - \widehat{f}_1(\theta) - \widehat{f}_2\left(\dot{\theta}\right) \tag{7.8}$$

This equation was used to estimate the active joint moment acting on the knee that produced the periodic joint motion. The result is plotted in Fig. 7.7.

7.4 Discussion

The NL-LTP method provided a good approach to identify the linear time periodic system which describes small deviations of the leg model about its full nonlinear periodic response. In particular, the highly time varying A(2,1) term that is associated with the state component θ was reasonably approximated. The pendulum leg model was simulated with a high level of structural damping, and although it is large compared to the damping in other systems that have been studied, the method showed some error when trying to estimate the A(2,2) term that is associated with system damping. This is consistent with previous work using with the NL-LTP method, and is the subject of further study. One approach that is being considered is to estimate the damping related terms in the system with an alternative approach, for example using Subspace Identification methods.

The terms in the LTP system state matrix were estimated well enough to integrate them and estimate a portion of the original nonlinear equations of motion. Figure 7.6 shows that this produced a result that matches quite well for the nonlinear restoring moments due to rotation θ. The curve in Fig. 7.6 shows that as theta follows its periodic orbit, the restoring moment is highly nonlinear in the first half of the response, when the muscle actuator is active. This is reasonable since the muscle actuator is a nonlinear function of θ. The discontinuity is the curve in Fig. 7.6, suggests a jump in the restoring moment when the periodic response is about to restart, but in reality there can be no such jump since the periodic response is smooth. This apparent discontinuity comes from neglecting the constant of integration that arises when each system state matrix term is integrated with respect to the corresponding periodic state component response. The muscle actuator causes the nonlinearity in this system to be highly non-symmetric (i.e., the nonlinear restoring moments do not go to zero when the state components go to zero, so the constant is not a simple offset). Nevertheless, the functions were still very well estimated with respect to a corresponding analytical result that also neglected the integration constant. The subject of the integration constants is under current investigation, and it is not prohibitive for applying the method to get some initial estimates of musculo-skeletal model system dynamics.

The most novel step in the methods of the paper includes the use of the estimated nonlinear equations of motion to obtain an estimate for the activation moment. The results in Fig. 7.7 suggest that this approach could indeed be used to estimate the net active moment applied to a joint by the muscles during periodic motion. The current results do suffer the same problem regarding the missing offsets due to the integration. This can be seen in curve for Fig. 7.7. Since the underlying nonlinear response is smooth and periodic, the net applied moment should be periodic, but the curve has a discontinuous jump between the starting and ending values. This would need to be accounted for the get a true representation of the net moment applied by the muscle. This is also under current investigation.

A final limitation of the current simulation is the lack of measurement noise. The NL-LTP method has been shown to be sensitive to noise in the response [1, 2]. Since the NL-LTP method is a frequency domain identification technique, one would need to employ some signal processing to ensure that the frequency domain response peaks of the measured signals can be resolved above the noise floor levels of those signals. When this is the case, the results of this work suggest the NL-LTP method can be a valid approach to identify models from real measurements of musculo-skeletal systems. This is subject of future efforts for this work.

7.5 Conclusions

This paper summarizes a method that can be used to identify nonlinear system dynamics from measurements of musculo-skeletal systems. A simple pendulum leg model with one muscle-tendon actuator was simulated in periodic and then periodic plus perturbation responses where the main forcing was due to a periodic muscle activation force. The simulated responses were then treated as "measurements", and the NL-LTP method was used to identify the linear time periodic system. The LTP system state coefficients were well estimated by the method, and were used to estimate the original nonlinear equations of motion as well as the joint activation moment due to the periodic muscle-tendon force. This last results suggests that the method could be useful for obtaining joint activation functions from clinical measurements of joint motion. In its current form, the method only provides an approximate form for the nonlinear equations of motion and active joint moment, differing from the complete estimates by term that is neglected in the calculation process. Current efforts are being made to account for all terms, thereby producing a complete estimate for the nonlinear dynamics that describe the equations motion of the system, as well as to apply the approach to real experimental measurements of joint motion. The identification method could then be used to produce patient specific models, or possibly baseline models which can be altered to iterate on pathologic parameters that affect a given musculo-skeletal systems' dynamics.

Appendix

The NL-LTP method was used to identify the eigenvalue and mode shape of the LTP system simulated in this work. As a part of the method, this mode has to be expanded in a Fourier series. Only the terms in the series that correspond to mode peaks in the frequency domain plot of the linear time periodic response should be kept and used to construct the LTP A(t) matrix (see [1, 2] for further detail). The Fourier terms that were kept for the mode in the results of this work are plotted below in the figure and tabulated for additional reference (Fig. 7.8, Table 7.1).

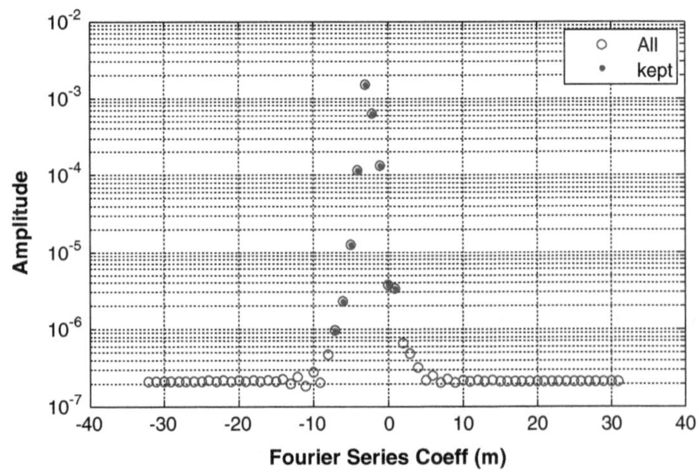

Fig. 7.8 Plot of the Fourier terms used in the Fourier description of the identified LTP mode

Table 7.1 Fourier expansion coefficients used to define the Fourier expansion of the mode identified with the NL-LTP method

Kept Fourier term, m	Fourier coefficient (magnitude)	Complex coefficient (complex valued)
−7	9.78E−07	$-4.8979e{-}09 + 1.4474e{-}08i$
−6	2.29E−06	$3.4225e{-}08 + 1.0340e{-}08i$
−5	1.28E−05	$-8.2993e{-}08 + 1.8120e{-}07i$
−4	1.18E−04	$1.6378e{-}06 + 8.3869e{-}07i$
−3	1.52E−03	$-1.0256e{-}05 + 2.1471e{-}05i$
−2	6.36E−04	$9.1710e{-}06 + 3.8285e{-}06i$
−1	1.35E−04	$-8.2563e{-}07 + 1.9453e{-}06i$
0	3.69E−06	$5.4632e{-}08 + 1.8471e{-}08i$
1	3.32E−06	$-1.9343e{-}08 + 4.8142e{-}08i$

References

1. Sracic M, Allen M (2014) Identifying parameters of multi-degree-of-freedom nonlinear structural dynamic systems using linear time periodic approximations. Mech Syst Signal Process 46(2):325–343
2. Sracic M, Allen M (2011) Method for identifying models of nonlinear systems using linear time periodic approximations. Mech Syst Signal Process 25(7):2705–2721
3. Erdemir A, McLean S, Herzog W, van den Bogert A (2007) Model-based estimation of muscle forces exerted during movements. Clin Biomech 22:131–154
4. Potluri C, Anugolu M, Schoen P, Naidu DS, Urfer A, Chiu S (2014) Hybrid fusion of linear, non-linear and spectral models for the dynamic modeling of sEMG and skeletal muscle force: an application to upper extremity amputation. Comput Biol Med 43:1815–1826
5. Eriten M, Dankowicz H (2007) A rigorous dynamical-systems-based analysis of the self-stabilizing influence of muscles. In: Proceedings of the ASME 2007 international design engineering technical conferences & computers and information in engineering conference IDETC/CIE, DETC2007-34469, Las Vegas, Nevada, pp 1–11
6. Dingwell J, Cusumano J, Cavanagh P, Sternad D (2001) Local dynamic stability versus kinematic variability of continuous overground and treadmill walking. J Biomech Eng 123:27–32
7. Dingwell J, Kang HG (2007) Differences between local and orbital dynamic stability during human walking. Trans ASME 129:586–593
8. Hurmuzlu Y, Basdogan C (1994) On the measurement of dynamic stability of human locomotion. J Biomech Eng 116:30–36
9. Dingwell J, Cusumano J (2000) Nonlinear time series analysis of normal and pathological human walking. Chaos 10(4):848–863
10. Hurmuzlu Y, Basdogan C, Stoianovici D (1996) Kinematic and dynamic stability of the locomotion of post-polio patients. J Biomech Eng 118:405–411
11. Huxley A (1957) Muscle structure and theories of contraction. Prog Biophys Biophys Chem 7:255–318
12. Zajac F (1989) Muscle and tendon: properties, models, scaling, and application to biomechanics and motor control. IEEE Crit Rev Biomed Eng 17(4):359–410
13. Delp S, Loan JP, Hoy M, Zajac F, Topp E, Rosen J (1990) An interactive graphics-based model of the lower extremity to study orthopaedic surgical procedures. IEEE Trans Biomed Eng 37(8):757–767
14. Thelen D, Anderson F (2006) Using computed muscle control to generate forward dynamic simulations of human walking from experimental data. J Biomech 39:1107–1115
15. Thelen D, Chumanov E, Best T, Swanson S, Heiderscheit B (2005) Simulation of biceps femoris musculotendon mechanics during the swing phase of sprinting. Off J Am Coll Sports Med 37(11):1931–1938
16. Siebert T, Rode C, Herzog W, Till O, Blickhan R (2005) Nonlinearities make a difference: comparison of two common Hill-type models with real muscle. Biol Cybern 98(11):133–143
17. Thelen D (2003) Adjustment of muscle mechanics model parameters to simulate dynamic contractions in older adults. Trans ASME 125:70–77
18. Sracic M, Allen M (2001) Numerical continuation of periodic orbits for harmonically forced nonlinear systems. Presented at the 29th international modal analysis conference (IMAC XXIX), Jacksonville

Chapter 8
On the Detectability of Femoral Neck Fractures with Vibration Measurements

Wolfgang Witteveen, Carina Wagner, Patrick Jachs, Stefan Froschauer, and Harald Schöffl

Abstract A suspicion of a femoral neck fracture is a frequently recurring situation, especially in nursing homes. For the clarification of such a suspicion normally imaging techniques are used. Such equipment is expensive and therefore is located in hospitals. In addition to the costs, a transport causes stress for the patient. This pilot study is devoted to the question whether the detection of a femoral neck fracture with vibration measurements is possible in principal. For this purpose vibration measurements on a dead body with intact, with partially fractured and with complete cut femoral neck have been performed. The frequency response function has been determined for all combinations on both sides of the body. It turned out that there is a difference in the frequency response functions of the fractured bone with respect to the intact bone when shaker testing is used.

Keywords Biomechanics • Vibrations • FRF • Femur • Femoral neck fracture

8.1 Introduction

The vibrations of a solid can be characterized by its resonant frequencies, see Maia [1] and Ewins [2]. The latter quantities are mainly system parameters which are dominated by a body's stiffness and mass distribution. A crack represents a change in stiffness which may lead to a change in the resonant frequency (=eigenfrequency). This observation is applicable to bones as well. This work is devoted to an answer on the question, whether a femoral neck fracture modifies the bones eigenfrequencies so much that it can be measured. If so, it would be possible to detect a femoral neck fracture based on vibration measurements.

A review on this approach from its beginning up to 1995 can be found in the review of Nokes [3]. In case of the tibia a lot of research confirms that the eigenfrequencies are significantly influenced by a fracture and the subsequent fracture-healing, see Cunningham et al. [4], Tower et al. [5], Nokes et al. [6], Benirschke et al. [7] and Nakatsuchi [8]. In most of the cases the frequency response function has been investigated which is a generalization of eigenfrequencies. Based on the same idea it has been investigated whether the status of integration of implants can be detected based on vibration date, see Cairns [9] and Cairns et al. [10]. In general it can be observed, that there was less focus on the femur as on the tibia. A femur related research can be found from Denker and Moberg [11] where the detectability of soft tissues between fractured bones has been investigated. Lappi et al. [12] determined some physical properties of the femur based on vibration measurements. Thomas et al. [13] observed in 1990 that the femurs eigenfrequencies depend on the amount of axial preload. Khalil et al. [14] performed vibration measurements on the embalmed human femur in order to document its vibrational characteristics. Based on the measurement data a mathematical model has been developed in order to compute the mode shape of a particular eigenfrequency. Investigations with respect to femoral neck fractures have been done by Misurya et al. [15] and Jawad, Odumala and Jones [16]. In the latter publication a vibrating tuning fork has been placed on the patella and the arrived vibrations at the anterior superior iliac spine (ASIS) have been measured. It has been tried to detect the fracture by a comparison of the vibration amplitudes. It turned out, that there was a significant amplitude reduction when comparing normal to fractured hips.

The objective of this work is similar as the one of Jawad, Obdumala and Jones [16]. In contrast to the latter publications frequency response functions have been used. It is known from mechanical engineering, that such functions are much more

W. Witteveen (✉) • C. Wagner
University of Applied Sciences Upper Austria– Wels Campus, Stelzhammerstr. 69, Wels 4600, Austria
e-mail: wolfgang.witteveen@fh-wels.at

P. Jachs • S. Froschauer • H. Schöffl
BioMed-zet Life Science GmbH, Linz, Austria

© The Society for Experimental Mechanics, Inc. 2015

R. Allemang (ed.), *Special Topics in Structural Dynamics, Volume 6*, Conference Proceedings of the Society for Experimental Mechanics Series, DOI 10.1007/978-3-319-15048-2_8

meaningful as time data, see Maia [1] and Ewins [2]. The goal of this pilot study is to see whether there is a significant change in the frequency response functions of a dead body's femur with and without neck fracture. Furthermore some investigations on the reproducibility of the FRFs will be presented.

8.2 Patients and Methods

An electromechanical vibrator is used in order to induce vibrations in the body. The vibrating part of the shaker is connected with the body as shown in Fig. 8.1. In this connection link a force sensor is mounted in order to get the time signal of the induced force. The connection link is fixed onto a piece of iron sheet which is pressed on the epicondylus lateralis via two cable retainer. For a better stability and repeatability a ground plate has been used. The shaker was fixed on the ground plate and the lower leg was fixated on that plate by means of a band. A sine sweep has been selected as excitation signal for the shaker. The sine sweep started at 40 Hz and the frequency has been continuously increased up to 800 Hz. Three FRFs have been determined because the patella, the epicondylus medialis and the ASIS have been considered as output locations. The third output location has been chosen in order to get a comparison with the work of Jawad, Obdumala and Jones [16]. The sensors have been applied by hand.

The software routines for LabView have been provided by the support of NI, Austria. The used hardware can be found in Table 8.1.

The dead body was provided by the Division of Clinical and Functional Anatomy of the Medical University in Innsbruck, Austria. It was a body of a male, died with 75 years due to a colon carcinoma. There was no abnormality and a bland anamnesis in the region of the femur. The first measurements have been done with the intact bone. For the next measurements the surrounding tissue has been opened and the femur was partially fractured by an osteotom (some kind of crosscut saw). The bone (femur) was partially sawed in approximately semi-diameter nearby femoral neck by the surgeon. This was done by experience and no measurements on the actual destruction have been performed. The final measurements have been performed on the totally cut through femoral neck. All the measurements (shaker and hammer testing) have been performed on the right and the left side. At each stage of fracture and with each method three FRFs have been measured.

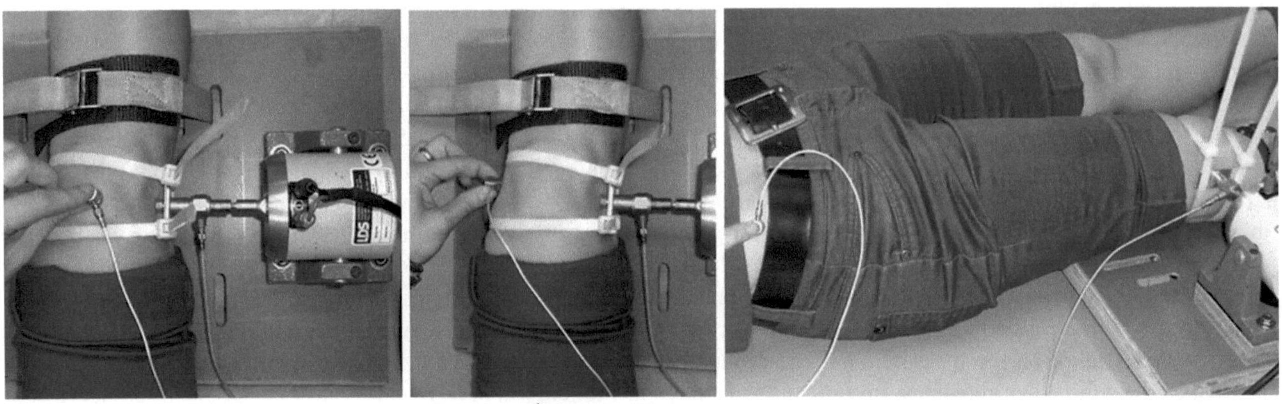

Fig. 8.1 Shaker testing, sensor at patella, at epicondylus medialis and at anterior superior iliac spine (left to right)

Table 8.1 Hardware

Description	Company	Model number
Shaker	LSD	V201/3-PA25E
Acceleration sensor	Kistler	8702B25
Force sensor	PCB	208C01
Data acquisition	NI	cDAQ-9178 with input module NI9234
Output (sine sweep for shaker)		cDAQ-9178 with output module NI9263

8.3 Ethics

The project with the registration number E-11-13(2.1.5) has been approved by the Upper Austrian Ethic Commission (http://www.ooe-ethikkommission.at). The project was permitted without any objections on the 29th of April 2013.

8.4 Results and Discussion

Figures 8.2, 8.3 and 8.4 contain the shaker testing FRFs. The black curves are the FRFs of the intact bone, the grey ones are the FRFs of the partially fractured bone and the dashed light grey curves hold FRFs of the totally cut through femoral neck.

The shaker FRFs (Figs. 8.2, 8.3 and 8.4) are considerably different in case of an intact or a broken femoral neck. For all three FRFs the differences become significant at 250 Hz and higher. It is interesting that the FRFs of the broken bone tend to higher values. That means that the mobility of the femur increases due to the fracture for frequencies higher as 250 Hz.

One can receive the impression that the FRFs obtained on the epicondylus medialis do have the best quality in terms of amplitude and noise. This is not surprisingly because both, input and output are close together and directly located on the surface of the femur itself. This is not true when the FRF is determined on the patella or on the ASIS.

There is no consent in the literature whether the properties of the left and the right leg should be similar. Khalil et al. [14] observed significant differences on the embalmed femur whereas Benirschke [7] did not observe a remarkable difference on the tibia in-vivo. In our case it is to report that the surgeon recognized a difference in the length of the left and the right femoral neck. There was no remark on that in the person's case history and there was no grain in the area of the femoral neck.

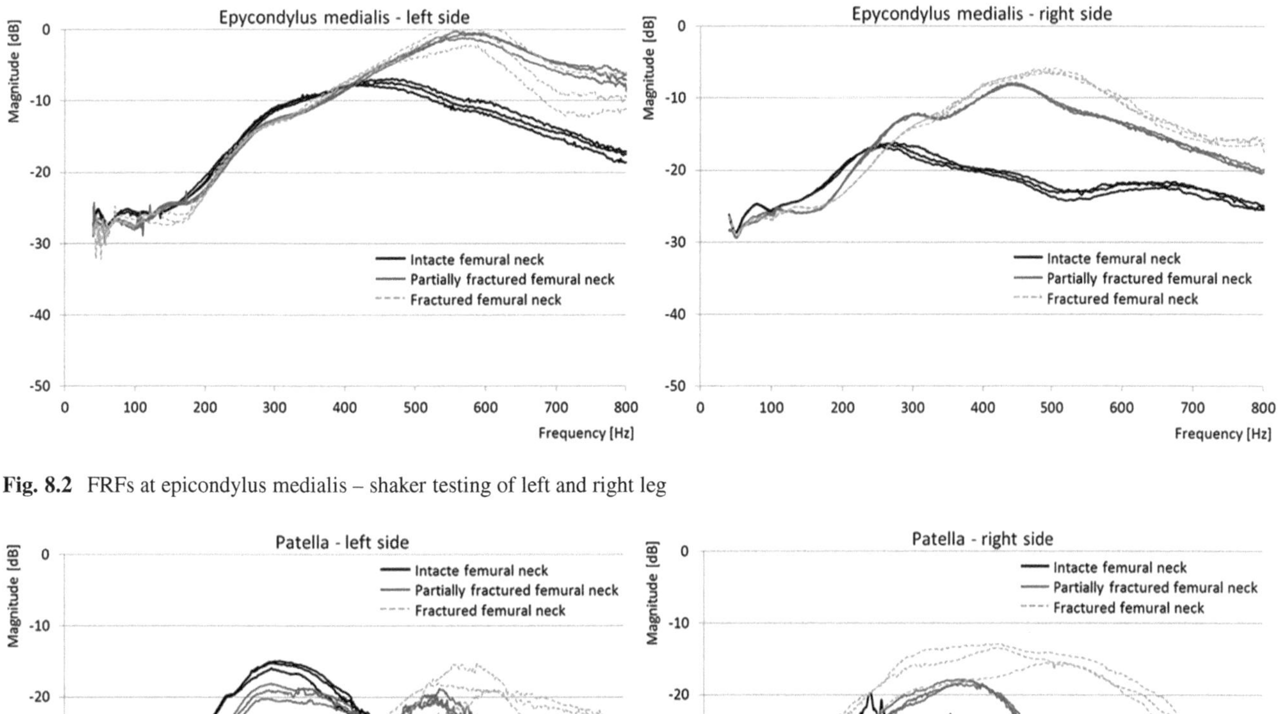

Fig. 8.2 FRFs at epicondylus medialis – shaker testing of left and right leg

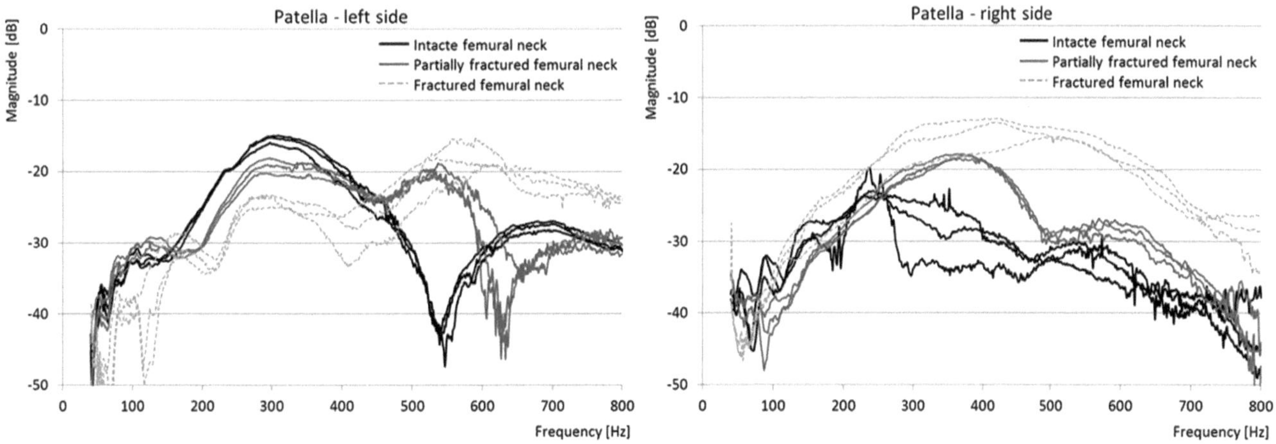

Fig. 8.3 FRFs at patella – shaker testing of left and right leg

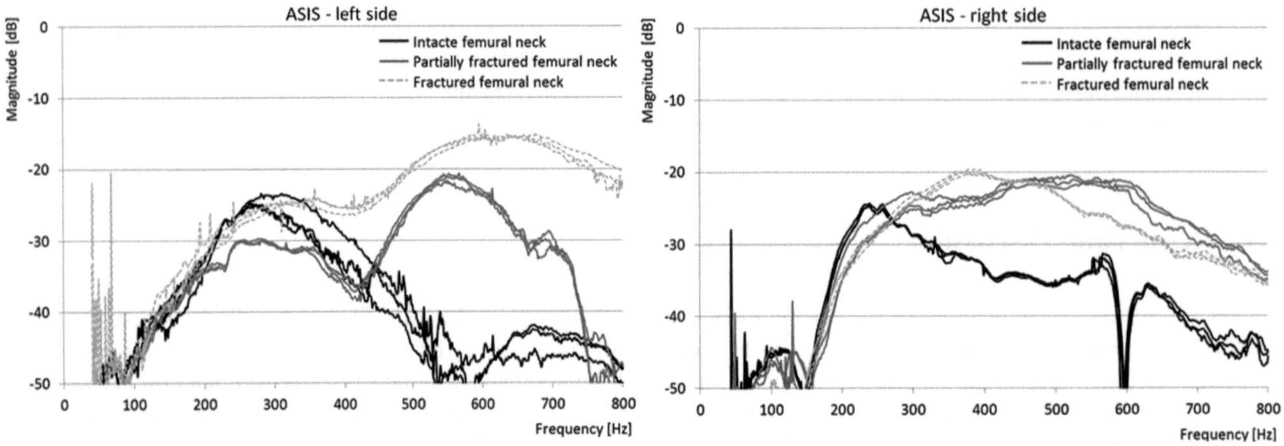

Fig. 8.4 FRFs at ASIS – shaker testing of left and right side

Maybe there was an unknown old fracture of the femoral neck. Consequently, the reason for this difference is not known and no general conclusion can be drawn on the symmetry of the measurements. But the conclusion that a fracture leads to significantly different FRFs still holds.

Figure 8.4 needs to be discussed in the context of the observations of Jawad, Obdumala and Jones [16]. There a vibrating tuning fork is pressed on the patella. The first eigenfrequency of this tuning folk was at 128 Hz. The thereby induced vibrations have been measured at the ASIS. It has been tried to detect the fracture by a comparison of the vibration amplitudes in case of a fractured and intact femoral neck. It turned out, that there was a significant amplitude reduction about a factor of 2 when comparing normal to fractured hips. Such a tuning folk is basically a shaker vibrating with one single frequency, even 128 Hz. One significant advantage of the FRF approach is that instead of one frequency an entire frequency band (40–800 Hz) is considered. This leads to much more information and Figs. 8.2, 8.3 and 8.4 indicate that frequencies higher as 250 Hz are better suited to detect a fracture. Another advantage of the FRF method is its objectivity. This is based on the fact, that the FRF is the quotient of output and input and not only the output, as it is the case in the publication of Jawad et al. [16]. If, for example, the input is twice as high as at another time, the FRF will lead to the same result whereas the output measurement only, will not. However, Jawad et al. [16] observed a reduction at about a factor of 2. This is equivalent to 6 dB when FRFs are used. The FRF value on the right leg of Fig. 8.4 with broken femoral neck is indeed smaller than the one with intact one. On the left leg, no significant difference in the FRFs can be observed at 128 Hz. All other FRFs have at 128 Hz a tendency to an amplitude reduction due to the fracture. Concluding it can be said, that the measurements indicate the most significant differences at higher frequencies and the shaker method is a generalization of the tuning fork method with significant more information.

Note, that the FRFs obtained by impact testing do not indicate a clear change due to the fractured femoral neck. The interested reader is referred to [17] for more comments on that observation.

8.5 Reproducibility of FRFs

A good repeatability of the FRFs is an essential characteristic for a clinical use. The literature indicates, that the repeatability is given for this kind of measurements, see Benirschke [7] for the tibia and Jurist [18] for the ulna. In order to verify this assumption four (healthy) test persons have been accompanied for several days in order to get several FRFs at different days and different times. Figure 8.5 contains a collection of nine FRFs of one test person measured during 3 days at different times at the day.

The FRFs of the other three persons are of the same quality. It turned out, that there is a remarkable reproducibility of the FRFs.

Acknowledgement The authors wish to thank the individual who donated his body and tissues for the advancement of education and research.

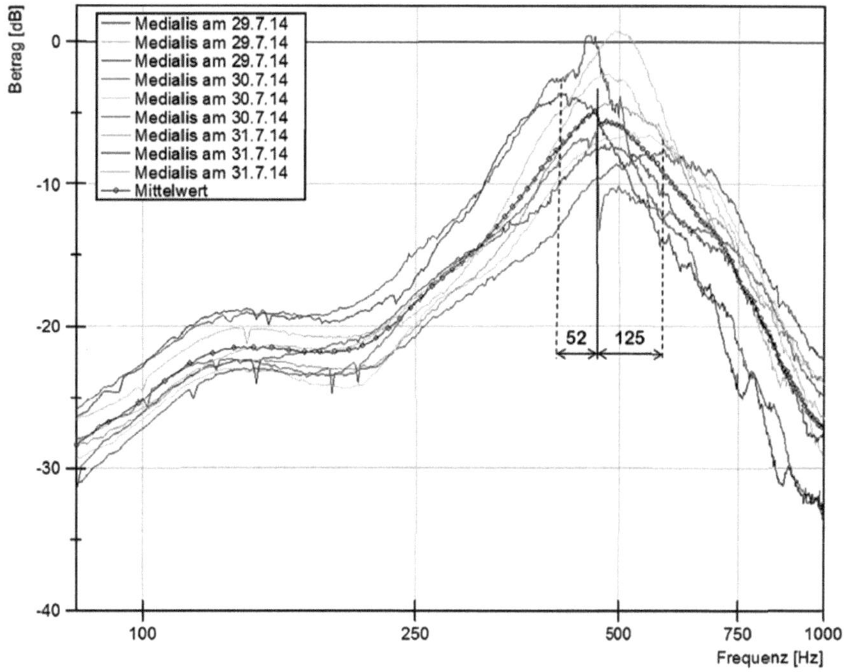

Fig. 8.5 FRFs of one test person at epicondylus medialis (3 days – 9 measurements)

References

1. Maia NMM (ed) (1998) Theoretical and experimental model analysis. Research Studies Press
2. Ewins DJ (2000) Modal testing, 2nd edn. Research Studies Press
3. Nokes L (1998) The use of low- frequency vibration measurement in orthopedics. Proc Instn Mech Engrs 213:271–289
4. Cunningham JL, Kenwright J, Kershaw CJ (1990) Biomechanical measurement of fracture healing. J Med Eng Technol 14(3):92–101
5. Tower SS, Rodney BK, Duwelius PJ (1993) Resonant frequency analysis of the tibia as a measure of fracture healing. J Orthop Trauma 7(6):552–557
6. Nokes L, Mintowt Czyzt WJ, Fairclought JA, Makie I, Williams J (1985) Vibration analysis in the assessment of conservatively managed tibial fractures. J Biomed Eng 7:40–44
7. Benirschke SK, Mirels H, Jones D, Tencer AF (1993) The use of resonant frequency measurements for the noninvasive assessment of mechanical stiffness of the healing tibia. J Orthop Trauma 7(1):64–71
8. Nakatsuchi Y, Tsuchikane A, Nomura A (1996) Assessment of fracture healing in the tibia using the impulse response method. J Orthop Trauma 10(1):50–62
9. Cairns NJ (2010) The feasibility of vibration analysis as a technique to detect osseointegration of transfemoral implants. PhD thesis, Faculty of Built Environment and Engineering, Queensland University of Technology, Brisbane
10. Cairns NJ, Adam CJ, Pearcy MJ, Smeathers JE (2011) Evaluation of modal analysis techniques using physical models to detect osseointegration of implants in transfemoral amputees. In: Proceedings of the 33rd annual international conference of the IEEE Engineering in Medicine and Biology Society. IEEE, Boston, pp 1600–1603
11. Dencker H, Moberg E (1968) Diagnosis of soft-tissue interposition in shaft fractures of the humerus and femur by measuring the conduction of vibration across the fracture. Acta Chir Scan 134:540–542
12. Lappi VG, King MS, May IL (1979) Determination of elastic constants for human femurs. J Biomech Eng 101:193–197
13. Thomas AMC, Luo DZ, Dunn JW (1991) Response of human femur to mechanical vibration. J Biomed Eng 13:58–60
14. Khalil TB, Viano DC, Taber LA (1981) Vibrational characteristics of the embalmed human femur. J Sound Vib 75(3):417–436
15. Misurya RK, Khare A, Mallick A, Sural A, Vishwakarma GK (1987) Use of tuning fork in diagnostic auscultation of fractures. Injury 18:63–64
16. Jawad Z, Odumala A, Jones M (2012) Objective sound wave amplitude measurement generated by a tuning fork. An analysis of its use as a diagnostic tool in suspected femoral neck fractures. Injury 43:835–837
17. Witteveen W, Wagner C, Jachs P, Froschauer S, Schöffl H (2014) Pilot study on the non-invasive detectability of femoral neck fractures with frequency response functions. World J Mech 4(7):210–216. http://dx.doi.org/10.4236/wjm.2014.47022
18. Jurist JM (1970) In vivo determination of the elastic response of bone – method of ulnar resonant frequency determination. Phys Med Biol 15(3):417–426

Chapter 9
Static Calibration of Microelectromechanical Systems (MEMS) Accelerometers for In-Situ Wind Turbine Blade Condition Monitoring

O.O. Esu, J.A. Flint, and S.J. Watson

Abstract As wind turbines get larger, operation and maintenance costs can be expected to rise unless reliability is improved. One important strategy for achieving this is through condition monitoring. Condition monitoring is a preventive type of maintenance based on the actual health of the wind turbine under observation. Its use allows an operational strategy to be employed based on information measured and provided by a condition monitoring system. This paper outlines the procedure for least square static calibration of inexpensive Microelectromechanical Systems accelerometers identified and instrumented for detecting and measuring changes in natural frequency (key information useful for condition monitoring) of a 4.5 m long wind turbine blade. The calibration procedure converts the local accelerometer coordinates to the global coordinate system of the blade and eliminates the accelerometer offsets from results. The objective is to enable easier comparison of results obtained from multiple accelerometers positioned arbitrarily along the blade, independent of the accelerometer orientation on the usually curved wind turbine blade surface.

Keywords Condition monitoring • Microelectromechanical systems (MEMS) accelerometers • Vibrations • Modal analysis • Wind turbine blades

9.1 Introduction

Wind energy is the second fastest growing form of renewable energy worldwide (after photovoltaic energy) and it is the fastest growing in the European Union [1]. Wind turbine systems harness the kinetic energy of the wind which can be used for mechanical work like pumping water (common in developing countries), or to turn a generator to produce electrical power [2].

The blades of wind turbines are made of different materials which contribute to a high-level of uncertainty in predicting their health [3–5]. Production defects (delamination, bond defects, imbalance, aerodynamic asymmetry and wrinkles), which typically occur during the manufacture and design stages or have been identified during experimental load testing of the blades, as well as operational defects which occur during the turbine lifetime due to accumulation of dirt, moisture, ice loading and insect contamination can cause cracks and surface damage to the blades. No detailed classification system exists for describing wind turbine blade failures and information on failures are not generally available. Turbine manufacturers and operators are not legally obliged to provide a detailed root-cause assessment in a publicly available forum; therefore, wind turbine blade defects and failures in the current paper, have been identified from studies conducted by various research groups [6–21].

There have been significant advances in wind turbine technology such as improved aerodynamics and improved structural dynamics with the primary purpose of increasing energy yield. A well-known example is the trend in turbine blade designs to larger and more flexible structures, which have been shown to increase energy capture efficiency. These enhancements have associated costs and unless reliability is improved through condition monitoring to manage maintenance, the high capital invested in wind energy projects will not be fully recovered. Condition monitoring is therefore necessary. It is an important and useful strategy which triggers maintenance actions, minimises breakdowns whilst avoiding periodic assessment and associated downtime [22, 23].

Various condition monitoring techniques exist and the current paper considers the vibration-based method. This method is based on well-established vibration principles of structures which suggest that the resonant properties of a system/structure such as its natural frequency, damping and mode shape are directly influenced by its physical properties. Therefore, any

O.O. Esu (✉) • J.A. Flint • S.J. Watson
School of Electronic, Electrical and Systems Engineering, Loughborough University, Loughborough, Leicestershire LE11 3TU, UK
e-mail: O.O.Esu@lboro.ac.uk

© The Society for Experimental Mechanics, Inc. 2015

R. Allemang (ed.), *Special Topics in Structural Dynamics, Volume 6*, Conference Proceedings
of the Society for Experimental Mechanics Series, DOI 10.1007/978-3-319-15048-2_9

changes or deviation in the physical properties of a structure such as cracks and erosion will cause a change in the modal parameters of the structure observed. Measuring the modal parameters of a wind turbine blade could potentially provide useful information on the blade health. Esu et al. [24–26] identified MEMS accelerometers as a potential alternative/replacement to existing sensors used.

MEMS accelerometers are miniaturised low-cost sensors with increasing applicability such as in navigation systems in smartphones and in the automotive industry for airbag deployment in vehicles [27, 28]. Their small size allows them to be placed in more obscure locations and on systems especially when weight variation is undesired. These characteristics make MEMS accelerometer suitable for vibration-based condition monitoring and the current paper introduces a calibration method for resolving the measured output from these accelerometers, while they are in use on a wind turbine blade for modal testing.

A significant problem with the use of accelerometers generally is the drift of the sensitivity and offset. Sensitivity, also referred to as scale factor, is the ratio of change in signal to change in acceleration and is ratiometric to the supply voltage in analogue-sensors. Offset, also known as the zero-g bias level, is the direct current (dc) output level of the accelerometer when it is not in motion or being acted upon by the earth's gravity [29, 30]. These two parameters (sensitivity and offset) of each axis on each sensor must be characterised to permit accurate conversion from voltage to acceleration. Thus, the purpose of calibrating the accelerometers is to limit the effects of these drifts. Static calibration where the accelerometer can be oriented freely in various positions is usually the method used. However, the current paper introduces an in-use static-calibration method where the accelerometers have constrained/restricted positioning.

Section 9.2 describes the MEMS accelerometers used in experiments and their positioning on a 4.5 m long 25 kW Carter wind turbine blade [31]. In Sect. 9.3, the in-use static calibration methodology is described and the results obtained from the calibration routine are also discussed. Section 9.4 offers conclusions on work conducted and ideas for future work to improve the campaign for cost-effective condition monitoring of wind turbine blades using low-cost MEMS accelerometers.

9.2 MEMS Accelerometers and Positioning

MEMS accelerometers are typically classified as piezoresistive or capacitive based accelerometers [28]. Capacitive based MEMS accelerometers generally offer more sensitivity and more resolution than the piezoresistive MEMS [32, 33]. The ADXL335 [29] capacitive based MEMS accelerometers were selected for use in the current paper. The ADXL335 accelerometer is a small, thin, low power device that can measure static acceleration due to gravity in tilt-sensing applications and dynamic acceleration resulting from motion, shock, or vibration in 3-axes (X_{out}, Y_{out} and Z_{out}) with a minimum full-scale range of ±3 g (where $g = 9.81\ ms^{-2}$). Figure 9.1a shows a picture of the ADXL335 accelerometer relative to a UK £1 coin and the accelerometer soldered to an expansion printed circuit board (PCB) to enable wires to be connected to the sensor. Figure 9.1b is a sketch of the ADXL335 device showing its measurement axes.

The analogue signals from these MEMS accelerometers are ratiometric to the supply voltage and vary by several hundred millivolts between axes. The ADXL335 has a nominal sensitivity of 300 mV/g of applied acceleration for a voltage supply of 3 V, centred on a +1.5 V offset per axis [29]. 0.01 μF capacitors were soldered to each of the output pins on each of the accelerometers to implement low-pass filtering with a bandwidth of 500Hz and to lower the noise floor whilst improving the resolution of the accelerometer measurements. A 0.1 μF capacitor was also soldered at each accelerometer supply pin to decouple the accelerometer from noise on the power supply.

A custom made mechanical support was constructed from steel box-section beams to mimic the hub of a real turbine blade and form a test fixture. This enabled a 4.5 m long 25 kW Carter wind turbine blade [31] to be suspended above the ground using its existing mechanical fixings and allowed the blade to be rotated manually about its axis. Five PCB mounted

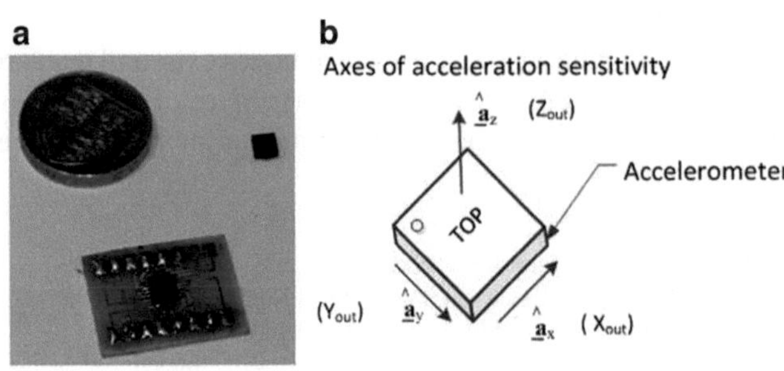

Fig. 9.1 ADXL335 MEMS accelerometer: (**a**) Stand alone and PCB mounted in relation to UK £1 coin and (**b**) Axes of acceleration sensitivity

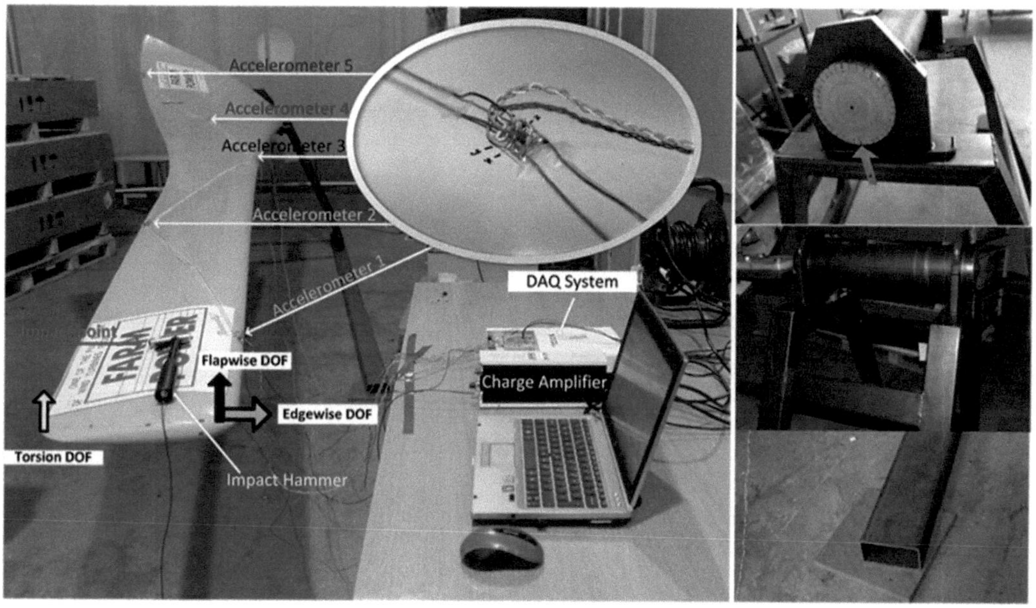

Fig. 9.2 A picture of the 4.5 m long wind turbine blade showing the positions of the accelerometers, the specially designed rig and anti-vibration pads used to support the test structure

ADXL335 MEMS accelerometers were glued to the blade as shown in Fig. 9.2, in order to measure movement within the three defined degrees of freedom (DOFs) of a wind turbine blade (flapwise, edgewise and torsion). This method was used to discretise the motion of a wind turbine blade as it would be impractical to deploy sensors throughout all material points of the blade to measure its motion. Flapwise DOF is the up and down deflection of the blade when in the horizontal position as shown in Fig. 9.2. Edgewise DOF describes the side deflections perpendicular to flapwise DOF. Torsion is assumed to be described by rotation of the chord about the pitch axis [34, 35]. Accelerometers were wired to a 16-channel NI USB-6,251 data acquisition system (DAQ) [36] set to read 5 s of data at a rate of 20 k samples/s using LabVIEW SignalExpress software [37]. MATLAB [38] was used to analyse measured data.

9.3 Static Calibration

In static calibration, the output voltage of a uniaxial accelerometer for instance, is a measure of the angle θ [rad] between the sensitive axis of the device and the direction of gravity. The parameters, sensitivity and offset of a uniaxial accelerometer can be obtained by applying two different angles to the device: thus, two equations with two unknowns are obtained. Therefore, a triaxial accelerometer could be calibrated by keeping each axis under two different known angles θ with respect to gravity, as shown in Fig. 9.3 and summarised in Table 9.1, thus obtaining six equations with six unknowns. This is the minimum required set of equations to determine the three different sensitivities S_x, S_y and S_z [V/g] and offsets O_x, O_y, O_z [V] of the device [39].

For wind turbine blades, it is time-consuming and inefficient to take measurements at all six different positions for each of the accelerometers positioned along the continuously varying blade surface profile. Maximum and minimum static acceleration will occur at different angles of orientation relative to gravity for each of the axes of the five accelerometers on the blade. Therefore, a global coordinate system was generated using the orientation of the wind turbine blade via the specially built and rotatable test rig. The blade was assumed to be fixed along the x-axis of the accelerometer and have no effect on the measured resultant acceleration as the blade is rotated. For this reason, positions θ_5 and θ_6 were ignored. θ_1 was set to 0° (when the blade lays flat as in Fig. 9.2), θ_2 was set to 90° (perpendicular blade position to θ_1), θ_3 to 180° (opposite blade position to θ_1) and θ_4 to 270°(opposite blade position to θ_3).

The relationship between the normalised accelerometer measurements (A_x, A_y and A_z) and the accelerometer raw measurements (V_x, V_y and V_z) [40, 41] can be expressed as;

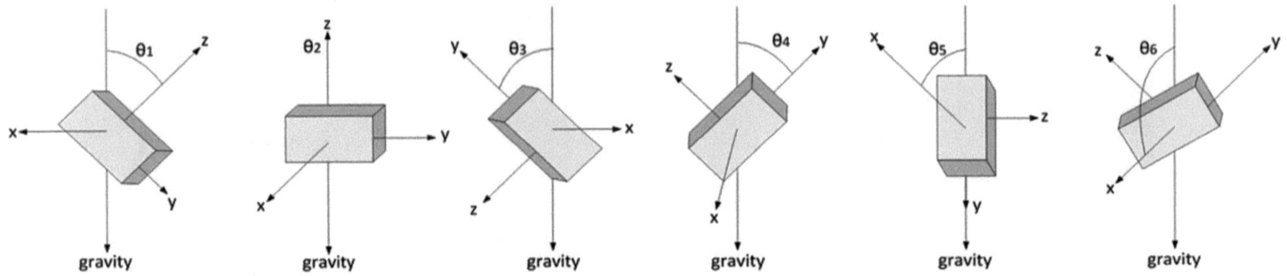

Fig. 9.3 Diagram showing the six possible orientations in which the accelerometer can be held in relation to gravity

Table 9.1 Accelerometer calibration positions from Fig. 9.4 and their corresponding normalised acceleration values (where A_x, A_y and A_z are in terms of g – acceleration due to gravity)

Stationary position	A_x	A_y	A_z
Z_{down} (θ_1)	0	0	+1 g
Z_{up} (θ_2)	0	0	−1 g
Y_{down} (θ_3)	0	+1 g	0
Y_{up} (θ_4)	0	−1 g	0
X_{down} (θ_5)	+1 g	0	0
X_{up} (θ_6)	−1 g	0	0

$$\begin{bmatrix} A_x \\ A_y \\ A_z \end{bmatrix} = [A_m]_{3\times3} \begin{bmatrix} {}^1/_{S_x} & 0 & 0 \\ 0 & {}^1/_{S_y} & 0 \\ 0 & 0 & {}^1/_{S_z} \end{bmatrix} \cdot \begin{bmatrix} V_x - O_x \\ V_y - O_y \\ V_z - O_z \end{bmatrix} \tag{9.1}$$

$$\begin{bmatrix} A_x \\ A_y \\ A_z \end{bmatrix} = \begin{bmatrix} ACC_{11} & ACC_{12} & ACC_{13} \\ ACC_{21} & ACC_{22} & ACC_{23} \\ ACC_{31} & ACC_{32} & ACC_{33} \end{bmatrix} \cdot \begin{bmatrix} V_x \\ V_y \\ V_z \end{bmatrix} + \begin{bmatrix} ACC_{10} \\ ACC_{20} \\ ACC_{30} \end{bmatrix} \tag{9.2}$$

Where;

- $[A_m]_{3\times3}$– is the 3 × 3 misalignment matrix between the accelerometer sensing axes and the wind turbine blade axes. Calculating the misalignment matrix compensates misalignment error.
- $S\,(S_x,\ S_y$ and $S_z)$ – is the sensitivity or scale factor of the accelerometer in terms of its output Volts per g (9.81 ms^{-2}) of acceleration.
- $O\,(O_x,\ O_y$ and $O_z)$ – is the offset or bias of the accelerometer in Volts.

The goal of accelerometer calibration is therefore to determine parameters from ACC_{10} to ACC_{33}, so that with any given raw measurements at arbitrary positions, the normalized values A_x, A_y and A_z can be obtained, resulting in;

$$|A| = \sqrt{{A_x}^2 + {A_y}^2 + {A_z}^2} = 1 \tag{9.3}$$

Where; $|A|$ is the resultant acceleration.

9.3.1 Methodology: Least Square Approximation

Using least square approximation in MATLAB [38], Eq. (9.2) was solved. The term *least squares*, describes a frequently used approach to solving over-determined or inexactly specified systems of equations in an approximate sense. Instead of solving the equations exactly, we seek only to minimize the sum of the squares of the residuals. The computation techniques for linear least squares problems make use of orthogonal matrix factorisations as well as simple calculus and linear algebra [42, 43].

Considering the accelerometer calibration at the selected stationary positions, Eq. (9.2) can be rewritten as;

$$\begin{bmatrix} A_x & A_y & A_z \end{bmatrix} = \begin{bmatrix} V_x & V_y & V_z & 1 \end{bmatrix} \cdot \begin{bmatrix} ACC_{11} & ACC_{21} & ACC_{31} \\ ACC_{12} & ACC_{22} & ACC_{32} \\ ACC_{13} & ACC_{23} & ACC_{33} \\ ACC_{10} & ACC_{20} & ACC_{30} \end{bmatrix} \tag{9.4}$$

Equation (9.4) can also be further simplified and represented as;

$$\boldsymbol{Y} = \boldsymbol{w} \cdot \boldsymbol{X} \tag{9.5}$$

Where,

Matrix $\boldsymbol{X}$ is the 12 calibration parameters that need to be determined. Matrix $\boldsymbol{w}$ is the accelerometer raw data collected at the stationary positions and Matrix $\boldsymbol{Y}$ is the known normalised earth gravity vector.

Now considering each stationary position,

- At Z_{down} position (θ_1), $\begin{bmatrix} A_x & A_y & A_z \end{bmatrix} = \begin{bmatrix} 0 & 0 & 1 \end{bmatrix}$ and assume that at Z_{down} position, *n1* sets of accelerometer raw data V_x, V_y and V_z have been collected. Then,

$$\boldsymbol{Y}_1 = \begin{bmatrix} 0 & 0 & 1 \end{bmatrix}_{n1\times3} \text{ and } \boldsymbol{w}_1 = \begin{bmatrix} A_{x\theta_1} & A_{y\theta_1} & A_{z\theta_1} & 1 \end{bmatrix}_{n1\times4} \tag{9.6}$$

Where, the matrix $\boldsymbol{Y}_1$ has the same row of [0 0 1] and matrix $\boldsymbol{w}_1$ contains raw data measured by the accelerometer.

- At Z_{up} position (θ_2), $\begin{bmatrix} A_x & A_y & A_z \end{bmatrix} = \begin{bmatrix} 0 & 0 & -1 \end{bmatrix}$ and assume that at Z_{up} position, *n2* sets of accelerometer raw data V_x, V_y and V_z have been collected. Then,

$$\boldsymbol{Y}_2 = \begin{bmatrix} 0 & 0 & -1 \end{bmatrix}_{n2\times3} \text{ and } \boldsymbol{w}_2 = \begin{bmatrix} A_{x\theta_2} & A_{y\theta_2} & A_{z\theta_2} & 1 \end{bmatrix}_{n2\times4} \tag{9.7}$$

- At Y_{down} position (θ_3), $\begin{bmatrix} A_x & A_y & A_z \end{bmatrix} = \begin{bmatrix} 0 & 1 & 0 \end{bmatrix}$ and assume that at Y_{down} position, *n3* sets of accelerometer raw data V_x, V_y and V_z have been collected. Then,

$$\boldsymbol{Y}_3 = \begin{bmatrix} 0 & 1 & 0 \end{bmatrix}_{n3\times3} \text{ and } \boldsymbol{w}_3 = \begin{bmatrix} A_{x\theta_3} & A_{y\theta_3} & A_{z\theta_3} & 1 \end{bmatrix}_{n3\times4} \tag{9.8}$$

- At Y_{up} position (θ_4), $\begin{bmatrix} A_x & A_y & A_z \end{bmatrix} = \begin{bmatrix} 0 & -1 & 0 \end{bmatrix}$ and assume that at Y_{up} position, *n4* sets of accelerometer raw data V_x, V_y and V_z have been collected. Then,

$$\boldsymbol{Y}_4 = \begin{bmatrix} 0 & -1 & 0 \end{bmatrix}_{n4\times3} \text{ and } \boldsymbol{w}_4 = \begin{bmatrix} A_{x\theta_4} & A_{y\theta_4} & A_{z\theta_4} & 1 \end{bmatrix}_{n4\times4} \tag{9.9}$$

Combining Eqs. (9.6–9.9) and assuming $n = n1 + n2 + n3 + n4$, then Eq. (9.5) becomes,

$$\boldsymbol{Y}_{n\times3} = \boldsymbol{w}_{n\times4} \cdot \boldsymbol{X}_{4\times3} \tag{9.10}$$

Where $\boldsymbol{Y} = \begin{bmatrix} \boldsymbol{Y}_1 \\ \boldsymbol{Y}_2 \\ \boldsymbol{Y}_3 \\ \boldsymbol{Y}_4 \end{bmatrix}_{n\times3}$ and $\boldsymbol{w} = \begin{bmatrix} \boldsymbol{w}_1 \\ \boldsymbol{w}_2 \\ \boldsymbol{w}_3 \\ \boldsymbol{w}_4 \end{bmatrix}_{n\times4}$

Therefore, the calibration parameter matrix $\boldsymbol{X}$ can be determined by the least squares method as,

$$\mathbf{X} = \left[\boldsymbol{w}^T \cdot \boldsymbol{w}\right]^{-1} \cdot \boldsymbol{w}^T \cdot \mathbf{Y} \tag{9.11}$$

Where; $\boldsymbol{w}^T$ means matrix transpose and $\left[\boldsymbol{w}^T \cdot \boldsymbol{w}\right]^{-1}$ means matrix inverse.

9.3.2 Results

The above equations were solved in MATLAB [38] using acceleration measurements taken when the blade was impact tested using a force hammer. Time domain plots in Fig. 9.4 show the direction in which the resultant acceleration of the blade acts. In Fig. 9.4a, it can be seen that the z-axis is the dominant direction of acceleration as the accelerometers measure a higher

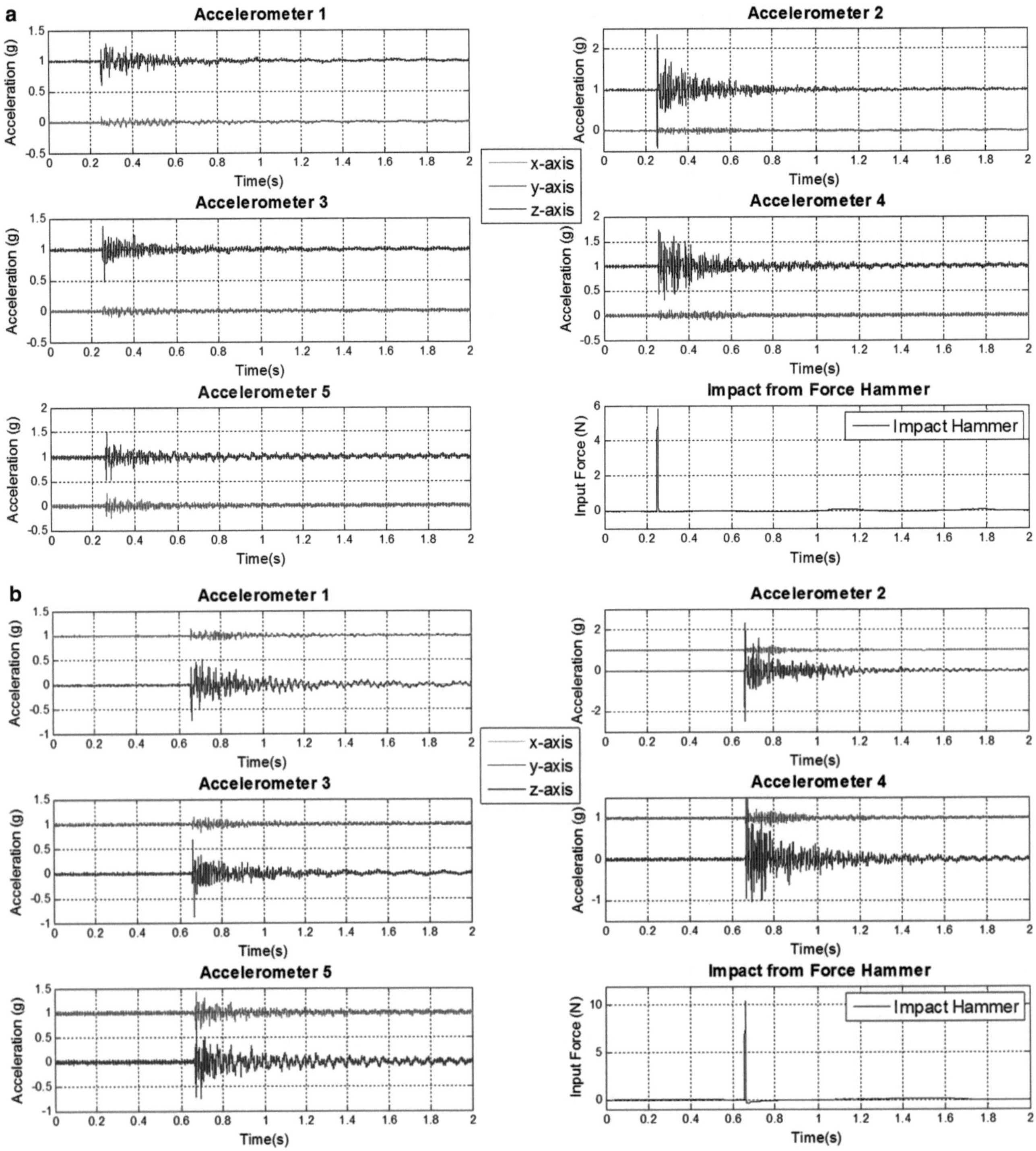

Fig. 9.4 Calibrated acceleration measurements when the turbine blade lies at (**a**) an angle of 0° and (**b**) an angle of 90° for 2 s of data sampled at a rate of 20 k samples/s (where x, y and z represent the axes of both the blade and accelerometers)

magnitude of acceleration at this position, normalised to 1 g when the blade is positioned at 0°. Similarly, in Fig. 9.4b, the y-axis is the direction in which the resultant acceleration is dominant when the blade is positioned at an angle of 90°. The x-axis is negligible in both plots as shown below because it was assumed to have no significant component of acceleration.

Accelerometer 2 measured the highest resultant acceleration magnitude in comparison to the other four accelerometers. This is because it was positioned along the trailing edge of the blade (the thinnest section of the blade) and vibrations along this area will generally be more visible. Accelerometers 1 and 3 measured smaller acceleration magnitudes as shown in Fig. 9.4 and this is also due to their positioning on the leading edge of the blade.

Comparing the results in Fig. 9.4a and b for accelerometer 5 to the picture of the blade in Fig. 9.2, it can be seen that accelerometer 5 which was placed at the curviest section of the blade is independent of its location. In general, all the accelerometers on the blade measure the blade vibrations independent of their position on the curved blade surface.

The local coordinates of all the accelerometers have therefore been converted to a global coordinate system of the wind turbine blade such that z is directly downwards towards gravity irrespective of the accelerometer orientation on the curved blade surface. Accurate comparisons between accelerometer measurements can therefore be made using the calibrated results. Also, faulty accelerometers such as damaged outputs can easily be spotted when this calibration is employed.

9.4 Conclusions

The high cost of the machinery and infrastructure of wind turbines combined with the difficulty of access to them by maintenance personnel, demands maintenance systems, if high reliability, availability, maintainability and safety are to be achieved. High reliability and condition monitoring benefit the maintenance management of wind turbine systems by reducing the operations and maintenance costs and by giving warnings of failures.

Instrumentation of MEMS accelerometers as described in this paper is a step forward towards achieving cost-effective wind turbine blade maintenance. MEMS accelerometers have shown potential for measuring useful vibration data from wind turbine blades. Calibration of these MEMS accelerometer to the global coordinate system of the wind turbine blade as outlined in this paper is useful for accurately interpreting vibration measurements of the blade needed for condition monitoring. This method can easily be applied to wind turbine blades with attached accelerometers while in operation. It also eliminates costs typically associated with the individual calibration of each accelerometer.

References

1. MSc CREST Flexible and Distance Learning Series. The wind turbine. In: MSc in Renewable Energy Systems Technology study notes for wind 1 2011/12, Loughborough, 2009, pp 2–9
2. Wroblewski D (2008) Wind turbines. Boston University College of Engineering. [Online]. http://people.bu.edu/dew11/turbinetypes.html. Accessed 25 Nov 2013
3. Borum KK, McGugan M, Brøndsted P (2006) Condition monitoring of wind turbine blades. In: 27th Risø international symposium on materials science: polymer composite materials for wind power turbines, Risø National Laboratory, Roskilde, Denmark, 2006, pp 139–145
4. McGowan JG, Hyers RW, Sullivan KL, Manwell JF, Nair SV, McNiff B, Syrett BC (2007) A review of materials degradation in utility scale wind turbines. Energy Mater Mater Sci Eng Energy Syst 2(1):41–64
5. Hayman B, Wedel-Heinen J, Brøndsted P (2008) Materials challenges in present and future wind energy. Mater Res Soc Bull 33(04):343–353
6. Musial W (2004) Wind turbine testing and certification. In: Wind turbine blade workshop. Sandia National Laboratory, Albuquerque, NM, USA, pp 1–23
7. Holmes JW, Sørensen BF, Brøndsted P (2007) Reliability of wind turbine blades: An overview of materials testing. In Proceedings. Chinese Renewable Energy Industry Association, Shanghai, pp 310–315
8. Risø DTU (National Laboratory for Sustainable Energy) (2010) Failure modes. [Online]. http://www.risoe.dtu.dk/research/sustainable_energy/wind_energy/projects/vea_structural_blade_design/failure_modes.aspx?sc_lang=en. Accessed 12 Dec 2011
9. Chou J-S, Chiu C-K, Huang I-K, Chi K-N (2013) Failure analysis of wind turbine blade under critical wind loads. Eng Fail Anal 27:99–118
10. Sagol E, Reggio M, Ilinca A (2013) Issues concerning roughness on wind turbine blades. Renew Sustain Energy Rev 23:514–525
11. Weiss P (2001) Insects in the wind lead to less power. In: Science news: magazine of the society for science & the public. [Online]. https://www.sciencenews.org/article/insects-wind-lead-less-power. Accessed 16 Apr 2014
12. Kelly Aerospace (2011) Wind turbine ice protection system. In: Thermal systems. [Online]. http://www.kellyaerospace.com/wind_turbine_deice.html. Accessed 16 Apr 2014
13. Soltani MR, Birjandi AH, Seddighi Moorani M (2011) Effect of surface contamination on the performance of a section of a wind turbine blade. Sci Iran 18(3):349–357
14. Branner K, Berring P, Berggreen C (2008) Buckling strength of thick composite panels in wind turbine blades – part II: effect of delamination. In: 4th international conference of composites testing & model identification. Dayton, OH, United States, 20–22 October, 2008

15. Astrom BT (1997) Manufacturing of polymer composites, Illustrate. CRC Press, London, pp 1–449
16. Ciang CC, Lee J-R, Bang H-J (2008) Structural health monitoring for a wind turbine system: a review of damage detection methods. Meas Sci Technol 19(12):122001
17. Sundaresan MJ, Schulz MJ, Ghoshal A (2002) Structural health monitoring static test of a wind turbine blade subcontract report NREL/SR-500-28719. Colorado
18. Corten GP, Veldkamp HF (2001) Insects can halve wind-turbine power. Nature 412:42–43, London
19. Ilinca A (2011) Analysis and Mitigation of icing effects on wind turbines. In: Al-Bahadly I (ed) Wind turbines. InTech, Rimouski, pp 177–215
20. Kraj AG, Bibeau EL (2010) Phases of icing on wind turbine blades characterized by ice accumulation. Renew Energy 35(5):966–972
21. Homola MC, Virk MS, Wallenius T, Nicklasson PJ, Sundsbø P a (2010) Effect of atmospheric temperature and droplet size variation on ice accretion of wind turbine blades. J Wind Eng Ind Aerodyn 98(12):724–729
22. Wiggelinkhuizen E, Verbruggen T, Braam H, Rademakers L, Xiang J, Watson S (2008) Assessment of condition monitoring techniques for offshore wind farms. J Sol Energy Eng 130(3):031004
23. McEwan JR, BHR Group Limited (1991) Condition monitoring. In: Proceedings of the 3rd international conference (Windsor, UK, 15–17 Oct 1990). Illustrate. Elsevier Applied Science, Essex, pp 3–35
24. Esu OO, Flint JA, Watson SJ (2013) Condition monitoring of wind turbine blades using MEMS accelerometers. In Proceedings Renewable Energy World Conference & Expo Europe: Keeping Europe's Power Flowing, Messe Wien, Vienna, Austria, pp 1–12. Published online by Pennwell corporation [pennwell.websds.net/2013/vienna/rewe/papers/t2s204-paper.pdf]
25. Esu OO, Flint JA, Watson SJ (2013) Integration of low-cost accelerometers for condition monitoring of wind turbine blades. In: Proceedings European wind energy association (EWEA) conference. Messe Wien, Vienna, Austria, 4–7 February, 2013, pp 1–4
26. Esu OO, Lloyd SD, Flint JA, Watson SJ (2014) Integration of low-cost consumer electronics for in-situ condition monitoring of wind turbine blades. In Renewable Power Generation Conference (RPG 2014), 3rd 24–25 September 2014, Naples, DOI 10.1049/cp.2014.0905IET[ieeexplore.ieee.org/xpl/articleDetails.jsp?tp=$amumber=6993298$searchWithin%3Dp_Authors%3A.QT.Flint%2C+J.%2FA%2F..QT.], pp 1–6
27. Ghosh A (2009) Introduction to measurements and instrumentation, 3rd edn. PHI Learning Private Limited, New Delhi, pp 1–810
28. Albarbar A, Badri A, Sinha JK (2009) Performance evaluation of MEMS accelerometers. J Meas 42(5):790–795
29. Analog-Devices (2010) ADXL335. [Online]. www.analog.com. Accessed 5 Dec 2011
30. Kitchin C (2009) Understanding accelerometer scale factor and offset adjustments. (Application Notes) AN-396 Published by ANALOG DEVICES, Norwood, Massachusetts
31. Carter Wind Energy. Carter wind turbines. [Online]. http://www.carterwindenergy.com/. Accessed 5 Nov 2012
32. Beliveau A, Spencer GT, Thomas K a, Roberson SL (1999) Evaluation of MEMS capacitive accelerometers. IEEE Des Test Comput 16(4):48–56
33. Minhang B (2005) Analysis and design principles of MEMS devices. Elsevier B. V, Amsterdam, pp 1–309
34. Larsen GC, Hansen MH, Baumgart A, Carlen I (2002) Modal analysis of wind turbine blades. Risø National Laboratory, Roskilde, Denmark, February 2002,
35. Pedersen HB, Kristensen OD (2003) Applied modal analysis of wind turbine blades. Risø National Laboratory, Roskilde, Denmark, 2003
36. National Instruments Corporation (2008) M Series user manual. [Online]. http://www.ni.com/pdf/manuals/371291h.pdf. Accessed 10 Jul 2012
37. National Instruments Corporation (2012) NI LabVIEW SignalExpress. [Online]. http://www.ni.com/labview/signalexpress/. Accessed 16 May 2012
38. MathWorks MATLAB: the language of technical computing. [Online]. http://www.mathworks.co.uk/products/matlab/. Accessed 1 Oct 2011
39. Lötters JC, Schipper J, Veltink PH, Olthuis W, Bergveld P (1998) Procedure for in-use calibration of triaxial accelerometers in medical applications. Sens Actuators A Phys 68(1–3):221–228
40. Sentera Technology Corporation Tri-axial accelerometer calibration. MEMS IMU Calibration Example. [Online]. http://motionsense.com/services/pdf/CalibrationReportExample.pdf. Accessed 3 Mar 2013
41. STMicroelectronics (2010) Tilt measurement using a low-g 3-axis accelerometer. AN3182, (Application Notes) STMicroelectronics, April, 2010
42. MathWorks Least squares. [Online]. http://www.mathworks.co.uk/moler/leastsquares.pdf. Accessed 23 Sep 2013
43. Miller SJ The method of least squares. [Online]. http://web.williams.edu/Mathematics/sjmiller/public_html/BrownClasses/54/handouts/MethodLeastSquares.pdf. Accessed 23 Sep 2013

Chapter 10
Predicting Full-Field Strain on a Wind Turbine for Arbitrary Excitation Using Displacements of Optical Targets Measured with Photogrammetry

Javad Baqersad, Peyman Poozesh, Christopher Niezrecki, and Peter Avitabile

Abstract Wind turbine blades and other structures are often subjected to dynamic loading that may not be predicted or measurable at critical locations of interest. Therefore, a non-contacting measurement technique that can provide information throughout an entire structure with the absence of instrumented sensors is desirable. Such an approach is particularly beneficial and relevant to operating rotor or wind turbine blades. In this paper, a three-bladed wind turbine placed in a semi-built-in boundary condition was subjected to a variety of different loadings. The turbine was excited using a sinusoidal excitation, a pluck test, arbitrary impacts on three blades, and random force excitations with a mechanical shaker. The response of the structure to these excitations at optical targets mounted to the blades was measured using three-dimensional point tracking. The limited set of measured displacement at the optical targets was expanded using a modal expansion algorithm. The expanded displacement was used in conjunction with a finite element model of the turbine to extract dynamic strain throughout the entire structure. The results from the technique were compared to instrumented strain gages and are shown be in close agreement. The predicted strain using the proposed approach is not limited to the locations of the optical targets or where the cameras have line of sight. This new technique may enable a new structural health-monitoring approach that has the ability to interrogate an entire structure, inside and outer surface.

Keywords Response identification • Point tracking • Modal expansion • Full-field strain • Wind turbine • Health monitoring • Blade

Nomenclature

$\{X_n\}$	Full-space displacement vector
$\{X_a\}$	Reduced-space displacement vector
$[M_a]$	Reduced mass matrix
$[M_n]$	Full-space mass matrix
$[K_a]$	Reduced stiffness matrix
$[K_n]$	Full-space stiffness matrix
$[U_a]$	Reduced mode shape matrix
$[U_a]^g$	Generalized inverse of reduced shape matrix
$[U_n]$	Full-space mode shape matrix
$[T]$	Transformation matrix
$[RTO_a]$	Real-time operating data at reduced-space
$[ERTO_n]$	Expanded real-time operating data
$\{X_i\}$ and $\{X_j\}$	Displacement vectors i and j
$\{RTO_i\}$ and $\{RTO_j\}$	Displacement time traces for points i and j

J. Baqersad (✉) • P. Poozesh • C. Niezrecki • P. Avitabile
Structural Dynamics and Acoustic Systems Laboratory, One University Avenue, Lowell, MA 01854, USA
e-mail: javad_baqersad@uml.edu

© The Society for Experimental Mechanics, Inc. 2015

R. Allemang (ed.), *Special Topics in Structural Dynamics, Volume 6*, Conference Proceedings of the Society for Experimental Mechanics Series, DOI 10.1007/978-3-319-15048-2_10

10.1 Introduction

Health monitoring of rotating structures (e.g. wind turbine or helicopter rotors) during operation are of particular interest due to cost, reliability, and safety issues. However, placing sensors on the rotating part of the structure is a challenge due to the wiring (that requires slip-rings for power and data transmission). Furthermore, for many wind turbines and complex structures mechanical failure may not be externally apparent and typically occurs at the interfaces between the structure's surface and the internal ribs or stiffening members where no sensor can be easily mounted. Understanding the dynamics of rotating structures to improve the design of these structures is highly desirable. For example, at the 2014 Sandia Wind Turbine Blade Workshop, it was identified that the edgewise-resonant vortex induced vibration of wind turbines and its effect on blades is currently poorly understood primarily because of a lack of appropriate measurement tools to quantify its influence and effect [1]. Therefore, a current area of interest to the wind and rotorcraft industries is how to accurately measure dynamics and monitor the condition of rotating blades using non-contacting sensors that have distributed sensing capability over a large area, while not adding significant cost or affecting the performance of the structure.

In the current work, a technique is proposed to extract dynamic strain on rotating structures using a non-contacting measurement approach and finite element (FE) modeling. The proposed approach uses photogrammetry along with a modal expansion technique. Therefore, these two techniques are disused next.

Three-dimensional (3D) digital image correlation (DIC) and three-dimensional point tracking (3DPT) are both non-contacting measurement approaches based on stereophotogrammetry principles and rely on a pair of digital cameras to capture images of the structure over a period of time. The 3D DIC technique is performed using a patterned area of interest and can identify the full-field dynamic displacement and strain over the visible surface of the structure. For 3DPT, high contrast optical targets are mounted to the structure and the displacements of those targets are measured.

Recently, researchers have started to use stereophotogrammetry to measure vibrations in structures [2–4]. The technique has been successfully applied to non-rotating wind turbine blades to measure displacement vectors and surface strain [5–8]. More recently, there have been some efforts to use the stereophotogrammetry technique on rotating structures. Helfrick et al. [9] used low speed cameras to perform a DIC measurement on a small rotating fan. In an effort to minimize measurement noise from DIC data, Wang et al. [10] used an image decomposition approach and extracted frequency response functions for several structures. These results were used to extract shape features [11] and eventually update finite element models [12, 13]. Due to the difficulties regarding surface preparations and high camera resolution needed for DIC measurements, it is far less common to find published literature that uses DIC system for rotating structures. On the other hand, 3DPT has more frequently been applied to the dynamic measurement of rotating structures. This technique has been used for measuring displacements of optical targets on small-scale wind turbines [14, 15] as well as utility-scale wind turbines. The 3DPT technique has also been successfully used to measure the dynamics of helicopter rotors [16–19]. Although measuring displacements of discrete targets mounted on the blades of rotating structures could improve the physical understanding of rotating structures, these measurements by themselves cannot provide detailed information about the levels of strain that the structure sees, which is important for some health monitoring algorithms. Therefore, establishing a comprehensive algorithm that enables full-field monitoring of structures using limited set of measurements is desirable.

Researchers have used Inverse Finite Element Method (IFEM) and expansion algorithms to extract full-field response of a structure using limited measurements. IFEM has generally been used for extracting full-field response of the structure using measured strain [20–22]. This method can be used for both static and dynamic loading conditions but the application of this approach for complicated structures have not been examined.

The modal expansion techniques can be used to expand real time measured data. Chipman et al. [23] showed that by expanding the limited set of data points measured with a laser Doppler vibrometer, an improved description of the real time operating response can be extracted. Pingle et al. [24] further expanded the method and applied it on several academic structures. Carr et al. [25] used the method to expand out-of-plane data for an aluminum beam and a blade placed in a cantilevered configuration and excited using a mechanical shaker. In their work, the loading condition was controlled by only applying either single or dual-sinusoidal inputs and may not necessarily replicate the operating condition of the structure. Iliopoulos et al. [26] expanded limited set of data measured with accelerometers and applied it to an analytical beam model to predict dynamic strain on a wind turbine tower. The expansion methods are usually used for linear expansions; however, it has also been successfully used to expand data for structures that include small nonlinearities [27, 28]. Using strain mode shapes is another approach that has been used by researchers to extract full-filed displacement and strain using strain measured at limited locations [29, 30]. Chierichetti et al. [31, 32] used a confluence algorithm to extract the full-field dynamic response of several structures by using limited set of measured data. Using iterative method can make correction for small nonlinearities; however, it makes the process computationally expensive. Although considerable research has been devoted to extract full-field response of structures, less attention has been paid to identify and validate an approach that can be used to extract full-field response for rotating structures subjected to arbitrary loading conditions.

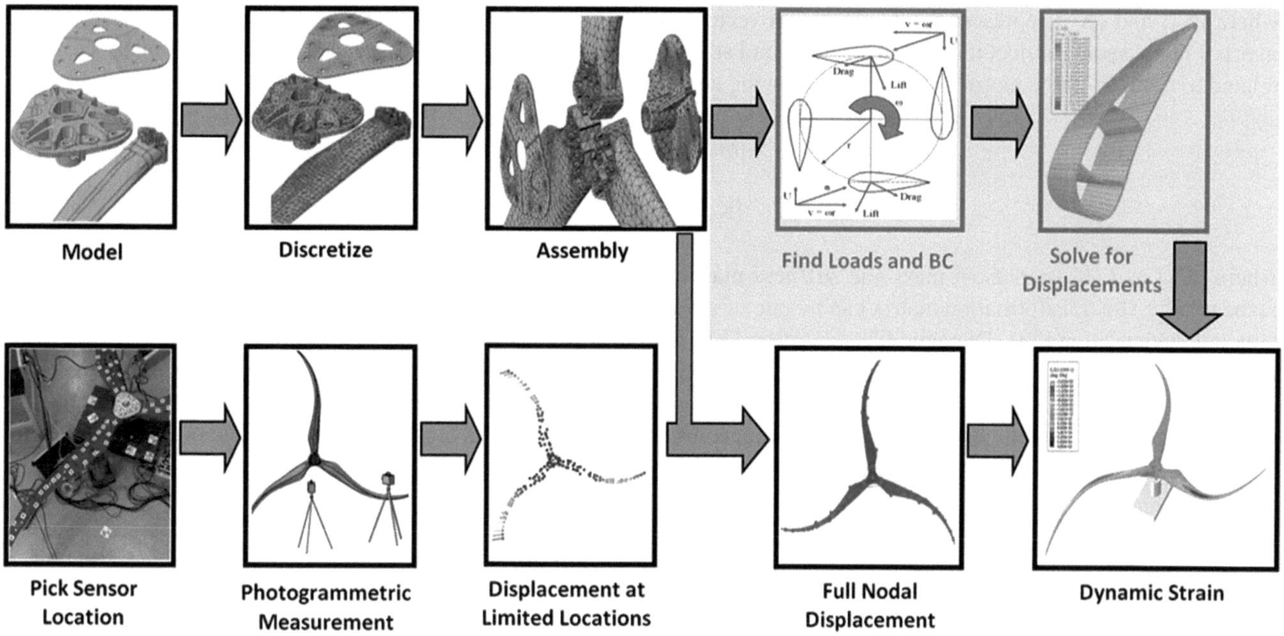

Fig. 10.1 A flow-chart of the expansion methodology showing the traditional approach (*shaded*) replaced with an alternative approach

Figure 10.1 shows a flow-chart of the conventional finite element approach replaced with the proposed approach. In the proposed approach, similar to the conventional finite element method, components are discretized and assembled. However, instead of using loads and boundary conditions (that is usually estimated using several assumptions), the actual displacement of points are applied to the structure.

Within this work, the optical three-dimensional point-tracking (3DPT) measurement approach is used in conjunction with a recently developed modal expansion technique. These two approaches (empirical and analytical) complement each other and enable the prediction of the full-field dynamic response on the surface of the structure as well as within the interior points for rotating and non-rotating structures. The current lack of distributed measurement capability for rotating structures compounded with the inability to accurately assess the deformation and global strain (or stress) has not been sufficiently addressed by the scientific community and provides the motivation for the research. Being able to fully characterize the global mechanics of the structure is highly desirable for condition based monitoring and for fatigue life prediction. Using such an approach is also beneficial to help assess the loading that helicopter rotors and wind turbine blades experience during operation; measurement of these loads is not currently possible.

10.2 Theory

The theoretical background of the expansion technique and correlations tools used in the current work is discussed herein.

10.2.1 Expansion

Model reduction/expansion approaches can be used to augment measurements and extract data for unmeasured degrees of freedom (DOF). In the current work, the modal expansion is used to augment the real time operating data at optical targets and extract the full-field dynamic displacement at all finite element degrees of freedom. For all modal expansion algorithms, the relationship between the full-space and reduced-space model can be written as:

$$\{X_n\} = [T]\{X_a\} \tag{10.1}$$

where $\{X_n\}$ and $\{X_a\}$ represent the displacement vectors in full-space and reduced-space, respectively. The transformation matrix [T] is used to project the full-space mass and stiffness matrices to the reduced-space. The full-space matrices can be related to reduced-space matrices using the following equations.

$$[M_a] = [T]^T [M_n] [T] \tag{10.2}$$

$$[K_a] = [T]^T [K_n] [T] \tag{10.3}$$

where $[M_n]$ and $[K_n]$ represent mass and stiffness matrices in the full-space and $[M_a]$ and $[K_a]$ are in the reduced-space, respectively. The transformation matrix can be one of several commonly used matrices in an expansion algorithm including Guyan Condensation [33], Dynamic Condensation [34], Improved Reduced System [35], and SEREP (System Equivalent Reduction and Expansion Process) [36], but because SEREP can produce an exact reduced system, it is used for all the work presented herein. The SEREP transformation matrix is developed by using mode shapes of the system at the full-space. The mode reduced by this method yields the exact frequencies and mode shapes as those obtained from the eigensolution of the full size matrices. The SEREP transformation is written as:

$$[T] = [U_n] \left([U_a]^T [U_a] \right)^{-1} [U_a]^T \tag{10.4}$$

$$[T] = [U_n] [U_a]^g \tag{10.5}$$

Where $[U_n]$ is the mode shapes of the structure at full-space and $[U_a]$ is the mode shapes at the *a* selected degrees of freedom. The superscript *g* refers to the general inverse of the matrix.

For the current work, the SEREP transformation matrix was developed using the finite element mode shapes. However, it can be developed using measured modal vectors such as the modes measured with Laser Doppler vibrometer in Chipman et al.'s work [23]. Equation 10.1 is used to expand measured real-time operating data $[RTO_a]$ to the full-space of the finite element model $[ERTO_n]$ and is written as:

$$[ERTO_n] = [T] [RTO_a] \tag{10.6}$$

10.2.2 Correlation Tools

The modal assurance criterion (MAC) [37] is a tool for quantifying the correlation between two vectors at all degrees of freedom. The MAC is given by:

$$MAC_{ij} = \frac{\left[\{X_i\}^T \{X_j\} \right]^2}{\left[\{X_i\}^T \{X_i\} \right] \left[\{X_j\}^T \{X_j\} \right]} \tag{10.7}$$

where X_i and X_j are displacement vectors (usually mode of the structure). The MAC values close to 1.0 indicate strong similarity between vectors, where values close to 0.0 indicate minimal or no similarity.

The TRAC is used in the current paper to find the correlation between two time traces. The TRAC can be written as:

$$TRAC_{ij} = \frac{\left[\{RTO_i\} \{RTO_j\}^T \right]^2}{\left[\{RTO_i\} \{RTO_i\}^T \right] \left[\{RTO_j\} \{RTO_j\}^T \right]} \tag{10.8}$$

where RTO_i and RTO_j represent time traces for point i and j. It should be noted that the TRAC only consider the correlation between the traces and not their amplitudes. Thus, if a time traces is scaled and compared to the original signal, the TRAC shows a strong correlation although the amplitudes of the traces are not equivalent.

Similar to MAC, TRAC values close to 1.0 indicate strong similarities, where values close to 0.0 indicate minimal or no similarity. The MAC and TRAC values in this paper are presented in percent (0–100 %).

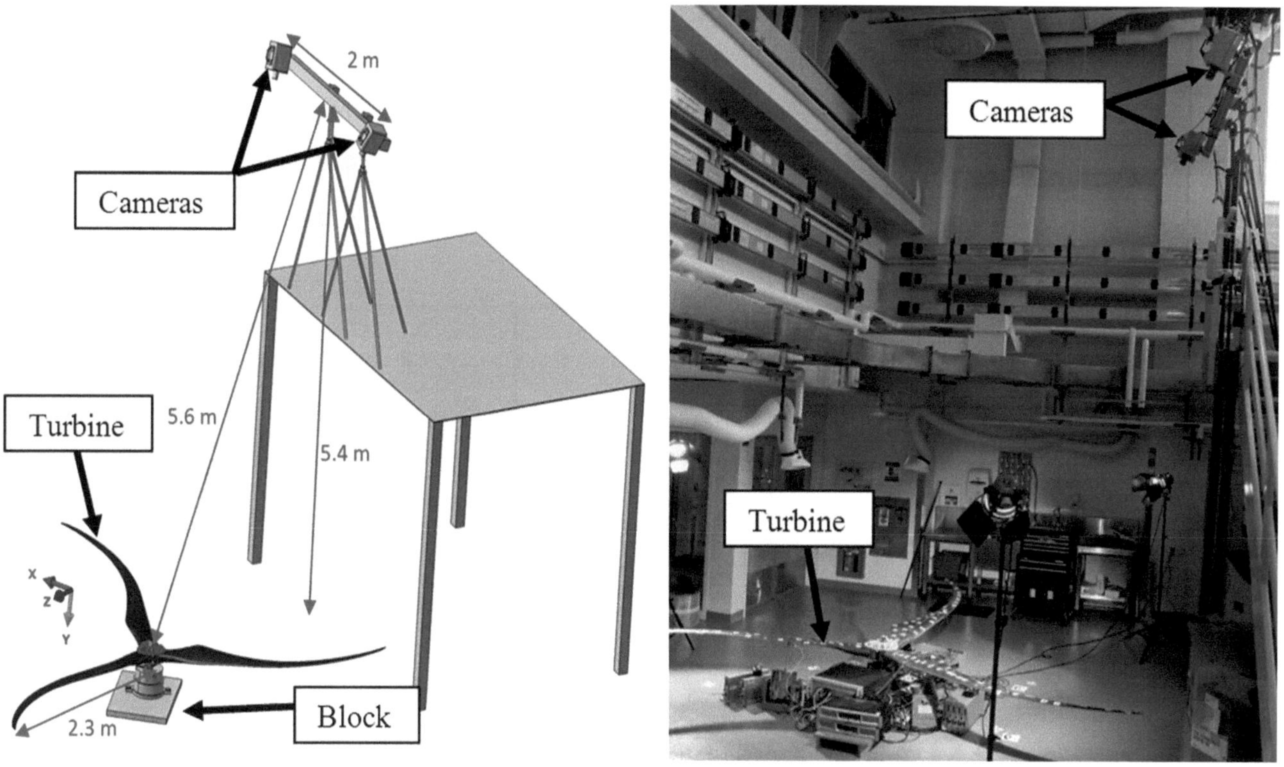

Fig. 10.2 A schematic of the set up of the cameras and turbine (*left*) and a photo of the actual test set-up (*right*)

10.3 Test Setup

A measurement to experimentally validate the proposed approach was performed on a Skystream 4.7 wind turbine with 2.3-m blades. In order to emulate the boundary conditions that a tower provides for the operating wind turbine, the turbine was mounted to a fixture that consisted of a 500-lb block and a machine tool chuck as shown in Fig. 10.2.

The camera setup is also shown in Fig. 10.2. A pair of Photron high-speed cameras (FASTCAM SA2) equipped with 24-mm lenses was used for the measurement. The calibration deviation for this test was calculated to be 0.0225 pixels. According to the PONTOS™ user manual [38], for a good calibration, the calibration deviation needs to be less than or equal to 0.04 pixels. The 4 mega-pixel cameras were mounted to a 2-m aluminum bar and placed approximately 5.6-m away from the turbine hub so they can observe the whole turbine area (see Fig. 10.2). Proper illumination was provided using several lights. A series of optical targets (~79) were mounted to the blades and the fixture to measure the in-plane and out-of-plane displacements of all the components. The cameras were configured to record images at a rate of 125 fps.

Eight strain-gages were also installed on the blades to measure dynamic strain at discrete locations. Strain measured with strain-gages was used for validation purposes.

10.4 Modeling and Data Processing

A validated solid finite element model of the Skystream 4.7™ turbine was needed for the expansion process. A finite element model the rotor including the blades, hub, and connecting plate was initially developed in Abaqus™ [39]. Each blade model consisted of approximately 15,000 quadratic tetrahedron elements. The fixture (the block and chuck) was modeled using rigid elements and the proper weight and inertial moment was applied to the model. After modeling the components, they were attached using rigid tie elements. In order to validate the FE mode shapes, an experimental modal test was performed on the wind turbine rotor in a semi-built-in configuration. In this test, the turbine was excited using three mechanical shakers and the response of the structure was measured at 36 points using triaxial accelerometers. A more detailed description of the experimental modal tests can be found in [40, 41]. A comparison between the FE and experimental mode shapes was

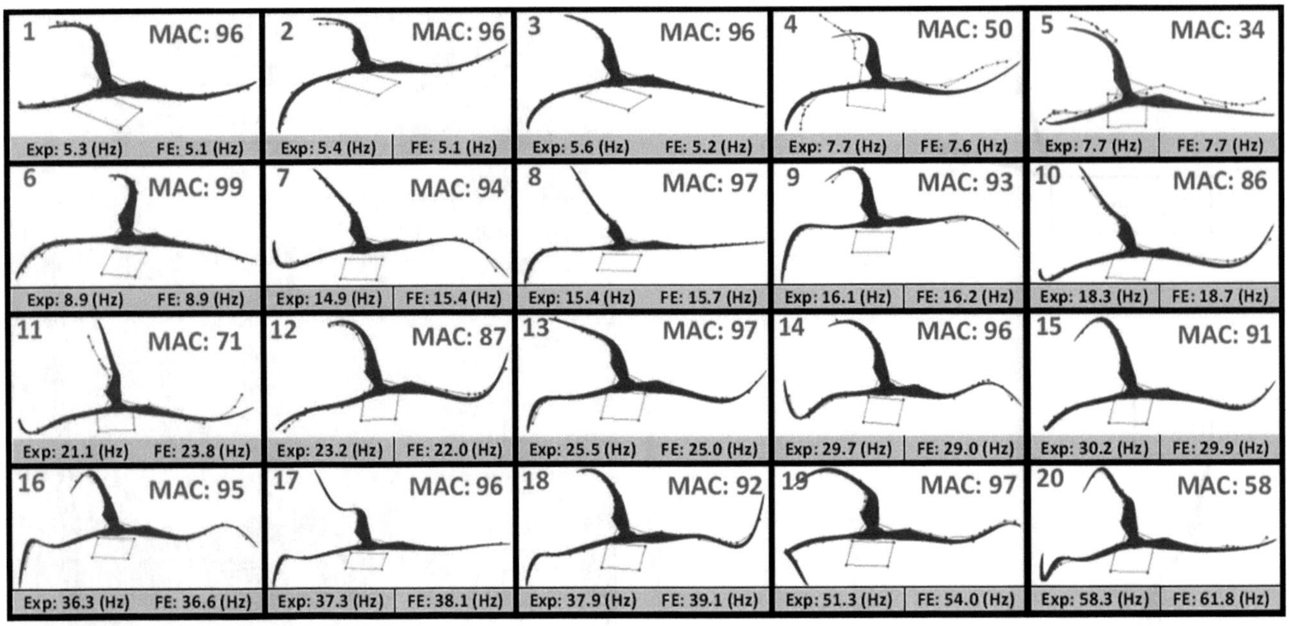

Fig. 10.3 The first 20 FE mode shapes of the turbine correlated with experimentally measured modes shapes using the modal assurance criterion

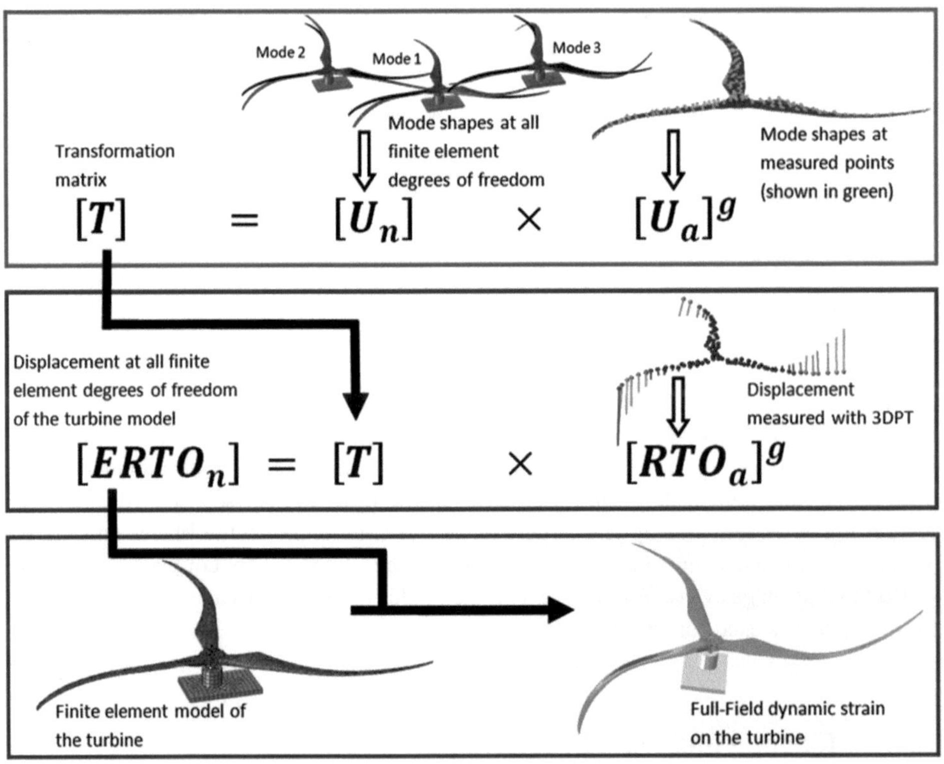

Fig. 10.4 A flowchart of the expansion methodology and 3DPT to predict full-field dynamic strain using modes of the wind turbine structure and optically measured displacement data for the expansion

performed using the modal assurance criterion. Strong similarities could be seen between the FE and test models for most of the modes as shown in Fig. 10.3. It should be noted that the modes that do not show strong correlations are generally related to the shapes that were generated due to the fixture modes (e.g. modes 4 and 5).

Figure 10.4 shows an overview of the proposed technique for dynamic strain extraction. A pair of stereophotogrammetric cameras was used to measure dynamic displacement of optical targets mounted on the structure. After extracting the dynamic displacement of optical targets by using the three-dimensional point tracking technique, the limited set of measured displacements were expanded to all FE degrees of freedom using the FE mode shapes. The calculated displacements were then applied to the FE model to extract the dynamic strain (or stress) at each stage.

The number of modes that are used in the model should be able to span the frequency range of the modes excited during operation. For example for a sinusoidal excitation at a resonant frequency of the turbine, using only the corresponding mode would be sufficient to expand the real time operating data. However, for a more complicated loading condition, using more modes for the expansion process is necessary. Because the arbitrary excitations used in this study may excite many modes, the first 20 modes of the structure were used to expand the measured displacement. Using that many modes assures that the expansion process will be able to cover the deformations that might occur in a frequency bandwidth from 0 to ~60 Hz.

10.5 Cases Studied

The wind turbine may experience different loading conditions during operation. In order to consider all possible conditions that a wind turbine may experience, different loading scenarios were examined in this paper. These loading scenarios include sine dwell excitation, pluck testing, arbitrary impact on all three blades, and random excitation with a shaker. The description of the loading scenarios, the work carried out for each case, and the results are now reviewed.

10.5.1 Sine Dwell Excitation

In the first test case, the turbine was excited using sine dwell excitations to verify the approach has merit. A sine dwell excitation is easier to characterize than arbitrary excitations, can validate the proposed approach, and allows the arbitrary excitations to be performed with more confidence. The sine dwell excitation test was performed by providing sinusoidal input to a mechanical shaker that was attached to one of the blades and could excite the entire turbine. In the current section, the predicted strain for the case when the blade is excited at its first resonant frequency is shown (see Fig. 10.5). It should be noted that the dynamics of an operating wind turbine is usually dominated by the modes at the low frequencies.

The predicted and measured strain at the strain-gage locations are compared when the turbine is excited by the shaker with a sinusoidal input at its first resonant frequency as shown in Fig. 10.5. It should be noted that the full-field strain was predicted for the entire turbine including the interior points; however, the results were validated only at the strain-gage locations. As can be seen, the predicted strain shows close agreement with the measured strain.

10.5.2 Pluck Test

In this case, a pluck test was performed on the blade by displacing the tip of a single blade using a dead-weight. As shown in Fig. 10.6, the weight imposed an initial displacement on the turbine blade that the mass was attached to. By cutting the connecting string, the wind turbine started to freely oscillate and the dynamic displacements of optical targets were measured during this oscillation using three-dimensional point tracking.

Similar to the previous case, measured displacements of the optical targets were expanded and applied to the FE model to extract the full-field dynamic strain on the structure. As shown in Fig. 10.7, the predicted strain is in close agreement with the measured strain and the graph shows that the strain is initially higher in the blade that the dead weight was attached to. After several oscillations, the two other blades also started to oscillate and approach a high level of strain.

10.5.3 Random Impact Excitations

In the second loading scenario, a highly arbitrary loading condition was considered. In this case, three impact hammers were used to simultaneously make random impact excitations that were spatially distributed along the three blades (see Fig. 10.8). This impacting input was completely random and impulsive and could replicate a very complicated loading condition for wind turbines.

The measured strain using strain-gages for the random impact excitations is compared to predicted strain in Fig. 10.9. The figure shows that although many modes were excited during this testing, the proposed approach could predict the dynamic strain accurately.

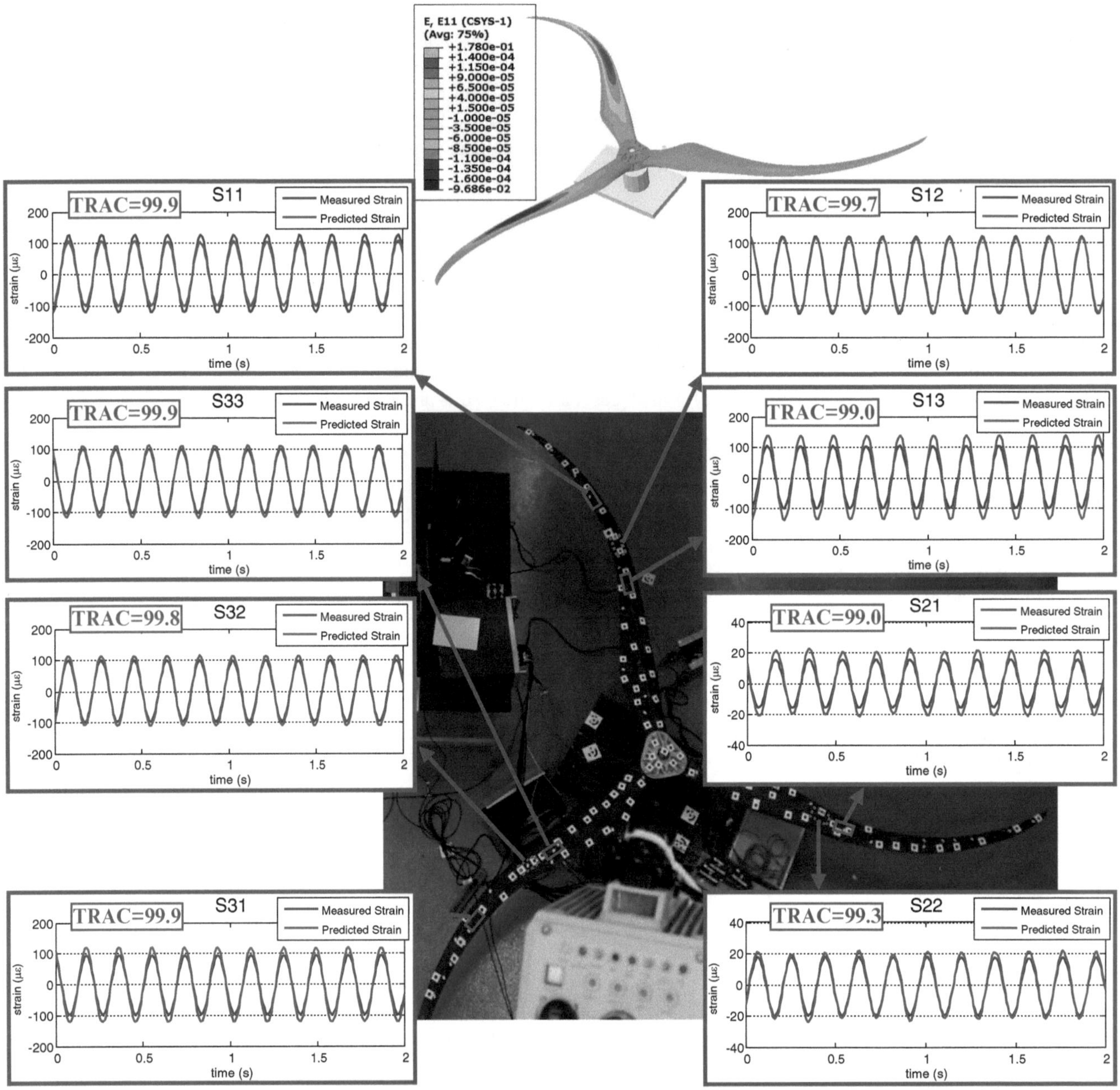

Fig. 10.5 Comparison of the strain measured with mounted strain-gages (*red boxes*) and the predicted strain using the proposed approach when the wind turbine is excited by a shaker with a sinusoidal input at the frequency of the first mode (5.3 Hz). The photo shows the location of the strain-gages where the comparison is performed. Sensors S12, S22, and S32 were mounted to the underside surface of the blades where no optical targets were placed or measured. The contour plot (*top*) shows the full-field predicted strain at one sample in time

10.5.4 Random Excitation with a Mechanical Shaker

In this case, the wind turbine was excited using a mechanical shaker with a random input that could excite all the modes below approximately 60 Hz.

A comparison between the measured and predicted strain for the random force excitation with the shaker is shown in Fig. 10.10. The results show very good agreement between the results for most of the strain-gages. It should be mentioned that for the strain-gages that do not have high TRAC values, the strain levels are very low and are below the noise floor of the proposed technique. In these cases there are slight variations and the predicted strain may differ from the measured strain. However, the amplitude of the predicted strain still correlates well with the measured strain.

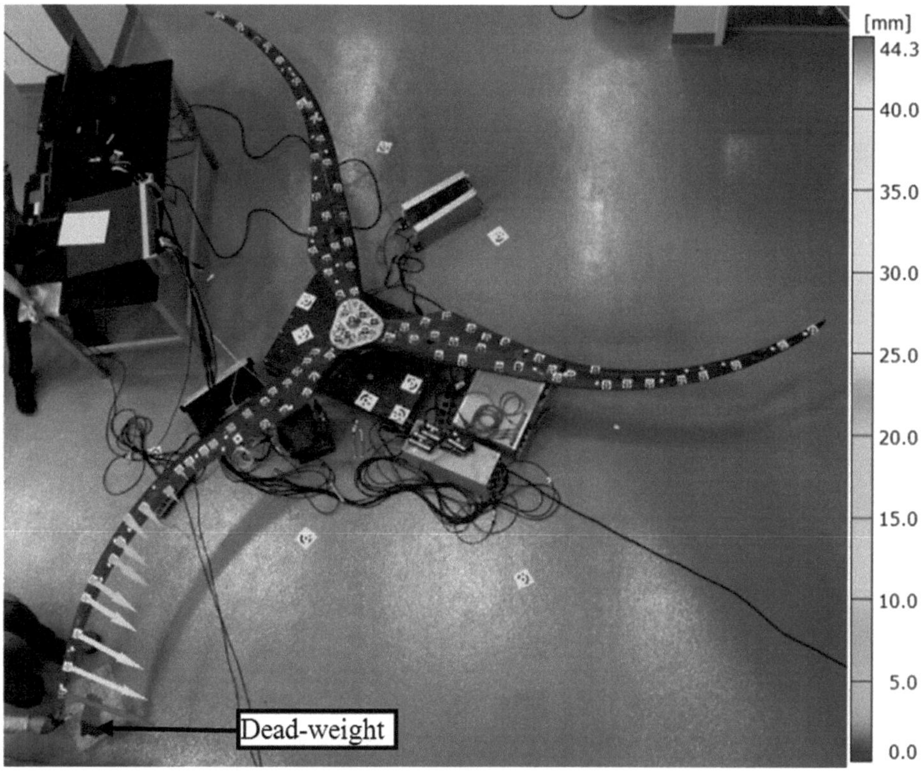

Fig. 10.6 A photo of the test set-up showing the dead-weight hung from the lower left blade and the associated displacement vectors of the optical targets

10.6 Observations

Performing the technique to expand real time operating data on a structure with several components and subjected to arbitrary excitation revealed several important observations.

The number of modes that were needed to expand the real time operating data on a wind turbine depends on the modes that were excited. In this paper, the measured strain showed that most of the modes that were excited were below 60 Hz. Therefore, all the modes below 60 Hz (20 modes) were used for the expansion. In the process of analyzing the number of modes required for the expansion process, it was found that in order to get good accuracy, the number of modes to be expanded must span the frequency range of the response produced by the external excitation. Using more modes than required to expand the measure displacement does not produce appreciably better results. The results of the paper revealed that although not all the FE modes had strong correlations to the experimental modes, the 20 modes that were used could effectively span the time trace of the operating data in the full-space model. Therefore, they could effectively show the dynamic strain with high accuracy.

The effects of choosing different sets of data points in SEREP to predict dynamic strain prediction for an academic structure was studied by Pingle et al. [24, 42]. It was shown that the method is very robust and can predict the strain with very few number of measurement points. It should be mentioned that not only the number of measurement points are important but also the locations of the points play an important role in the data expansion. The minimum number of required sensors also depends on the type of the excitation and the number of modes excited during the measurement. Studying these effects is beyond the scope of the current work and will be investigated in future work.

To identify the accuracy of the technique in predicting dynamic strain, a noise floor measurement was performed on the turbine in a stationary state. In this measurement, the wind turbine was not excited by any external sources. The measured optical displacement error (noise floor) in this test was augmented using the expansion algorithm. The expanded noise floor displacement was applied to the finite element model. The results of this test showed the noise level in the predicted strain. As can be seen in Fig. 10.11, even when the turbine is in stationary state, the predicted strain shows a variation of strain between -10 and 6 micro-strain. These results help to quantify the accuracy of the technique in predicting measured strain.

It should also be noted that the proposed method can be used to extract dynamic strain not only on the visible surface of the structure but also on non-visible parts. For the results in Figs. 10.7, 10.9, and 10.10, three strain-gages (S12, S22, and S32) were mounted on the underside surface of the structure; still strong correlations between the measured and predicted strain could be seen. As shown in Fig. 10.12, the strain could also be extracted inside the blades where no-other sensor could be mounted.

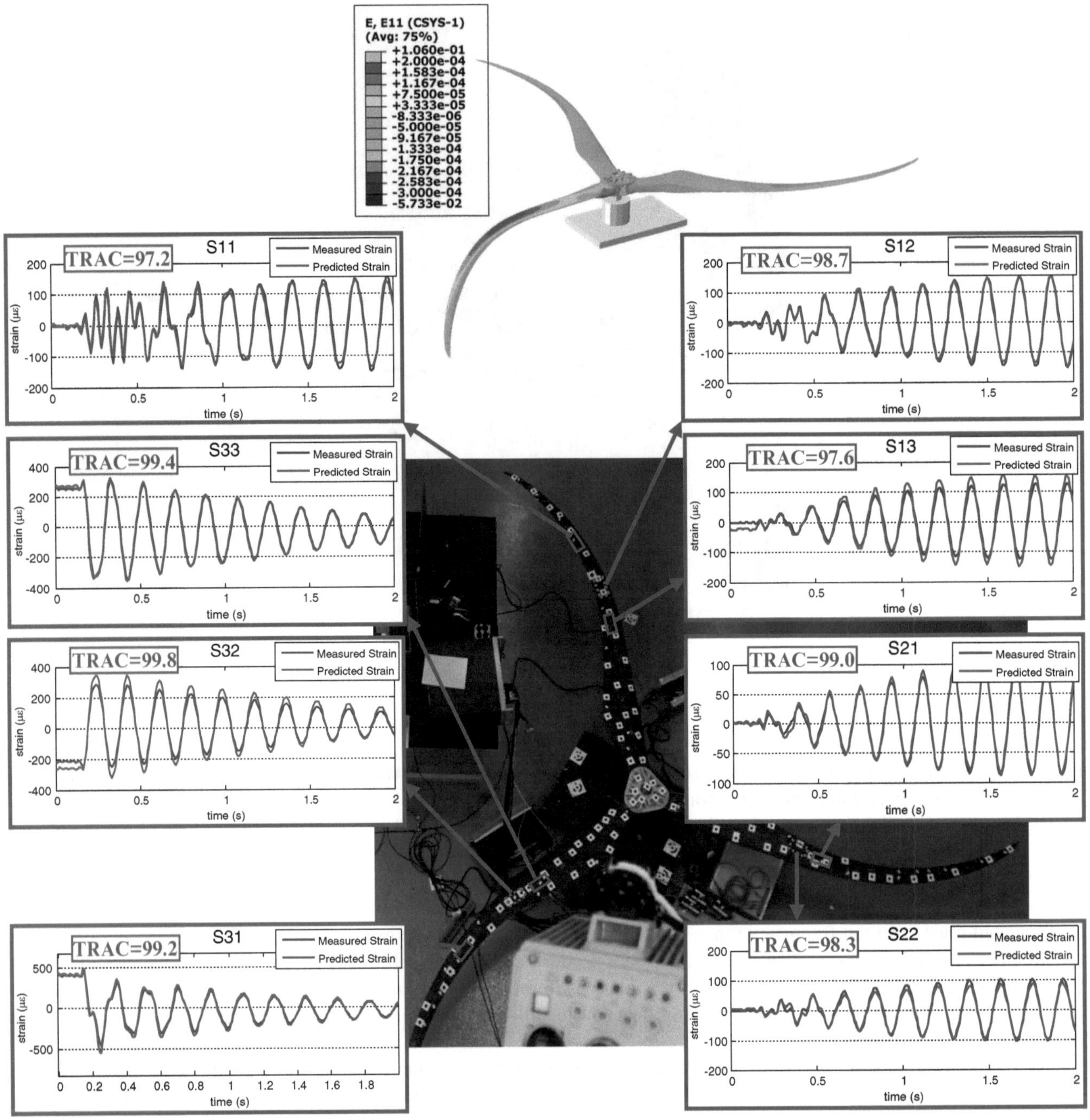

Fig. 10.7 Comparison of the strain measured with mounted strain-gages (*red boxes*) and the predicted strain using the proposed approach during the pluck test. The photo shows the location of the strain-gages where the comparison is performed. Sensors S12, S22, and S32 were mounted to the underside surface of the blades where no optical targets were placed or measured. The contour plot (*top*) shows the full-field predicted strain at one sample in time

The results of the paper are very encouraging with the proposed technique having a higher accuracy than compared to measured strain by using digital image correlation. The accuracy of the digital image correlation that is frequently used for measuring full-field surface strain is reported in the literature is approximately 30–100 microstrains (depending on camera resolution, facet size, facet step, computation size, and filters used). For instance, the high-speed cameras used in the current work were previously used for a DIC measurement on a small field of view ($\sim$1 m^2) of a single blade [6, 7]. It was seen that the accuracy of the DIC was approximately 50–100 microstrains. In the current test, the field of view ($\sim$16 m^2) was several time larger than compared to the single blade DIC case ($\sim$1 m^2) in the previous work. Generally, it is expected that the accuracy decreases with the extending the field of view. However, the results of the paper showed that the proposed

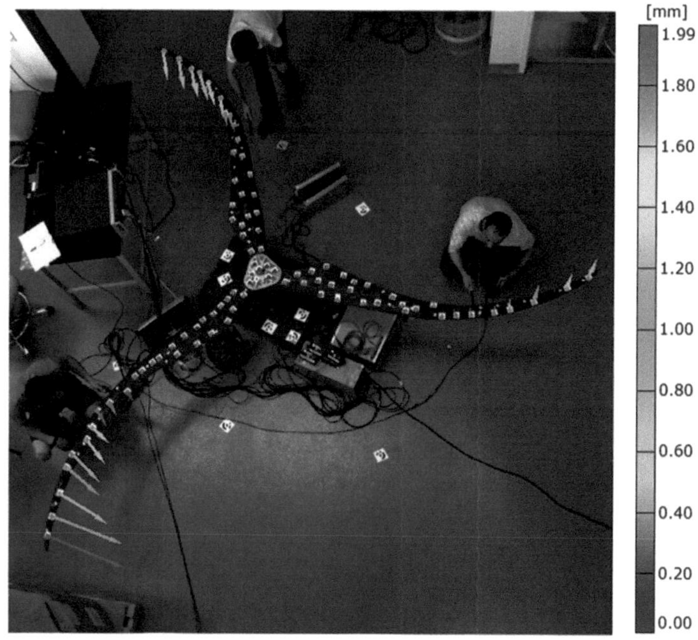

Fig. 10.8 A photo of the test set-up showing arbitrary impacts made on the three blades using three impact hammers and the associated displacement vectors of optical targets due to the impacts at one instant in time

method generated strain with an accuracy up to 5 microstrain even for the large field of view. Furthermore, the technique does not need the distributed pattern surface preparation that is required for DIC measurement. This paper has demonstrated the strong potential of the proposed technique for full-field strain extraction by using displacements of optical targets.

10.7 Conclusions

In this paper, an algorithm to monitor the dynamic response of operating wind turbines is proposed and experimentally verified. This algorithm combines a non-contacting measurement approach and an expansion technique that can be used to extract full-field dynamic strain on rotating structures. It was shown that the proposed approach could predict dynamic strain with accuracy higher that digital image correlation. Furthermore, the proposed approach is not limited to the line of sight of the cameras and can predict strain on locations where no other sensor can be mounted. The results of the paper indicate the strong potential of the approach to be used to measure the dynamics of rotating structure and assess their structural health during operation. The predicted full-field response can also be used to promote the understanding of the loading conditions that rotating structures experience and improve the performance of these structures.

Acknowledgements This research presented in this paper is partly supported by the National Science Foundation under Grant Number 1230884 (Achieving a Sustainable Energy Pathway for Wind Turbine Blade Manufacturing). Any opinions, findings, and conclusions or recommendations expressed in this material are those of the author(s) and do not necessarily reflect the views of the National Science Foundation.

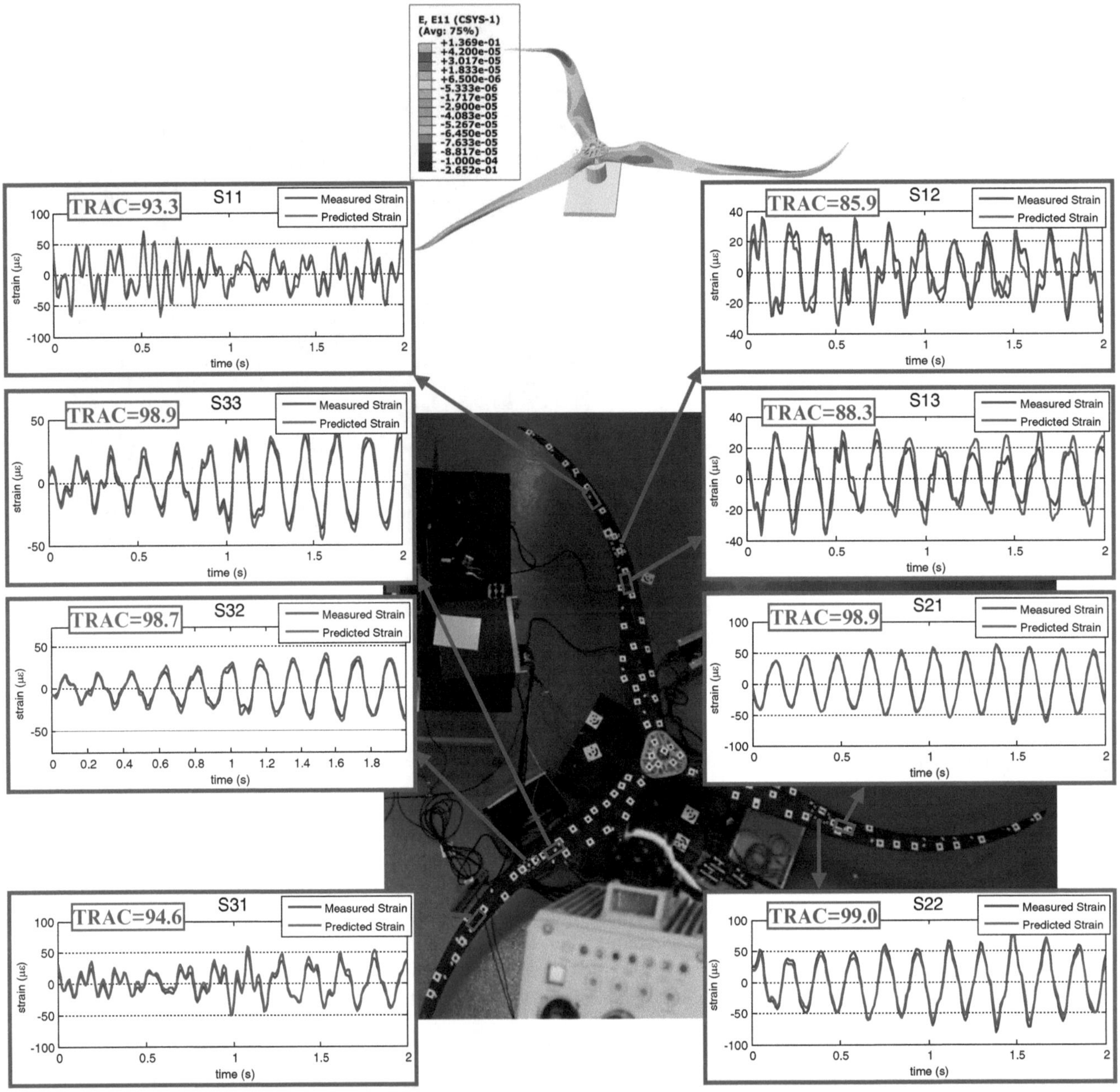

Fig. 10.9 Comparison of the strain measured with mounted strain-gages (*red boxes*) and the predicted strain using the proposed approach when the turbine was impacted with three impact hammers simultaneously. The photo shows the location of the strain-gages where the comparison is performed. Sensors S12, S22, and S32 were mounted to the underside surface of the blades where no optical targets were placed or measured. The contour plot (*top*) shows the full-field predicted strain at one sample in time

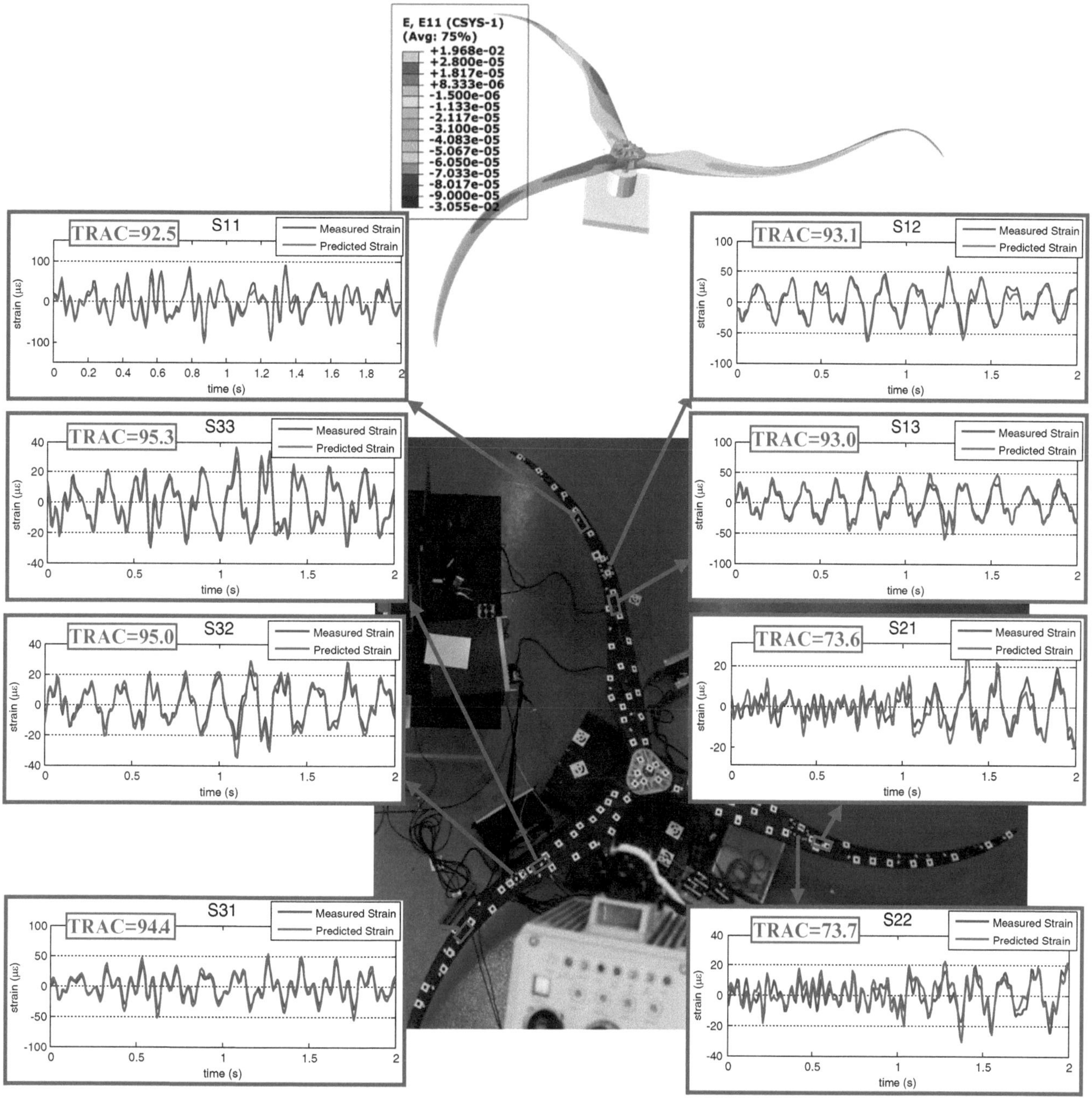

Fig. 10.10 Comparison of the strain measured with mounted strain-gages (*red boxes*) and the predicted strain when the wind turbine was excited by a shaker with a burst random excitation. The photo shows the location of the strain-gages where the comparison is performed. Sensors S12, S22, and S32 were mounted to the underside surface of the blades where no optical targets were placed or measured. The contour plot (*top*) shows the full-field predicted strain at one sample in time

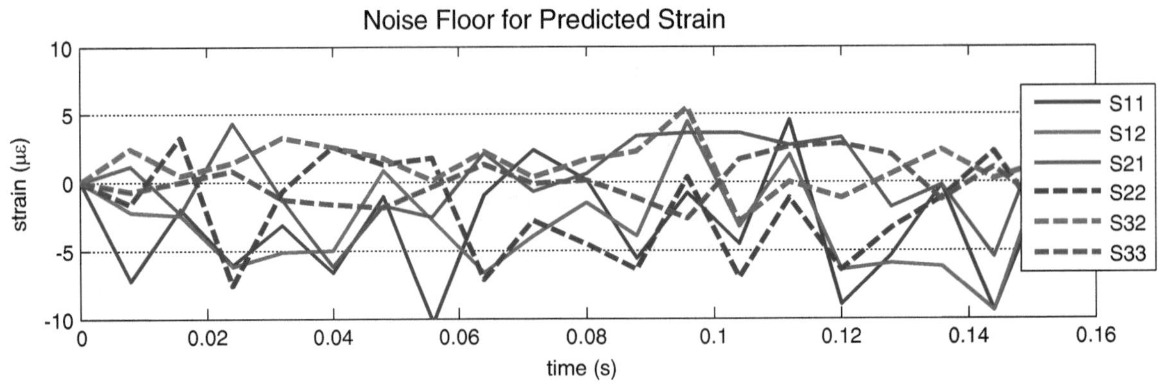

Fig. 10.11 Noise level of the predicted strain at the locations of physical strain-gages

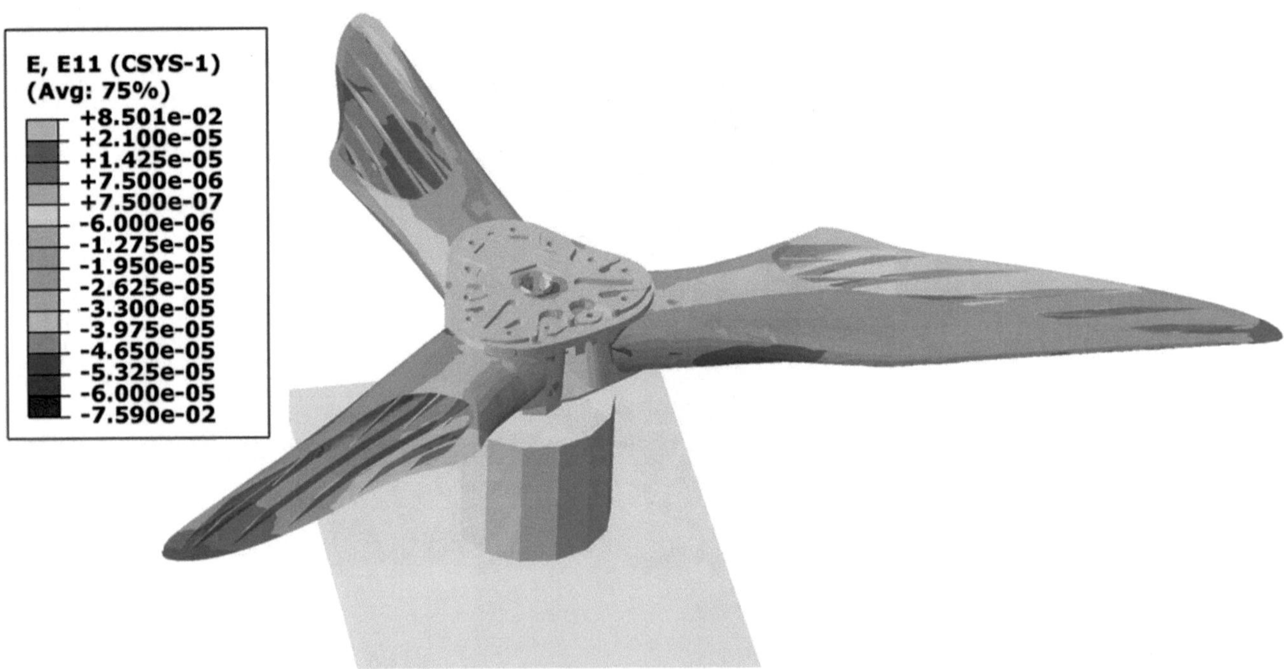

Fig. 10.12 A section view of the turbine showing the strain distribution inside the blade at one instant of time

References

1. Malkin M (2014) Reliability trends. In: Sandia National Laboratories wind turbine blade workshop, Albuquerque
2. Helfrick MN, Niezrecki C, Avitabile P, Schmidt T (2011) 3D digital image correlation methods for full-field vibration measurement. Mech Syst Signal Process 25:917–927. doi:10.1016/j.ymssp.2010.08.013
3. Siebert T, Crompton MJ (2010) Application of high speed digital image correlation for vibration mode shape analysis. In: SEM annual conference and exposition on experimental and applied mechanics 2010, Indianapolis, pp 1776–1783
4. Warren C, Niezrecki C, Avitabile P, Pingle P (2011) Comparison of FRF measurements and mode shapes determined using optically image based, laser, and accelerometer measurements. Mech Syst Signal Process 25:2191–2202. doi:10.1016/j.ymssp.2011.01.018
5. Baqersad J, Carr J, Lundstrom T, Niezrecki C, Avitabile P, Slattery M (2012) Dynamic characteristics of a wind turbine blade using 3D digital image correlation. In: Health monitoring of structural and biological systems 2012, San Diego. doi:10.1117/12.915377
6. Carr J, Baqersad J, Niezrecki C, Avitabile P, Slattery M (2012) Dynamic stress–strain on turbine blades using digital image correlation techniques part 2: dynamic measurements. In: 30th IMAC, a conference on structural dynamics 2012, Jacksonville, pp 221–226. doi:10.1007/978-1-4614-2422-2_21
7. Carr J, Baqersad J, Niezrecki C, Avitabile P, Slattery M (2012) Dynamic stress–strain on turbine blade using digital image correlation techniques part 1: static load and calibration. In: 30th IMAC, a conference on structural dynamics 2012, Jacksonville, pp 215–220. doi:10.1007/978-1-4614-2422-2_20
8. Poozesh P, Baqersad J, Niezrecki C, Harvey E, Yarala R (2014) Full field inspection of a utility scale wind turbine blade using digital image correlation. CAMX, Orlando
9. Helfrick MN, Pingle P, Niezrecki C, Avitabile P (2009) Optical non-contacting vibration measurement of rotating turbine blades. In: 27th conference and exposition on structural dynamics 2009, IMAC XXVII, Orlando
10. Wang W, Mottershead JE, Siebert T, Pipino A (2012) Frequency response functions of shape features from full-field vibration measurements using digital image correlation. Mech Syst Signal Process 28:333–347. doi:10.1016/j.ymssp.2011.11.023
11. Wang W, Mottershead JE (2013) Adaptive moment descriptors for full-field strain and displacement measurements. J Strain Anal Eng Des 48:16–35
12. Wang W, Mottershead JE, Ihle A, Siebert T, Reinhard Schubach H (2011) Finite element model updating from full-field vibration measurement using digital image correlation. J Sound Vib 330:1599–1620
13. Wang W, Mottershead JE, Patterson E, Siebert T, Ihle A (2013) Model updating using shape descriptors from full-field images. In: Conference proceedings of the Society for Experimental Mechanics series, pp 425–436
14. Lundstrom T, Baqersad J, Niezrecki C, Avitabile P (2012) Using high-speed stereophotogrammetry techniques to extract shape information from wind turbine/rotor operating data. In: 30th IMAC, a conference on structural dynamics 2012, Jacksonville, pp 269–275. doi:10.1007/978-1-4614-2419-2_26
15. Warren C, Niezrecki C, Avitabile P (2011) Determination of wind turbine operating deflection shapes using full-field 3D point-tracking. In: 29th IMAC, a conference on structural dynamics 2011, Jacksonville, pp 217–226
16. Lundstrom T, Baqersad J, Niezrecki C (2013) Using high-speed stereophotogrammetry to collect operating data on a Robinson R44 Helicopter. In: 31st international modal analysis conference on structural dynamics, IMAC 2013, Garden Grove, pp 401–410. doi:10.1007/978-1-4614-6546-1_44
17. Olson LE, Abrego AI, Barrows DA, Burner AW (2010) Blade deflection measurements of a full-scale UH-60A rotor system. In: AHS aeromechanics specialists conference 2010, San Francisco, pp 738–747
18. Schneider O (2005) Analysis of SPR measurements from HART II. Aerosp Sci Technol 9:409–420. doi:10.1016/j.ast.2005.01.013
19. Schneider O, Van der Wall BG, Pengel K (2003) HART II blade motion measured by stereo pattern recognition (SPR). In: American Helicopter Society 59th annual forum, Phoenix
20. Tessler A, Spangler JL (2003) A variational principle for reconstruction of elastic deformations in shear deformable plates and shells. National Aeronautics and Space Administration, Langley Research Center, Hampton
21. Tessler A (2007) Structural analysis methods for structural health management of future aerospace vehicles. Key Eng Mater 347:57–66
22. Tessler A, Spangler JL (2005) A least-squares variational method for full-field reconstruction of elastic deformations in shear-deformable plates and shells. Comput Methods Appl Mech Eng, Uxbridge, England, 194:327–339. doi:10.1016/j.cma.2004.03.015
23. Chipman C, Avitabile P (2012) Expansion of transient operating data. Mech Syst Signal Process 31:1–12. doi:10.1016/j.ymssp.2012.04.013
24. Pingle P (2010) Prediction of full-field dynamic stress–strain from limited sets of measured displacement data. In: Mechanical Engineering Departmentz,Ph.D. Thesis, University of Massachusetts Lowell, Lowell, MA
25. Carr J, Baqersad J, Niezrecki C, Avitabile P, Slattery M (2013) Predicting dynamic strain on wind turbine blade using digital image correlation techniques in conjunction with analytical expansion methodologies. In: 31st international modal analysis conference on structural dynamics, IMAC 2013, Garden Grove, CA, pp 295–302. doi:10.1007/978-1-4614-6546-1_31
26. Iliopoulos AN, Devriendt C, Iliopoulos SN, Van Hemelrijck D (2014) Continuous fatigue assessment of offshore wind turbines using a stress prediction technique. In: SPIE/NDE, San Diego, pp 90640S-90640S-90611
27. Avitabile P, Obando SE, Truong K (2014) Full field dynamic deflection and strain for linear components connected with nonlinear connectors. In: International conference on noise and vibration engineering (ISMA2014), Leuven, Belgium
28. Harvie J, Obando S, Avitabile P (2013) Reduced order system model nonlinear response and expansion for full field results. In: Eleventh international conference on recent advances in structural dynamics, Pisa
29. Bogert PB, Haugse E, Gehrki RE (2003) Structural shape identification from experimental strains using a modal transformation technique. In: 44th AIAA/ASME/ASCE/AHS/ASC structures, structural dynamics, and materials conference
30. Derkevorkian A, Masri SF, Alvarenga J, Boussalis H, Bakalyar J, Lance Richards W (2013) Strain-based deformation shape-estimation algorithm for control and monitoring applications. AIAA J 51:2231–2240. doi:10.2514/1.10275
31. Chierichetti M, Ruzzene M (2012) Dynamic displacement field reconstruction through a limited set of measurements: application to plates. J Sound Vib 331:4713–4728

32. Chierichetti M (2014) Load and response identification for a nonlinear flexible structure subject to harmonic loads. J Comput Nonlinear Dyn 9:011009. doi:10.1115/1.4025505
33. Guyan RJ (1965) Reduction of stiffness and mass matrices. AIAA J 3:380–380
34. Kidder RL (1973) Reduction of structural frequency equations. AIAA J 11:892–892
35. O'Callahan JC (1989) A procedure for an improved reduced system (IRS) model. In: Proceedings of the 7th international modal analysis conference. Union College Press, Schenectady, pp 17–21
36. O'Callahan J, Avitabile P, Riemer R (1989) System equivalent reduction expansion process (SEREP). In: 7th IMAC, a conference on structural dynamics 1989, pp 29–37
37. Allemang RJ, Brown DL (1989) Correlation coefficient for modal vector analysis. In: Proceedings of the 1st international modal analysis conference & exhibit. Union College, Orlando, pp 110–116
38. (2011) PONTOS v6.3. GOM mbH, Braunschweig
39. (2010) Abaqus/CAE 6.10-2. Dassault System
40. Baqersad J, Poozesh P, Niezrecki C, Avitabile P (2014) Comparison of modal parameters extracted using MIMO, SIMO, and impact hammer tests on a three-bladed wind Turbine. In: 32th IMAC, a conference on structural dynamics, Orlando. doi:10.1007/978-3-319-04774-4_19
41. Baqersad J, Niezrecki C, Avitabile P, Slattery M (2013) Dynamic characterization of a free-freewind turbine blade assembly. In: 31st international modal analysis conference on structural dynamics, IMAC 2013, Garden Grove, pp 303–312. doi:10.1007/978-1-4614-6546-1_32
42. Pingle P, Avitabile P (2011) Full field dynamic stress/strain from limited sets of measured data. In: 29th IMAC, a conference on structural dynamics 2011, Jacksonville, pp 187–200

Chapter 11
Predicting the Vibration Response in Subcomponent Testing of Wind Turbine Blades

Mohamad Eydani Asl, Christopher Niezrecki, James Sherwood, and Peter Avitabile

Abstract Currently new wind turbine blade materials are certified by starting with coupon testing for initial strength and fatigue analysis, followed by full-scale blade testing as a final quality control to assess material characteristics. Subcomponent testing has been proposed as a supplement to the structural analysis and material characterization, bridging the gap between coupon and full-scale tests. In this study, similitude theory is applied to a simply-supported rectangular plate that is representative of a wind turbine blade spar cap with the goal of designing a validated scaled-down subcomponent. The vibration of a specially orthotropic rectangular laminated plate is analyzed to extract the scaling laws based on direct use of the field equations. The accuracy of the derived scaling laws is analyzed as a model validation criteria by mapping the first natural frequency of the variant subcomponents to the full-scale plate. The effect of the ply stack up scheme and size of the subcomponents in predicting accuracy of the scaling laws are then investigated by applying partial and complete similarity conditions. According to the results, subcomponents with modified ply stack up could be found that have a good accuracy in predicting the first natural frequency of the full-scale plate. However, picking an appropriate aspect ratio is critical to the success of the prediction of full scale plate response as shown in the cases studied.

Keywords Wind turbine blade • Similitude • Sub-component • Vibrations • Plate

11.1 Introduction

The rapidly increasing size of wind turbine blades in the last two decades is making full-scale blade testing more challenging than ever before. These large sized test structures are expensive to build, difficult to ship and manipulate during test, require several months of testing for evaluation, and are very expensive to construct and test. Coupon testing of the materials and full-scale blade testing are both required for certification of the blade. As coupon testing does not always reveal the durability of the blade, a final full-scale testing is required to certify a blade design. If accurate and meaningful subcomponent tests could be designed, they could potentially be used to facilitate blade certification by expediting the evaluation rate of newly released materials.

Numerous studies have been carried on to investigate and design small-scale subcomponents emulating a wind turbine blade. Mandell et al. [1] fabricated and tested composite I-beams with shear webs and flanges representative of the cross section of a wind turbine blade. Resulting measurements from four point bending test of the composite I-beams were in good agreement with strain and stiffness predicted by finite element model and simple beam theory. The root section of the blade was studied by Cairns et al. [2] in which the specimens emulating a single insert of a blade to hub joint were evaluated by using a pull out test. A significant amount of static and fatigue strength data were generated, however the primary focus of the study was manufacturing. Mandell et al. [3] conducted a study on sandwich panel terminations and skin-stiffener intersections. The objective was to estimate skin-stiffener fracture loads and evaluate the performance at transition location between sandwich panel and normal laminate by using tension tests.

There are a few studies focusing on the performance of adhesive joints and bond lines in a wind turbine blade using the subcomponent testing concept. Within these tests a complex stress field was applied to the components to investigate the integrity of adhesive bond lines. An asymmetric three-point bending test performed on what is called a Henkel beam was proposed by Sayer et al. [4]. The goal was to design an I-beam with a comparable combination of bending moment and shear forces as a three-point bending test and reducing the stress concentrations at the clamped end. The specimens constructed using the Henkel design were tested for a parametric study to investigate the influence of the design and manufacturing

M. Eydani Asl (✉) • C. Niezrecki • J. Sherwood • P. Avitabile
Structural Dynamics and Acoustic Systems Laboratory, University of Massachusetts Lowell, One University Avenue, Lowell, MA 01854, USA
e-mail: mohamad_eydaniasl@student.uml.edu

© The Society for Experimental Mechanics, Inc. 2015

R. Allemang (ed.), *Special Topics in Structural Dynamics, Volume 6*, Conference Proceedings of the Society for Experimental Mechanics Series, DOI 10.1007/978-3-319-15048-2_11

variables on the shear web to spar caps adhesive joints [5]. Zarouchas et al. [6] proposed a static four-point bending test on two symmetric composite I-beams. The I-beams represented a scaled down load carrying connection between the spar caps to the shear web in a blade.

Although subcomponent testing concept is usually considered a laboratory-scale test rig, a few mid-level blade test approaches have been proposed which fall into the subcomponent category. A comprehensive study of the structural static strength and failure mechanisms of a box girder of a wind turbine blade, loaded in the flap-wise direction, was conducted by Jensen [7]. In order to reduce the total test time required for fatigue evaluation of the blade, White et al. [8] developed a dual-axis test setup on a truncated wind turbine blade that combined resonance excitation with forced hydraulic loading.

The subcomponent testing approaches developed to date have aimed to explore the characterization of the materials and failure mechanisms of the substructures used in the wind turbine blades. Some of the models designed as a subcomponent of the blade are fairly simple, small, and easy to manufacture and some are well detailed and close to an original utility-scale blade. The remaining question is how simple can a subcomponent test be and still provide meaningful results for the intricacies of a utility-scale blade? It is important to know if the data from the subcomponent tests are able to accurately characterize and estimate the behavior of its corresponding component in the full-scale blade, because the subcomponent specimens are not necessarily a geometrically scaled-down version of the original full-scale blade. In other words, it is important to verify that the designed subcomponent is able to predict the behavior of its corresponding full-scale component and data from the subcomponent test can be used to predict or estimate the response of the full-scale component. This possible connection between subcomponent (referred to as "model") and its corresponding full-scale component (referred to as "prototype") must be based on the existence of the same field equations and structural parameters that govern the response of both model and prototype when subjected to different load conditions. If such structural parameters could be found, a scaled down model (subcomponent) can be built that duplicates the behavior of its prototype but on a much smaller scale. The descriptor which describes this relationship is "structural similarity".

The necessary and sufficient similarity conditions between two systems are described by similitude theory. These similarity conditions between two systems may be derived directly from the governing equations of the system that lead to more specific similarity conditions than dimensional analysis. Simitses and Rezaeepazhand [9] proposed a technique to derive the scaling laws by means of the governing equation of a system directly. The response of the prototype was then predicted by projecting the response of its associate models using derived scaling laws. As a benchmark, the buckling response of an orthotropic and symmetric cross-ply laminated plate was investigated in that study. They also analyzed the vibration of scaled laminated rectangular plates in later studies [10] and investigated the effect of axial and shear loads on stability of scaled laminated rectangular plates [11, 12]. Later, this method was extensively used in their works regarding the prediction of vibration response of laminated shells [13, 14]. According to their results, the scaling laws that are obtained directly from the governing equations yield perfect accuracy for regular cross-ply laminates, while for the angle-ply laminates, the scaling laws did not show good accuracy.

The accuracy of derived scaling laws highly depends on the ply stack-up scheme of the prototype. Hence, derived scaling laws and suggested models for a specific laminated prototype are only valid for that case and may not be used for the prototypes with a different ply stack up. In this study, accuracy of the derived scaling laws is evaluated for fundamental frequency prediction of a specific case of a layup scheme used in a wind turbine blade spar cap near the maximum chord of the blade.

11.2 Governing Equations

In this section, the governing equations for the vibration of a simply-supported specially orthotropic laminated plate are introduced and the Navier method is applied to derive the closed form solution. Then, scaling laws are derived and their applicability is discussed.

11.2.1 *Vibration of Specially Orthotropic Laminated Plate*

The objective is to predict the fundamental vibration frequency of the prototype which in this case is a simply-supported laminated plate with a layup scheme similar to a spar cap of a wind turbine blade near its maximum chord. The simply-supported boundary conditions were selected to get an exact solution for equations. For this study, the static bending in the absence of thermal effects and in-plane forces for a specially orthotropic laminated rectangular plate ($B_{ij} = D_{16} = D_{26} = 0$, where B and D are coupling and bending stiffness matrices) is considered. Neglecting the secondary rotary inertia, the equation governing the free vibration of the plate is given by [15]:

$$D_{11}\frac{\partial^4 w_0}{\partial x^4} + 2\left(D_{12} + 2D_{66}\right)\frac{\partial^4 w_0}{\partial x^2 \partial y^2} + D_{22}\frac{\partial^4 w_0}{\partial y^4} + I_0\frac{\partial^2 w_0}{\partial t^2} = 0 \tag{11.1a}$$

$$I_0 = \sum_{k=1}^{L} \rho_0{}^{(k)}\left(z_{k+1} - z_k\right) \tag{11.1b}$$

The simply-supported boundary conditions on all four edges of the rectangular plate can be expressed as:

$$w_0(x,0) = 0, \quad w_0(x,b) = 0, \quad w_0(0,y) = 0, \quad w_0(a,y) = 0$$

$$M_{xx}(0,y) = 0, \quad M_{xx}(a,y) = 0, \quad M_{yy}(x,0) = 0, \quad M_{yy}(x,b) = 0 \tag{11.1c}$$

where D_{ij} are bending stiffnesses, w_0 is transverse deflection, I_0 represents mass moment of inertia, ρ_0 indicates mass per volume, L is the total number of plies and z-axis is taken positive upward from the mid-plane. M_{ij} describe moments and a and b are the in-plane dimensions along the x and y coordinates. The origin of coordinate system is chosen to be taken at the lower left corner of the mid-plane as shown in Fig. 11.6.

Following the standard Navier method, a periodic solution is assumed in the form of:

$$W_{mn}(t) = W_{mn}^{0} e^{i\omega t} \tag{11.2}$$

Where $i = \sqrt{-1}$ and ω is the natural frequency. Substitution of Eq. (11.2) in Eq. (11.1a) and neglecting the effect of rotary inertia, yields the natural frequency ω_{mn} for different values of m and n which is described as:

$$\omega_{mn}^2 = \frac{\pi^4}{I_0 b^4}\left[D_{11}m^4\left(\frac{b}{a}\right)^4 + 2\left(D_{12} + 2D_{66}\right)m^2 n^2\left(\frac{b}{a}\right)^2 + D_{22}n^4\right] \tag{11.3}$$

which fundamental frequency occurs at $m = n = 1$.

11.3 Complete Similarity

The fundamental frequency of the prototype and its models (subcomponents) are all described by Eq. (11.3). To establish similitude theory and derive the scaling laws it is assumed that all variables of the governing equation of the model (x_m) are connected to their corresponding values in prototype (x_p) by one to one mapping. Then the scale factor may be defined as $\lambda_x = x_p/x_m$ and standard similitude procedure [9] can be applied to the Eq. (11.3) that results in the following scaling laws:

$$\lambda_{D_{22}}\lambda_n{}^4 = \lambda_{D_{11}}\left(\frac{\lambda_m\lambda_b}{\lambda_a}\right)^4 = \lambda_{\overline{D}_{12}}\left(\frac{\lambda_n\lambda_m\lambda_b}{\lambda_a}\right)^2 \tag{11.4}$$

$$\lambda_\omega^2 = \frac{\lambda_{D_{22}}\lambda_n{}^4}{\lambda_b{}^4\lambda_\rho} \tag{11.5a}$$

$$\lambda_\omega^2 = \frac{\lambda_{\overline{D}_{12}}\lambda_n{}^2\lambda_m{}^2}{\lambda_b{}^2\lambda_a{}^2\lambda_\rho} \tag{11.5b}$$

$$\lambda_\omega^2 = \frac{\lambda_{D_{11}}\lambda_m{}^4}{\lambda_a{}^4\lambda_\rho} \tag{11.5c}$$

where $D_{12} + 2D_{66} = \overline{D}_{12}$ in Eq. (11.4).

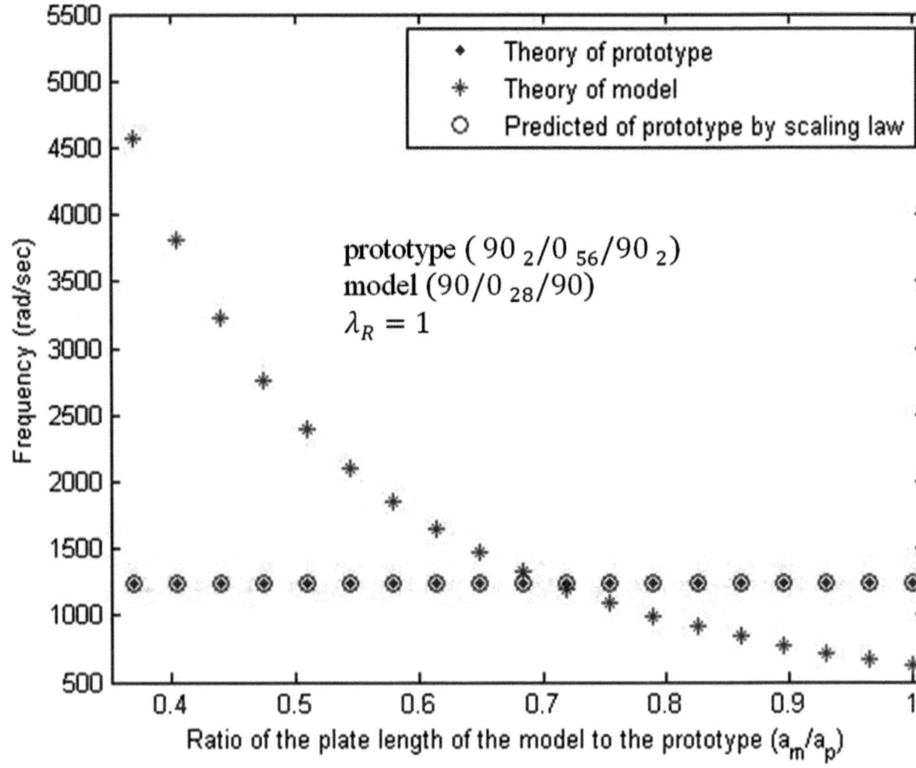

Fig. 11.1 Predicted and theoretical fundamental frequency of the prototype ($90_2/0_{56}/90_2$) when models are ($90/0_{28}/90$) with $a_p \neq a_m$ and $\lambda_R = 1$

To achieve the complete similarity between prototype and model, all terms of Eq. (11.4) must be satisfied simultaneously. Then the frequency scale factor (λ_ω) may be obtained exactly using Eqs. (11.5a), (11.5b), and (11.5c).

To satisfy Eq. (11.4) flexural stiffness ratios $\left(\lambda_{D_{22}}, \lambda_{\overline{D}_{22}} \text{ and } \lambda_{D_{11}}\right)$ must be equal and there should be geometrical symmetry ($\lambda_a = \lambda_b$) between model and prototype. The first condition may be satisfied using ply-level scaling method $[(+\Theta_n/-\Theta_n)_s,\ n = 1, 2, 3, \ldots]$ which the angle of plies in the layup scheme of the laminate is denoted by Θ [14]. The latter condition is then satisfied by choosing an aspect ratio for the model the same as the prototype ($\lambda_R = 1,\ R = b/a$).

The prototype geometry and lay up scheme considered in this study emulate the spar cap flanges of a 5 MW wind turbine blade [16]. The prototype dimensions are $600 \times 300 \times 30\ mm^3$ made of Carbon/Epoxy layers ($90_2/0_{56}/90_2$) and the thickness of the plies are $t = 0.5$ mm. The geometry and layup scheme of the prototype is kept fixed in all case studies while comparing to the developed models with various dimensions and ply stack up.

In the complete similarity case, the fundamental vibration frequency of the prototype with simply supported boundary conditions was compared to the models with ply-level scaled down thickness and overall dimensions with an aspect ratio same as the prototype. Theoretical values of the models and the prototype were obtained directly by Eq. (11.3), and then predicted values were projected by applying derived response scaling laws Eqs. (11.5a), (11.5b), and (11.5c) to the theoretical values of the models. In the complete similarity case, all response scaling laws predict the frequency of the prototype with excellent and same accuracy because the complete similarity assumptions in Eq. (11.4) were satisfied simultaneously with specific design considered for the models (ply-level scaling and preserving aspect ratio of the prototype).

Figure 11.1 shows the accuracy of scaling laws for the complete similarity case. The thicknesses of the models ($90/0_{28}/90$) have been scaled down using ply-level scaling and overall dimensions are varying $\left(a_p \neq a_m\right)$ while preserving same aspect ratio as the prototype (See Fig. 11.7). According to the results, theoretical values from the models were projected by using the scaling laws to those of the prototype with excellent agreement.

11.4 Partial Similarity

In the previous section, it was shown that complete similarity can be achieved in the special case of ply-level scaling while keeping the same aspect ratio for the prototype and its models. To extend the applicability of scaled down models, partial similarity with distortions in ply stack up and aspect ratios of the models are investigated. In these cases, necessary conditions

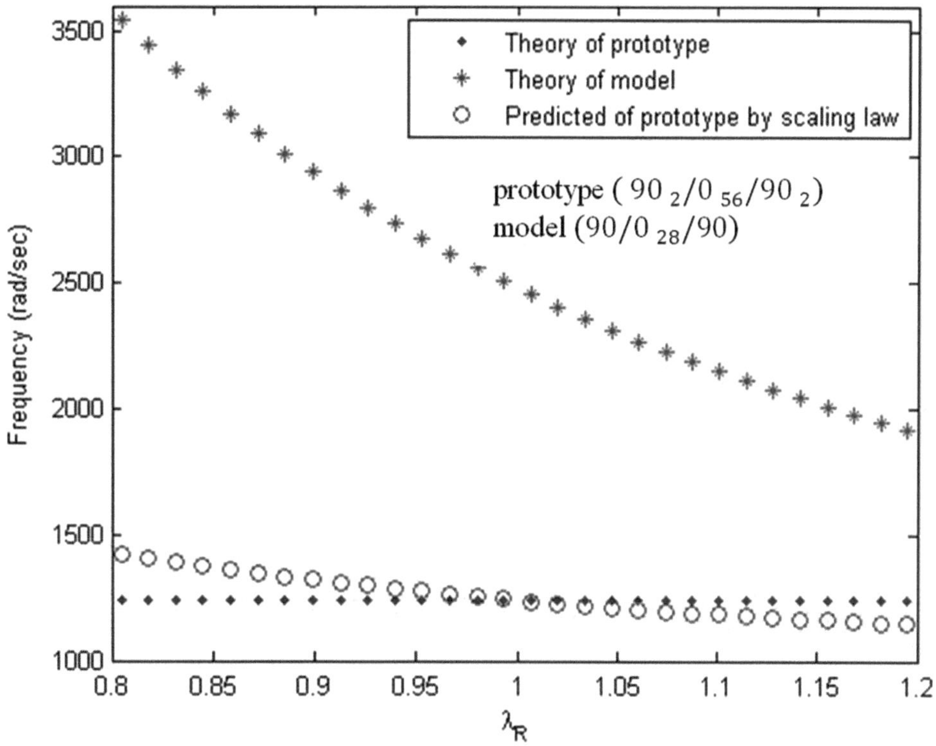

Fig. 11.2 Predicted and theoretical fundamental frequency of the prototype ($90_2/0_{56}/90_2$) when models are $(90/0_{28}/90)$ with $a_p = 2a_m$ and $\lambda_R \neq 1$

for complete similarity in Eq. (11.4) are not fully satisfied. Discrepancy in Eq. (11.4) yields inaccuracy in the response scaling laws defined in Eqs. (11.5a), (11.5b), and (11.5c). The level of inaccuracy for various distortions is investigated and the most accurate response scaling law is selected to project the data from the model to prototype.

11.4.1 Distortion in Aspect Ratio

In this case, models are still a ply-level scaled down of the prototype in the thickness direction but have different aspect ratios compared to the prototype (See Fig. 11.8). While flexural stiffness ratios are equal $\left(\lambda_{D_{22}} = \lambda_{\overline{D}_{12}} = \lambda_{D_{11}}\right)$, the aspect ratio scale factor has values other than unity ($0.8 < \lambda_R < 1.2$). Figure 11.2 represents the theoretical and predicted fundamental frequency of the prototype and scaled down models while changing the model aspect ratio. Models with different aspect ratio were selected to predict the prototype response. The response scaling laws in Eqs. (11.5a), (11.5b), and (11.5c) are very sensitive to distortion of the aspect ratio. The scaling law presented by Eq. (11.5c) yields the best prediction in this case. When distortions in the aspect ratio are being considered, it is not a good candidate for designing models, because the response scaling laws are unable to predict the prototype response accurately using response scaling laws.

11.4.2 Distortion in Ply Stack Up

In this case, it is assumed that models have the same aspect ratio as the prototype but ply-level scaling is not considered, so models may hold different flexural stiffness ratios ($\lambda_{D_{22}} \neq \lambda_{\overline{D}_{12}} \neq \lambda_{D_{11}}$). Basically ply-level scaling is not applicable when laminate thickness is supposed to be scaled-down below the range in which there is no multiple lamina left in the stack up and changing the lamina thickness is not an option. Considering the prototype with the ply stack up ($90_2/0_{56}/90_2$), it is not possible to scale down the thickness to one quarter of the prototype thickness using ply-level scaling. As the only constraint to achieve complete similarity in scaling down the prototype thickness is obtained by satisfying Eq. (11.4), any ply scheme for the model that approximately satisfies Eq. (11.4) may be considered a solution. Based on the prototype layup scheme, an algorithm is proposed to search for the potential model ply schemes with an overall thickness less than half of the prototype thickness and approximately same flexural stiffness ratios.

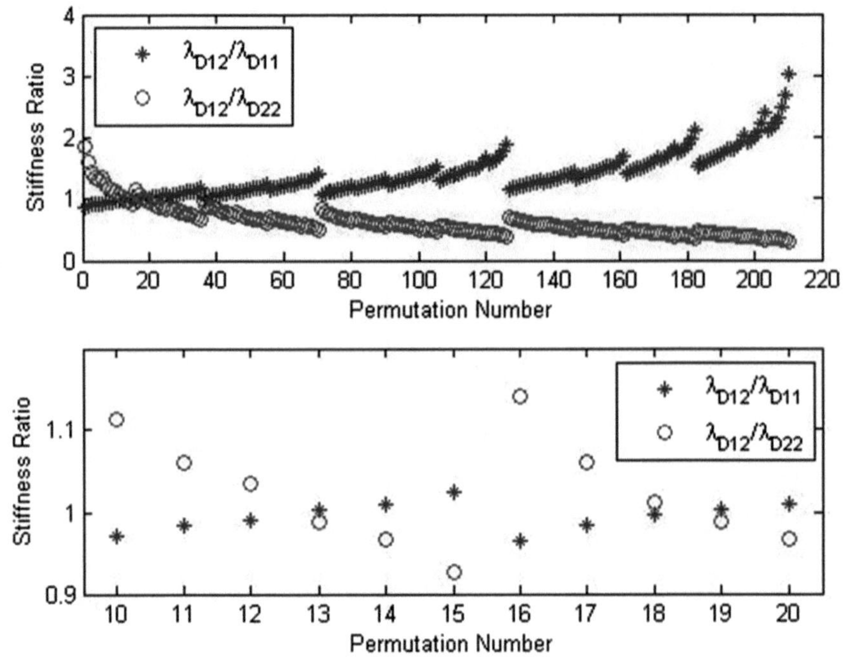

Fig. 11.3 Evaluating Eq. (11.6) for the prototype ($90_2/0_{56}/90_2$) when models are all possible cases with 20 plies and include eight 90 degree plies. The top graph shows all 210 possible cases and the bottom graph shows cases 13, 18 and 19 have the least deviation with less than 1 % error in the stiffness ratio law

Rearranging the terms in Eq. (11.4) assuming that $\lambda_m = \lambda_n = \lambda_R = 1$ yields:

$$\frac{\lambda_{D_{22}}}{\lambda_{D_{11}}} = 1, \quad \frac{\lambda_{\overline{D}_{12}}}{\lambda_{D_{11}}} = 1 \tag{11.6}$$

This is the necessary and sufficient condition for scaling down the thickness that results in complete similarity. The goal is then to find ply schemes for the models that approximately satisfy Eq. (11.6) and show values close to unity $\left(\lambda_{D_{22}}/\lambda_{D_{11}} \simeq 1,\ \lambda_{\overline{D}_{12}}/\lambda_{D_{11}} \simeq 1\right)$.

To develop scaled down models with approximate flexural stiffness ratios, a permutation algorithm was developed to search the ply schemes with a small number of plies that approximately satisfy Eq. (11.6). As the prototype ply stack up ($90_2/0_{56}/90_2$) has only cross ply layers, models with symmetric cross ply layups were searched.

Figure 11.3 shows all possible ply schemes for the models with a total of 20 layers that include 12 zero degree and 8 ninety degree plies. Among the 210 possible ply scheme cases, the case with permutation number 13 and layup $[0_4/\ 90_3/0_2/90]_s$, the case with permutation number 18 and layup $[0_3/\ 90/0_2/90_2/0/90]_s$, and the case with permutation number 19 and layup $[0_3/\ 90/0_2/90_2/90/0]_s$ were found with less than 2 % error $\left(\lambda_{D_{22}}/\lambda_{D_{11}} < 1.02,\ \lambda_{\overline{D}_{12}}/\lambda_{D_{11}} < 1.02\right)$ in stiffness ratio law.

Figure 11.4 shows all possible ply schemes for the models with a total of 20 layers which include 16 zero degree and 4 ninety degree plies. Among all 45 possible ply scheme cases, the case with permutation number 20 and layup $[0_3/90/0/90/\ 0_4]_s$ and the case with permutation number 24 and layup $[0_2/\ 90/0_4/90/0_2]_s$ were found to have with less than 4 % error $\left(\lambda_{D_{22}}/\lambda_{D_{11}} < 1.04,\ \lambda_{\overline{D}_{12}}/\lambda_{D_{11}} < 1.04\right)$ in stiffness ratio law.

Figure 11.5 shows all possible ply schemes for the models with total 14 layers which include 8 zero degree and 6 ninety degree plies. Among all 35 possible ply scheme cases, the case with permutation number 4 and layup $[0_3/90_3/0]_s$ was found to have less than 1.5 % error $\left(\lambda_{D_{22}}/\lambda_{D_{11}} < 1.015,\ \lambda_{\overline{D}_{12}}/\lambda_{D_{11}} < 1.015\right)$ in stiffness ratio law.

11.5 Discussion

Based on the presented results, Eq. (11.4) is always valid for ply-level scaled down models that have the same aspect ratio as the prototype and response scaling laws Eqs. (11.5a), (11.5b), and (11.5c) yield excellent accuracy in predicting the fundamental frequency. In partial similarity, developed ply-level scaled down models with a different aspect ratios do not satisfy Eq. (11.4) and results in high sensitivity in response scaling laws and discrepancies for predicted frequencies.

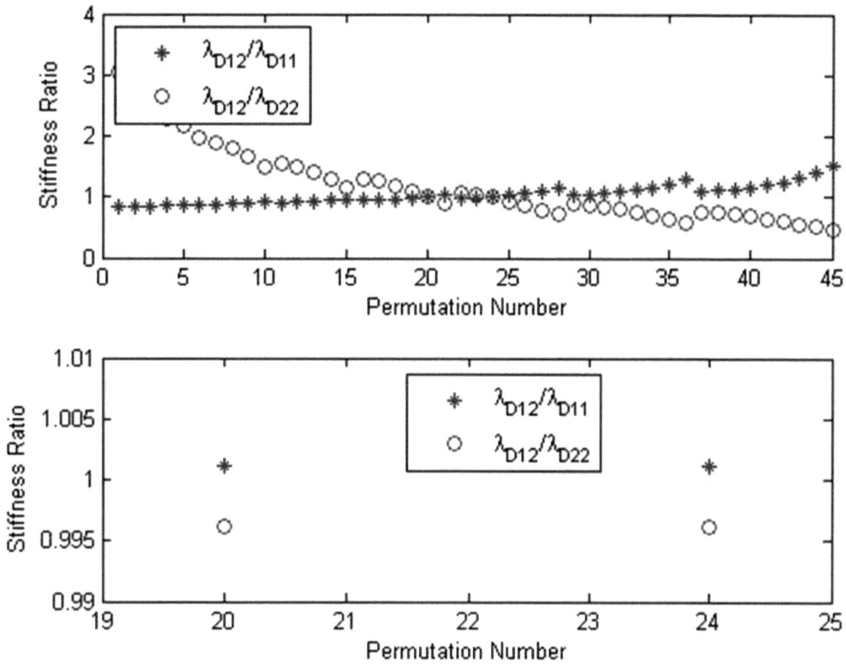

Fig. 11.4 Evaluating Eq. (11.6) for the prototype ($90_2/0_{56}/90_2$) when models are all possible cases with 20 plies and include eight 90 degree plies. The top graph shows all 45 possible cases and the bottom graph shows cases 20 and 24 have the least deviation with less than 1 % error in stiffness ratio law

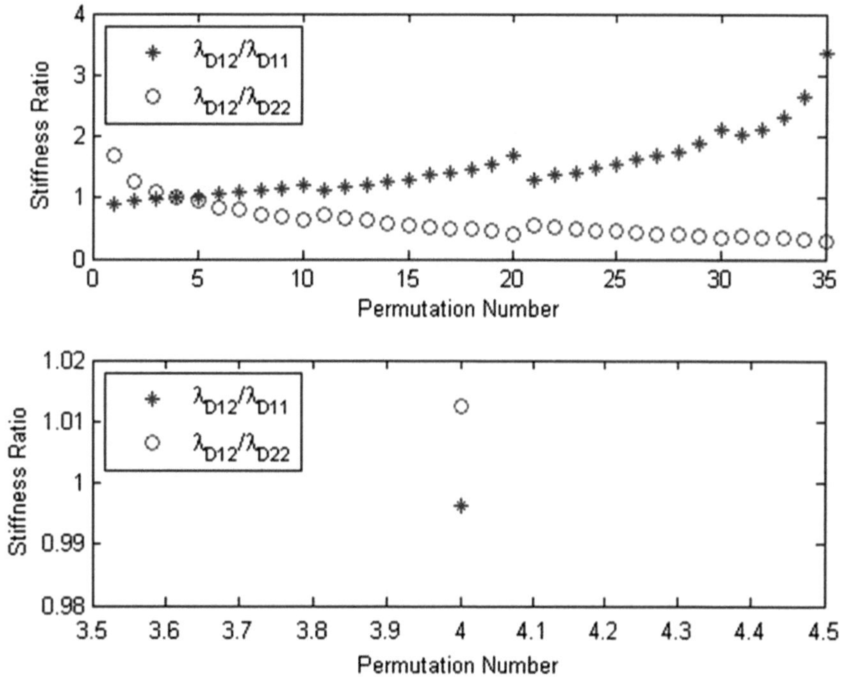

Fig. 11.5 Evaluating Eq. (11.6) for the prototype ($90_2/0_{56}/90_2$) when models are all possible cases with 14 plies and include three 90 degree plies. The top graph shows all 35 possible cases and the bottom shows permutation number 4 that has the least deviation with less than 1.5 % error in stiffness ratio law

Applying the ply-level scaling approach to the prototype results in a model with an overall thickness equal to half of the prototype thickness. To scale down the thickness of the model more than half, models with total layers less than 30 should be searched that satisfy the specially orthotropic laminates condition ($B_{ij} = D_{16} = D_{26} = 0$). Based on results shown in Figs. 11.1, 11.2, and 11.3, symmetric cross ply laminates were found as models with a total number of layers (e.g. 20 and 14) which satisfy Eq. (11.6) with negligible error. Selected models using this approach may be used with very good accuracy to predict the prototype frequency depending on their error in satisfying Eq. (11.6).

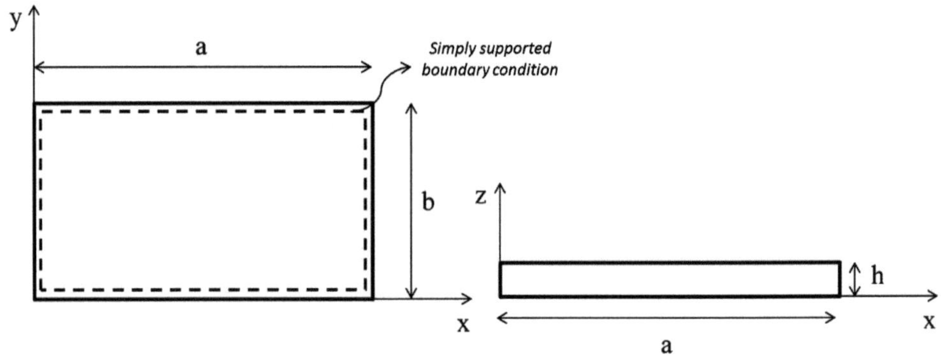

Fig. 11.6 Geometry, coordinate system and simply supported boundary condition for the rectangular plate [15]

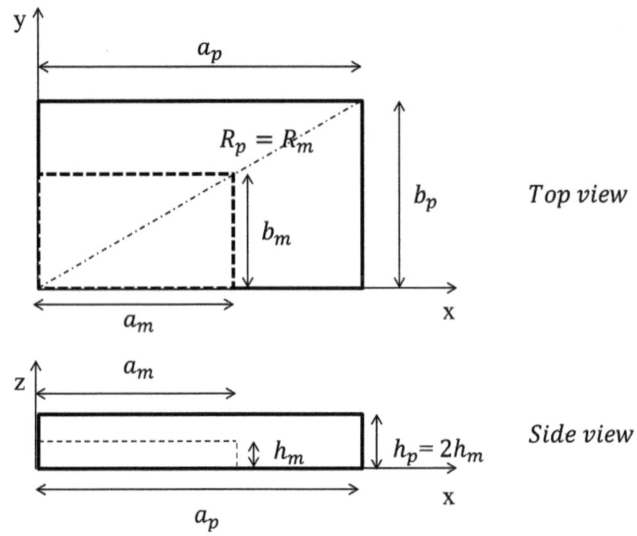

Fig. 11.7 Models have the same aspect ratio as the prototype ($\lambda_a = \lambda_b$) while changing the in plane dimensions and keeping the thicknesses fixed

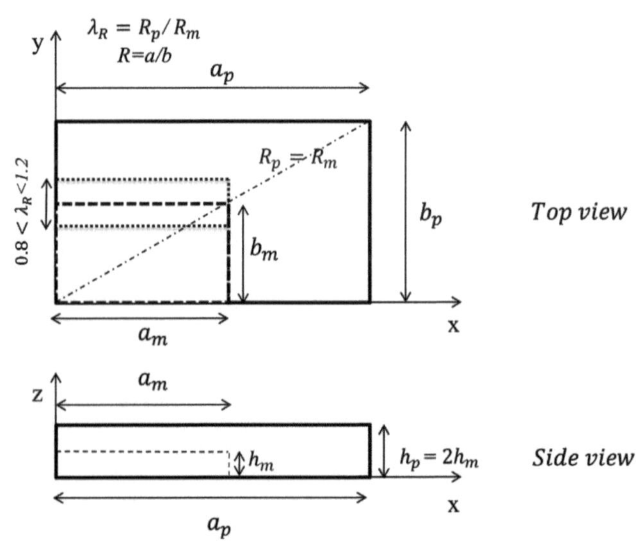

Fig. 11.8 Models have the same flexural stiffness ratios due to ply-level scaling while holding different aspect ratios than the prototype

11.6 Conclusion

In order to expedite the certification and testing of new materials for wind turbine blades, a scaled down model may be developed. To evaluate applicability of scaled down models in testing new materials, an analytical study was conducted. Similitude theory was applied to the governing equations of vibration of a specially orthotropic rectangular laminated plate emulating a spar cap of a wind turbine blade. Derived scaling laws were treated as design criteria to propose scaled down models. The effect of different design parameters on the accuracy of scaling laws was investigated and possible partial similarity scenarios were evaluated. Then the scaling laws were evaluated for various conditions of both complete and partial similarity. A permutation algorithm was developed to search for scaled down models when ply-level scaling in not available. Applicability of this method to design the scaled down models, for predicting the fundamental natural frequency of the prototype, was established based on an analytical investigation. According to the results, established scaling laws can be applied with excellent accuracy in complete similarity conditions and very good accuracy for partial similarity to predict the fundamental frequency of a specially orthotropic laminated rectangular plate.

Acknowledgments This material is based upon work supported by the National Science Foundation under Grant Number 1230884 (Achieving a Sustainable Energy Pathway for Wind Turbine Blade Manufacturing). Any opinions, findings, and conclusions or recommendations expressed in this material are those of the author(s) and do not necessarily reflect the views of the National Science Foundation.

References

1. Mandell JF, Combs D, Samborsky DD (1995) Fatigue of fiberglass beam substructures. Wind Energy 16:99
2. Cairns DS, Skramstad JD, Mandell JF (2001) Evaluation of hand lay-up and resin transfer molding in composite wind turbine blade structures. In 20th 2001 ASME Wind Energy Symposium
3. Mandell JF, Creed R Jr, Pan Q, Combs DW, Shrinivas M (1994) Fatigue of fiberglass generic materials and substructures. Wind Energy 15:207
4. Sayer F, Post N, Van Wingerde A, Busmann H, Kleiner F, Fleischmann W, Ganso M (2009) Testing of adhesive joints in the wind industry. In European Wind Energy Conference and Exhibition 2009, EWEC 2009, pp 288–315
5. Sayer F, Antoniou A, Van Wingerde A (2012) Investigation of structural bond lines in wind turbine blades by sub-component tests. Int J Adhes Adhes 37:129–135
6. Zarouchas DS, Makris AA, Sayer F, Van Hemelrijck D, Van Wingerde AM (2012) Investigations on the mechanical behavior of a wind rotor blade subcomponent. Compos Part B 43:647–654
7. Jensen FM, Falzon BG, Ankersen J, Stang H (2006) Structural testing and numerical simulation of a 34 m composite wind turbine blade. Compos Struct 76:52–61
8. White D, Musial W, Engberg S (2005) Evaluation of the B-REX fatigue testing system for multi-megawatt wind turbine blades. In Collection of the 2005 ASME Wind Energy Symposium Technical Papers at the 43rd AIAA Aerospace Sciences Meeting and Exhibit, pp 52–65
9. Simitses GJ, Rezaeepazhand J (1993) Structural similitude for laminated structures. Compos Eng 3:751–765
10. Rezaeepazhand J, Simitses GJ (1995) Use of scaled-down models for predicting vibration response of laminated plates. Compos Struct 30: 419–426
11. Rezaeepazhand J, Simitses GJ, Starnes JH Jr (1995) Design of scaled down models for stability of laminated plates. AIAA J 33:515–519
12. Simitses GJ, Rezaeepazhand J (1995) Structural similitude and scaling laws for buckling of cross-ply laminated plates. J Thermoplast Compos Mater 8:240–251
13. Rezaeepazhand J, Simitses GJ, Starnes JH Jr (1996) Design of scaled down models for predicting shell vibration response. J Sound Vib 195:301–311
14. Rezaeepazhand J, Simitses GJ (1997) Structural similitude for vibration response of laminated cylindrical shells with double curvature. Compos Part B 28:195–200
15. Reddy JN (2004) Mechanics of laminated composite plates and shells: theory and analysis. CRC press, Boca Raton
16. Resor BR (2013) Definition of a 5MW/61.5 m wind turbine blade reference model. Sandia National Laboratories, Albuquerque, SAND2013-2569 2013

Chapter 12
Linear Modal Analysis of a Horizontal-Axis Wind Turbine Blade

Gizem Acar and Brian F. Feeny

Abstract In this work, mode shapes and modal frequencies of a horizontal axis wind turbine blade are found. The blade is modeled as a pre-twisted beam under bend-bend-twist deformations, and the strain, gravitational potential and kinetic energy expressions are found. These expressions are then simplified for linear theory, and further by using the assumed modes method in which assumed modes are chosen independently for flapwise, edgewise, and twist deflections from a cantilevered uniform beam. Lagrange's equations of motion are applied to the assumed modal coordinates, and coupled linear equations are thus found. Modal analysis is applied to the assumed modal equations to find the resonance frequencies, and the resulting mode shapes are recombined to express the mode shapes in the bend-bend-twist coordinates. The approach was applied to a uniform twisted rectangular beam, and the first three mode shapes are found to be combinations of the in-plane and out-of-plane bending. The results of the modal analysis are compared with results from a finite element analysis, and the mode shapes and frequencies are consistent. The method is also applied to NREL's 5 MW wind turbine blade to find its bend-bend-twist mode shapes and natural frequencies.

Keywords Wind turbines • Bend-bend-twist coupled vibrations • Assumed modes • Finite element analysis • Position dependent modal properties

12.1 Introduction

Horizontal-axis wind turbines are subjected to dynamic loading due to wind shear, cyclic changes in the transverse and axial components of gravity, and tower shadowing. In most wind turbines, failure takes place in the hub, and the gearbox. The aim of this project is to investigate the dynamics of the blades and to identify the components of loading on hub induced by blades.

Dynamics of rotor blades have been studied by many researchers [1–4]. Much of the wind-turbine blade analysis address either the in-plane (edgewise) deformations, or the out-of-plane (flapwise) deformations [5, 6]. In reality, for a twisted blade, in-plane and out-of-plane motions are coupled, and true modes will involve both sets of deformations.

A model for a helicopter blade under bend-bend-twist loading was derived by Crespo Da Silva [3]. This model was then extended to blades with cross sections having a shear center different than the center of mass by Avramov et al. [4]. The work presented here is based on Crespo Da Silva's and Avramov's models. A beam having a structural twist and a varying cross section, under bend-bend-twist coupled deformation is studied. First the kinetic energy and the potential energy of the beam are found, and then the equations of motion are derived by using energy methods together with an assumed modes method. Aeroelastic effects are not included, and rotational effects, such as cyclic changes in gravity and centrifugal effects, are taken into account in the equations of motion. However, the modal analysis is applied to a quasistatic blade. For example the difference in modal behavior between an upright blade and a horizontal blade is studied for a stationary blade. Rotational effects include cyclic loading and cyclic stiffening. Treatment of the dynamic parametric stiffening will be considered in the future. One way to do this is by using Floquet modes, as done by Allen et al. [7].

G. Acar (✉) • B.F. Feeny
Michigan State University, East Lansing, MI, USA
e-mail: dilbergi@egr.msu.edu

© The Society for Experimental Mechanics, Inc. 2015

R. Allemang (ed.), *Special Topics in Structural Dynamics, Volume 6*, Conference Proceedings of the Society for Experimental Mechanics Series, DOI 10.1007/978-3-319-15048-2_12

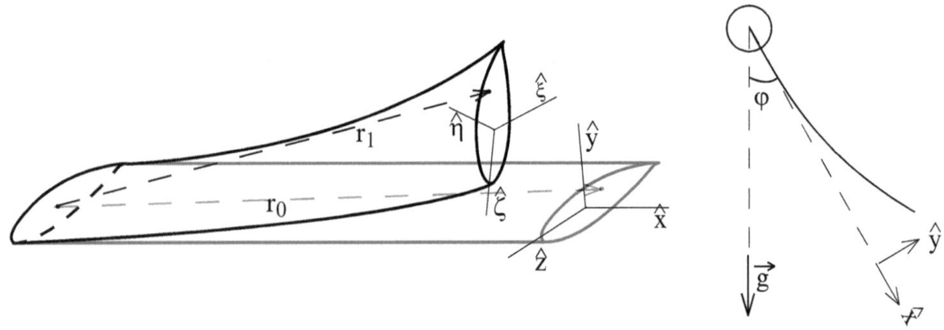

Fig. 12.1 Coordinate systems in deformed, and undeformed blade

12.2 Model Definition

The wind turbine blade is modeled as a beam having a structural twist and a varying cross section. For airfoil cross sections, the shear center is usually different than the mass center, and this difference induces coupling between torsion and bending. Besides, in-plane and out-of-plane bending deformations are also coupled because of the pre-twist. Therefore the vibrational modes are expected to be a combination of in-plane bending, out-of-plane bending and torsional degrees of freedom.

In this work, two different coordinate systems are used as shown in Fig. 12.1. The $\hat{x}\hat{y}\hat{z}$ coordinate system is fixed to the undeformed blade, and it rotates with the rotor, whereas $\hat{\xi}\hat{\eta}\hat{\zeta}$ is fixed to the deformed blade. The combined motion is modeled as three translations, u, v, w and three rotations $\theta_x, \theta_y, \theta_z$, in and about the $\hat{x}, \hat{y}, \hat{z}$ directions, respectively. θ_x is the torsion angle, v and w are in-plane and out-of-plane bending deflections, and θ_y and θ_z are the corresponding rotations. The beam is assumed to be inextensible, therefore the axial deflection, u, is affected only by the foreshortening due to bending, and it can be written in terms of v and w, as given by

$$u' = -\sqrt{1 - v'^2 - w'^2}. \tag{12.1}$$

To find the strain energy, strain and stress distributions are found first. The strain distribution is obtained by using Green's formula, as given by

$$2\,(\mathrm{d}x \quad \mathrm{d}\eta \quad \mathrm{d}\zeta)\,(\varepsilon)\begin{pmatrix}\mathrm{d}x\\ \mathrm{d}\eta\\ \mathrm{d}\zeta\end{pmatrix} = \mathrm{d}\overrightarrow{\mathbf{r_1}}\cdot\mathrm{d}\overrightarrow{\mathbf{r_1}} - \mathrm{d}\overrightarrow{\mathbf{r_0}}\cdot\mathrm{d}\overrightarrow{\mathbf{r_0}}, \tag{12.2}$$

where $\overrightarrow{\mathbf{r_0}}$ and $\overrightarrow{\mathbf{r_1}}$ are positions of a point, before and after deformation, respectively [8]. The stress distribution is then found by Hooke's law, and the strain energy is given by the volume integral

$$U = \frac{1}{2}\int_V \left(\sigma^T\varepsilon\right)\mathrm{d}V. \tag{12.3}$$

The kinetic energy of the blade is given by a volume integral, as

$$T = \frac{1}{2}\int_V \left(\dot{\overrightarrow{\mathbf{r_1}}}\cdot\dot{\overrightarrow{\mathbf{r_1}}} + \dot{\phi}^2\overrightarrow{\mathbf{r_1}}\cdot\overrightarrow{\mathbf{r_1}}\right)\mathrm{d}m. \tag{12.4}$$

Here $\dot{\overrightarrow{\mathbf{r_1}}}$ is the time derivative of $\overrightarrow{\mathbf{r_1}}$ in $\hat{x}\hat{y}\hat{z}$ coordinate system, and the rigid body rotation of the rotor is included with the $\dot{\phi}^2\overrightarrow{\mathbf{r_1}}.\overrightarrow{\mathbf{r_1}}$ term, where ϕ is the rotation angle of the blade, as shown in Fig. 12.1.

The gravitational potential energy is given by

$$V_g = \int_V -\overrightarrow{\mathbf{g}}\cdot\overrightarrow{\mathbf{r_1}}\mathrm{d}m, \tag{12.5}$$

where both gravity vector, $\overrightarrow{\mathbf{g}}$, and $\overrightarrow{\mathbf{r_1}}$ are written in $\hat{x}\hat{y}\hat{z}$ coordinates. Since this coordinate system rotates with the blade, $\overrightarrow{\mathbf{g}}$ has cyclicly changing components.

Following Crepo Da Silva's work, these volume integrals are then separated into integrals on the cross sectional area inside an integral on the length of the blade, as

$$
\begin{aligned}
U &= \frac{1}{2}\int_0^L \left(\int_A \left(\sigma^T \varepsilon\right) \mathrm{d}A\right) \mathrm{d}x \\
T &= \frac{1}{2}\int_0^L \left(\int_A \left(\dot{\vec{\mathbf{r}}}_1 \cdot \dot{\vec{\mathbf{r}}}_1 + \dot{\phi}^2 \vec{\mathbf{r}}_1 \cdot \vec{\mathbf{r}}_1\right) \mathrm{d}A\right) \mathrm{d}x \\
V_g &= \int_0^L \left(\int_A \left(-\vec{\mathbf{g}} \cdot \vec{\mathbf{r}}_1\right) \mathrm{d}A\right) \mathrm{d}x.
\end{aligned}
\tag{12.6}
$$

Then the terms in the inner integral are identified in terms of familiar quantities such as the mass per unit length, mass moments of inertia per unit length, and area moments of inertia. This results in expressions of energy densities integrated along the length of the beam, written in terms of the distributed coordinates $v(x,t), w(x,t)$, and $\theta_x(x,t)$, and their partial derivatives, which can be used with the extended Hamilton's principle, or the Lagrange method on assumed modal coordinates, to get the equations of motion.

12.3 Equations of Motion

The assumed modes method is used together with Lagrange's method to find the equations of motion. In each direction (v, w, and θ_x), n cantilever beam modes are assumed. Then v, w, and θ_x, given as

$$
\begin{aligned}
v(x,t) &= \sum_{i=1}^{n} \phi_{vi}(x)\, q_{vi}(t) \\
w(x,t) &= \sum_{i=1}^{n} \phi_{wi}(x)\, q_{wi}(t) \\
\theta_x(x,t) &= \sum_{i=1}^{n} \phi_{\theta i}(x)\, q_{\theta i}(t)
\end{aligned}
\tag{12.7}
$$

are plugged into the energy expressions, and Lagrange's method is applied to find the equations of motion.

The linearized equations of motion for n are

$$
\begin{bmatrix} \mathrm{M_{vv}} & \mathrm{M_{vw}} & 0 \\ \mathrm{M_{vw}} & \mathrm{M_{ww}} & 0 \\ 0 & 0 & \mathrm{M_{\theta\theta}} \end{bmatrix}
\begin{pmatrix} \ddot{\underline{q}}_v \\ \ddot{\underline{q}}_w \\ \ddot{\underline{q}}_\theta \end{pmatrix}
+
\begin{bmatrix} \mathrm{K_{vv}} & \mathrm{K_{vw}} & \mathrm{K_{v\theta}} \\ \mathrm{K_{vw}} & \mathrm{K_{ww}} & \mathrm{K_{w\theta}} \\ \mathrm{K_{v\theta}} & \mathrm{K_{w\theta}} & \mathrm{K_{\theta\theta}} \end{bmatrix}
\begin{pmatrix} \underline{q}_v \\ \underline{q}_w \\ \underline{q}_\theta \end{pmatrix}
=
\begin{pmatrix} \underline{Q}_v \\ \underline{Q}_w \\ \underline{Q}_\theta \end{pmatrix},
\tag{12.8}
$$

where M_{ij} and K_{ij} are block matrices denoting coupling between coordinates i and j.

Here, $\mathrm{K}_{v\theta}$ and $\mathrm{K}_{w\theta}$ are non zero only when the shear center is different than the center of mass. Therefore the transverse modes are not coupled with the torsional mode when the shear center is on the cross sectional centroid. Also, K_{ij} matrices are cyclic (ϕ dependent), but they will be analyzed for fixed ϕ.

12.4 Case Studies

12.4.1 A Hollow Rectangular Beam with a Structural Twist

A pre-twisted beam with a hollow rectangular cross-section, as shown in Fig. 12.2, was analyzed. Since the cross section has two axes of symmetry, the cross sectional center of mass and shear center are the same. This greatly simplifies the model, and eliminates the coupling between torsion and bending. The beam has dimensions 3 m × 1.5 m × 50 m with 0.1 m

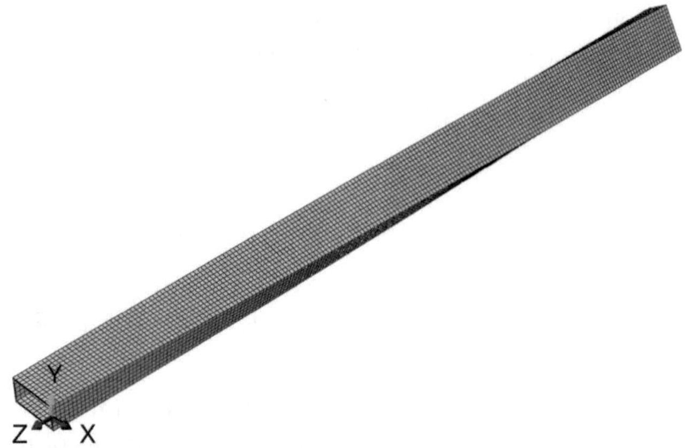

Fig. 12.2 *Hollow rectangular* pre twisted beam

Table 12.1 Natural frequencies of the rectangular pre-twisted beam, $\Omega = 0$

		$n = 1$	$n = 2$	$n = 3$	FEA
First mode	$\phi = 0$	0.4073	0.4010	0.4008	0.4037
	$\phi = \frac{\pi}{2}$	0.4025	0.3961	0.3959	0.3936
	$\phi = \pi$	0.3976	0.3911	0.3910	0.3839
Second mode	$\phi = 0$	0.6782	0.6678	0.6667	0.6649
	$\phi = \frac{\pi}{2}$	0.6753	0.6648	0.6637	0.6589
	$\phi = \pi$	0.6724	0.6618	0.6607	0.6530
Third mode	$\phi = 0$	8.7706	2.6170	2.5524	2.4862
	$\phi = \frac{\pi}{2}$	8.7706	2.6128	2.5482	2.4769
	$\phi = \pi$	8,7706	2.6086	2.5440	2.4693

thickness, and structural twist angle changing linearly with axial position, given by the relation $\beta(x) = \frac{\pi x}{3l}$. The material is assumed to be linearly elastic, isotropic, having Young's modulus $E = 20\,\text{GPa}$, Poisson's ratio $\nu = 0.3$, and mass density $\rho = 2,500\,\text{kg/m}^3$. The modal analysis is applied to beams in horizontal and vertical positions, by assuming static conditions, $\Omega = 0, \phi = \phi_0$.

The first three natural frequencies and mode shapes were found by assuming up to three modes in each direction, Table 12.1. As we increased the number of assumed modes, the natural frequencies converged to their final values. For two and three assumed modes, all first three modes are coupled bending modes, and they do not involve torsion. Yet, for only one assumed mode in each distributed coordinate, the third mode is a torsional one. This is expected, because we assumed one mode in each direction (v, w, θ_x), so we have to have at least one mode consisting of torsion. Therefore it is not sufficient to predict the first three modes correctly by assuming only one mode in each direction ($n = 1$). On the other hand, $n = 2$ resulted in coupled transverse modes for the first three, and the natural frequencies have nearly converged. Adding one more assumed mode ($n = 3$) does not change the first three natural frequencies by more than 5 %.

The mode shapes for $n = 2$, are given as

$$\underline{q}_1 = \begin{pmatrix} 0.9806 \\ 0.0064 \\ 0.1950 \\ -0.0192 \\ 0 \\ 0 \end{pmatrix} \quad \underline{q}_2 = \begin{pmatrix} -0.2187 \\ -0.0940 \\ 0.9705 \\ -0.0386 \\ 0 \\ 0 \end{pmatrix} \quad \underline{q}_3 = \begin{pmatrix} 0.1761 \\ 0.8931 \\ 0.1914 \\ 0.3669 \\ 0 \\ 0 \end{pmatrix}, \tag{12.9}$$

where $\underline{q}_j = \begin{pmatrix} q_{v1j} & q_{v2j} & q_{w1j} & q_{w2j} & q_{\theta 1j} & q_{\theta 2j} \end{pmatrix}^{\mathrm{T}}$.

For the first mode, the first in-plane bending assumed modal coordinate is dominant, whereas for the second mode the first out-of-plane bending, and for the third mode the second in-plane bending, are the dominant assumed modal coordinates.

The beam was also analyzed with a finite-element analysis (FEA) software, ABAQUS, as a cantilever beam, using rectangular solid elements. The number of elements was increased until the frequencies converged to their final values.

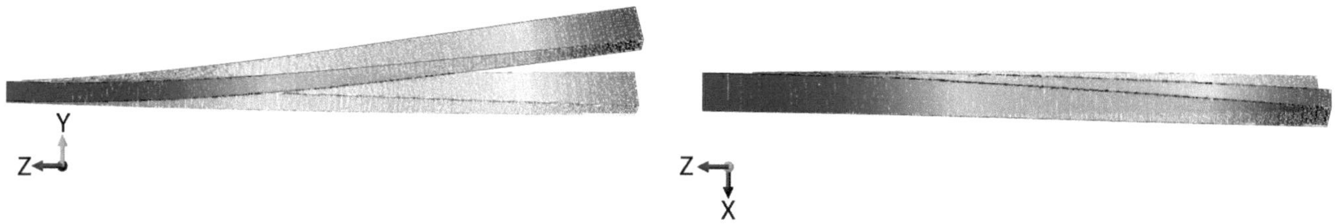

Fig. 12.3 Mode 1, $\phi = \pi$ and $\Omega = 0$, *right view* and *top view*

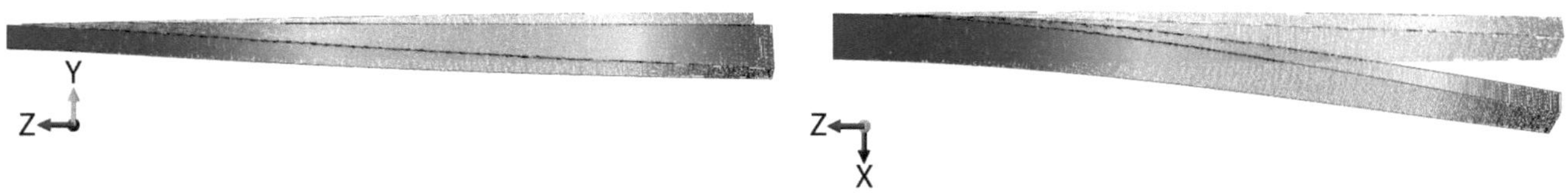

Fig. 12.4 Mode 2, $\phi = \pi$ and $\Omega = 0$, *right view* and *top view*

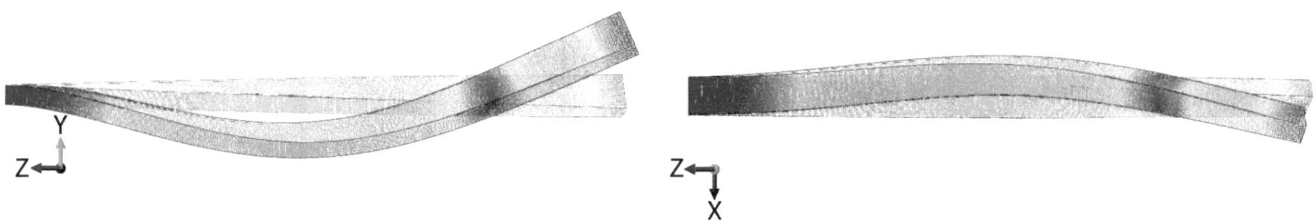

Fig. 12.5 Mode 3, $\phi = \pi$ and $\Omega = 0$, *right view* and *top view*

Table 12.2 Natural frequencies of NREL's 5 MW turbine blade, $\Omega = 0, \phi = \pi$

	$n = 1$	$n = 2$	$n = 3$	NREL's
Edgewise	1.2922	1.0933	1.0933	1.0898
Flapwise	1.1459	0.7408	0.6953	0.6664

The results of the finite element analysis are consistent with those obtained from assumed modes method, as shown in Table 12.1. The mode shapes obtained from FEA are given in Figs. 12.3, 12.4, and 12.5. The shapes are consistent with the findings of the analytical calculations.

12.4.2 NREL's 5 MW Turbine Blade

As a case study, we worked on NREL's 5 MW turbine blade, which is adopted from DOWEC's study [9]. This 62m blade has a varying cross section, the rotor has a precone of 2.5 °, and the plane of rotation is tilted by 5 °. For 50 different positions, the distributed properties, such as cross sectional area and moment of inertia, are tabulated in Jonkman's study [6]. The assumed modes method was applied, and the potential and kinetic energy densities were numerically integrated for each section. Then these qualities were summed up to find the total kinetic and potential energy of the blade. The linearized equations of motion were found by Lagrange's method, and the modal analysis was applied to find the natural frequencies.

Jonkman et al., provided the natural frequencies for the stationary blade in edgewise and flapwise directions separately [6]. To compare our results with the given frequency data, we first assumed in-plane only ($\phi_{wi}(x) = 0, \phi_{\theta i}(x) = 0$), and out-of-plane only ($\phi_{vi}(x) = 0, \phi_{\theta i}(x) = 0$) vibrational modes, and found the corresponding natural frequencies. With three assumed modes, the natural frequencies were found to be at most 4 % different than NREL's results, as given in Table 12.2.

We then applied our method with up to four modes in each direction in order to find the natural frequencies for the combined deformation. The results are given in Table 12.3.

Table 12.3 Coupled motion frequencies for NREL's 5 MW turbine blade, $\Omega = 0$

		$n=1$	$n=2$	$n=3$	$n=4$
First mode	$\phi = 0$	1.1454	0.7466	0.7091	0.6945
	$\phi = \frac{\pi}{2}$	1.1420	0.7380	0.6961	0.6791
	$\phi = \pi$	1.1387	0.7287	0.6826	0.6628
Second mode	$\phi = 0$	1.3045	1.1113	1.1092	1.1026
	$\phi = \frac{\pi}{2}$	1.3016	1.1055	1.1040	1.0972
	$\phi = \pi$	1.2986	1.1003	1.0988	1.0917
Third mode	$\phi = 0$	12.082	4.4518	2.2639	2.0291
	$\phi = \frac{\pi}{2}$	12.082	4.4555	2.2507	2.0091
	$\phi = \pi$	12.082	4.4520	2.2376	1.9890

For $n = 4$, the first three mode shapes are given as

$$\underline{q}_1 = \begin{pmatrix} -0.9579 \\ 0.2496 \\ -0.0728 \\ 0.0133 \\ 0.1186 \\ -0.0220 \\ 0.0033 \\ -0.0005 \\ 0.0001 \\ -0.0001 \\ 0.0000 \\ 0.0000 \end{pmatrix} \quad \underline{q}_2 = \begin{pmatrix} -0.1284 \\ 0.0466 \\ -0.0209 \\ 0.0041 \\ -0.9855 \\ 0.0978 \\ -0.0063 \\ 0.0048 \\ 0.0000 \\ 0.0000 \\ 0.0000 \\ 0.0000 \end{pmatrix} \quad \underline{q}_3 = \begin{pmatrix} -0.0643 \\ 0.8654 \\ -0.4811 \\ 0.0888 \\ 0.0395 \\ -0.0692 \\ 0.0335 \\ -0.0033 \\ 0.0005 \\ -0.0001 \\ 0.0001 \\ 0.0000 \end{pmatrix} \tag{12.10}$$

The first mode is dominated by the first and second flapwise assumed modal coordinates coupled with out-of-plane modes, whereas in the second mode the first edgewise assumed modal coordinate is dominant. The third mode is composed of flapwise and edgewise coordinates, and the dominant one is the second flapwise coordinate. In all three modes, torsional coupling is negligible. This is because in NREL's blade, the difference between the shear center and the mass center is very small. For different positions of the blade ($\phi = 0$, $\phi = \pi/2$ and $\phi = \pi$), the gravity affects the stiffness. The change in stiffness alters the natural frequencies of the blade in different positions.

12.5 Conclusions

Bend-bend-twist coupled equations of motion for a turbine blade were found by using energy methods and applying assumed modes. Then, these equations were linearized, and natural frequencies and mode shapes were found by modal analysis. The method was applied to NREL's 5 MW turbine blade, and the primary modes are found to be dominantly of either in-plane or out-of-plane motion. Yet, the in-plane and out-of-plane coordinates are coupled in the mode shapes, and the coupling is not negligible. The modes described by assumed modal coordinates can be used for reduced-order modeling, which will be useful in the analysis of forced, parametrically excited, nonlinear, and coupled blades. This method will also be applied to three blade turbines, by introducing the coupling between the blades through the rotor hub and shaft.

Acknowledgements This project is funded by the National Science Foundation, under grant CMMI-1335177.

References

1. Hodges DH, Dowell EH (1974) Nonlinear equations of motion for the elastic bending and torsion of twisted nonuniform rotor blades. National aeronautics and space administration, 1974
2. Kallesøe BS (2007) Equations of motion for a rotor blade, including gravity, pitch action and rotor speed variations. Wind Energy 10(3):209–230

3. Crespo Da Silva MRM (1988) Nonlinear flexural-flexural-torsional-extensional dynamics of beams. i. formulation. Int J Solids Struct 24(12):1225–1234
4. Avramov KV, Pierre C, Shyriaieva N (2007) Flexural-flexural-torsional nonlinear vibrations of pre-twisted rotating beams with asymmetric cross-sections. J Vib Control 13(4):329–364
5. Cheng J, Xu H, Yan A (2006) Frequency analysis of a rotating cantilever beam using assumed mode method with coupling effect. Mech Based Des Struct Mach 34(1):25–47
6. Jonkman JM, Butterfield S, Musial W, Scott G (2009) Definition of a 5-MW reference wind turbine for offshore system development. National Renewable Energy Laboratory Golden, CO, 2009
7. Allen MS, Sracic MW, Chauhan S, Hansen MH (2011) Output-only modal analysis of linear time-periodic systems with application to wind turbine simulation data. Mech Syst Signal Process 25(4):1174–1191
8. Shames IH (1985) Energy and finite element methods in structural mechanics. CRC Press, Boca Raton
9. Lindenburg C (2002) Aeroelastic modelling of the lmh64-5 blade. ECN, Petten, December, 2002

Chapter 13
Reduced-Order Modeling of Turbine Bladed Discs by 1D Elements

Luigi Carassale, Mirko Maurici, and Laura Traversone

Abstract The dynamic behavior of turbine bladed discs is deeply influenced by blades geometry, which is determined as a compromise between fluid-dynamic and mechanical needs. One of the most important issue from the mechanical side is the tailoring of the dynamic characteristics to allow a sufficient separation between the natural frequencies and the harmonics of the rotational speed. The most common approach to predict the natural frequencies of bladed discs involves the realization of 3D FE models using solid elements. This approach is highly demanding in terms of time for the realization of the models as well as for the computation, even when only one blade if explicitly modeled and a cyclic symmetry constraint is employed.

In this work, we propose a reduced order model based on mono-dimensional finite elements. The proposed element is based on the Timoshenko beam formulated for non-symmetrical cross sections. The shape functions are based on the exact solution of the beam equation to avoid shear-locking. Torsional stiffness includes the effect of warping. The proposed model is compared with accurate 3D FE models in terms of modal properties. Blades of different shape and slenderness taken from turbine and compressor of a large gas turbine are considered.

Keywords Turbine blades • Reduced order model • Finite element • Beam

13.1 Introduction

The dynamic behavior of turbine bladed discs is deeply influenced by blades geometry, which is determined as a compromise between fluid-dynamic and mechanical needs. One of the most important issues from the mechanical side is the tailoring of the dynamic characteristics to allow a sufficient separation between the natural frequencies and the harmonics of the rotational speed. The most common approach to predict the natural frequencies of bladed discs involves the realization of 3-D Finite Element (FE) models using solid elements [1]. This approach is highly demanding in terms of time for the realization of the models as well as for the computation, even when only one blade if explicitly modeled and a cyclic symmetry constraint is employed. The situation obviously becomes even more complicated when the whole disc must be explicitly modelled to assess the effect of random or intentional blade detuning [2].

The use of the mentioned modeling approach in the preliminary stage of a blade design in very inefficient as a relatively large ensemble of blade candidates is usually delivered by the fluid-dynamic analyses. This situation clearly requires for the development of fast computational tools able to predict with sufficient accuracy the natural frequency of the first few vibration modes of blades and blade assemblages. In this work, we propose a Reduced Order Model (ROM) based on mono-dimensional finite elements based on the Timoshenko beam theory [3]. The parameters of the ROM are obtained on the basis of the geometry of the blade cross sections without the knowledge of a reference 3D solution. The proposed model is compared with accurate 3D FE models in terms of modal properties. The case of a compressor blade of a large gas turbine is considered.

L. Carassale (✉) • M. Maurici
Department of Civil, Chemical and Environmental Engineering, University of Genova, Via Montallegro 1, Genoa 16145, Italy
e-mail: luigi.carassale@unige.it; mirko.maurici@edu.unige.it

L. Traversone
Ansaldo Sviluppo Energia, Via Lorenzi 8, Genoa 16152, Italy

© The Society for Experimental Mechanics, Inc. 2015

R. Allemang (ed.), *Special Topics in Structural Dynamics, Volume 6*, Conference Proceedings of the Society for Experimental Mechanics Series, DOI 10.1007/978-3-319-15048-2_13

13.2 1-D Modelling of a Turbine Blade

This section discusses a strategy for the realization of a ROM of a turbine blade. Since the ROM is intended to be predictive, its parameters should be obtained from the sole blade geometry and material properties, without relying on an existing 3-D reference solution. A possible choice consists of a model composed by 1-D finite elements, whose properties are evaluated from the blade geometry. In practice the blade geometry is provided through a series of closed lines representing a number of cross-sections. These lines are not necessarily plane and are described by sets of points in the 3-D space.

The definition of a 1-D ROM based on the classic beam theory requires (1) the identification of the beam axis and of a family of plane cross-sections orthogonal to the axis, (2) the definition of the FEs in terms of stiffness and mass matrices, (3) the assemblage of the FEs into a blade model, (4) the calculation of the static and dynamic response with pertinent load and constraint conditions. The use of the mentioned modeling approach in the preliminary stage of a blade design in very inefficient as a relatively large ensemble.

13.3 Cross-Section

The definition of a blade cross section starts from a closed curve defined by the position $\mathbf{X}(s)$, s being a curvilinear abscissa measuring the curve starting from an arbitrary point. Since the input geometry is given as a set of points, the function $\mathbf{X}(s)$ is conveniently parametrized through cubic splines. The position $\mathbf{X}$ is referred to a blade reference system having origin somewhere at the clamping point. The orientation of the blade reference system is defined by the unit vectors $\mathbf{e}_x^{(0)}$, $\mathbf{e}_y^{(0)}$ and $\mathbf{e}_z^{(0)}$ with $\mathbf{e}_x^{(0)}$ having radial direction and $\mathbf{e}_y^{(0)}$ being parallel to the machine axis.

Let $\mathbf{X}_G$ be the Center of Gravity (CG) of the cross section and $\mathbf{a}$ be the unit vector defining its normal direction. Both this quantities are unknown, however they can be used to define, at least in principle, four orthogonal reference systems that are proper of the cross section. The reference system SYS0 is parallel to the blade reference system, but centered in the CG of the cross section. In SYS0, the border of the cross-section has coordinates

$$\mathbf{x}_0(s) = \mathbf{X}(s) - \mathbf{X}_G \tag{13.1}$$

The reference system SYS1 is obtained rotating SYS0 about $\mathbf{e}_z^{(0)}$ of an angle $\alpha_z^{(0)}$. The orientation of SYS1 is defined through the unit vectors $\mathbf{e}_x^{(1)}$, $\mathbf{e}_y^{(1)}$ and $\mathbf{e}_z^{(1)} = \mathbf{e}_z^{(0)}$.

The reference system SYS2 is obtained by rotating SYS1 around the direction $\mathbf{e}_y^{(1)}$ of an angle $\alpha_y^{(1)}$. The orientation of SYS2 is defined through the unit vectors $\mathbf{e}_x^{(2)}$, $\mathbf{e}_y^{(2)} = \mathbf{e}_y^{(1)}$ and $\mathbf{e}_z^{(2)}$.

The reference system SYS3 is the is obtained by rotating SYS2 about $\mathbf{e}_x^{(2)}$ of an angle $\alpha_2^{(2)}$. The orientation of SYS2 is defined through the unit vectors $\mathbf{e}_x^{(3)} = \mathbf{e}_x^{(2)}$, $\mathbf{e}_y^{(3)}$ and $\mathbf{e}_z^{(3)}$, with $\mathbf{e}_y^{(3)}$ and $\mathbf{e}_z^{(3)}$ being inertial principal axes of the cross section. The angles $\alpha_z^{(0)}$, $\alpha_y^{(1)}$ and $\alpha_x^{(2)}$ are usually referred to as Tait-Bryan angles and define completely the orientation of the cross-section.

The angles $\alpha_z^{(0)}$ and $\alpha_y^{(1)}$ can be calculated on the basis of the normal vector $\mathbf{a}$ as

$$\tan\alpha_{z_0} = \frac{a_y}{a_x}; \quad \tan\alpha_{y_1} = -\frac{a_z}{a_{xy}} \tag{13.2}$$

where a_x, a_y and a_z are the components of $\mathbf{a}$ in SYS0 and $a_{xy}{}^2 = a_x{}^2 + a_y{}^2$.

The plane ($\mathbf{e}_y^{(2)}$, $\mathbf{e}_z^{(2)}$) contains the orthogonal cross-section of the blade, but does not contain, necessarily, the given profile $\mathbf{x}_0(s)$. A plane profile contained in ($\mathbf{e}_y^{(2)}$, $\mathbf{e}_z^{(2)}$) can be obtained by orthogonal projection as (Fig. 13.1):

$$\bar{\mathbf{x}}_0 = \mathbf{x}_0 - \mathbf{E}_2^{\mathrm{T}}\mathbf{x}_0\mathbf{E}_2 \tag{13.3}$$

where

$$\mathbf{E}_2 = \left[\mathbf{e}_x^{(2)}, \mathbf{e}_y^{(2)}, \mathbf{e}_z^{(2)}\right] = \begin{bmatrix} a_x & -\frac{a_y}{a_{xy}} & -\frac{a_x a_z}{a_{xy}} \\ a_y & \frac{a_x}{a_{xy}} & -\frac{a_y a_z}{a_{xy}} \\ a_z & 0 & a_{xy} \end{bmatrix} \tag{13.4}$$

Fig. 13.1 Reference systems

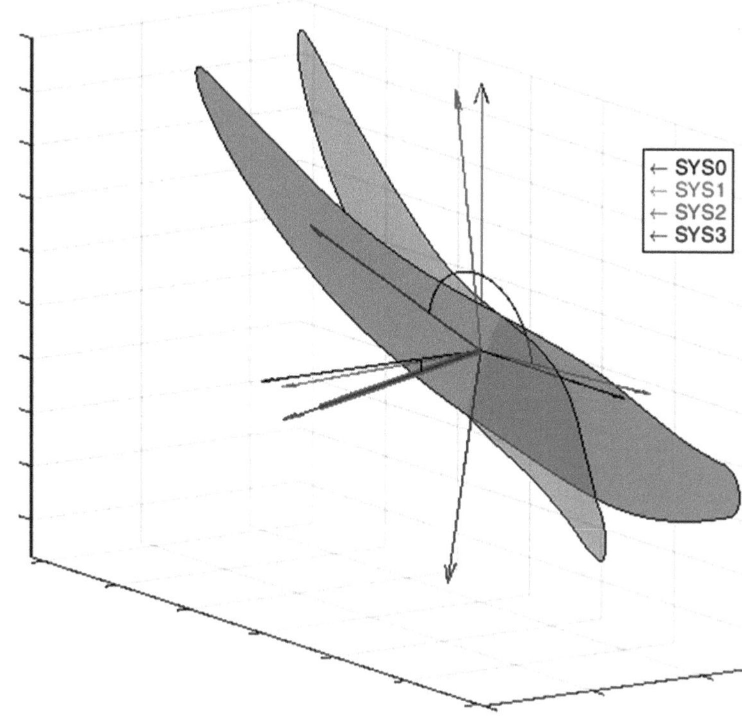

The function $\bar{\mathbf{x}}_0(s)$ defines the border of the reference (plane and orthogonal to the blade axis) cross section that is used for the calculation of the mechanical properties of the beam model. The position is defined in SYS0, however is can be translated into SYS2 through the relationship:

$$\mathbf{x}_2 = \mathbf{R}_y^{(1)}\mathbf{R}_z^{(0)}\bar{\mathbf{x}}_0 \tag{13.5}$$

where $\mathbf{R}_z^{(0)}$ and $\mathbf{R}_y^{(1)}$ are rotation matrices defined as

$$\mathbf{R}_z^{(0)} = \begin{bmatrix} \cos\alpha_z^{(0)} & \sin\alpha_z^{(0)} & 0 \\ -\sin\alpha_z^{(0)} & \cos\alpha_z^{(0)} & 0 \\ 0 & 0 & 1 \end{bmatrix}; \quad \mathbf{R}_y^{(1)} = \begin{bmatrix} \cos\alpha_y^{(1)} & 0 & -\sin\alpha_y^{(1)} \\ 0 & 1 & 0 \\ \sin\alpha_y^{(1)} & 0 & \cos\alpha_y^{(1)} \end{bmatrix} \tag{13.6}$$

In SYS2 the first component of the position vector identifying the cross-section border is identically zero, i.e. $\boldsymbol{x}_2 = [0, y, z]^{\mathrm{T}}$. The functions $y(s)$ and $z(s)$ define completely the profile and can be used for the evaluation of the cross-section area A, first moments $S_y{}^{(2)}$ and $S_z{}^{(2)}$, as well as second moments $I_y{}^{(2)}$, $I_z{}^{(2)}$, $I_{yz}{}^{(2)}$. These quantities are defined through integrals over the cross-section area, however they can be conveniently translated into line integrals along the cross section border using the Gauss-Green formula:

$$\begin{aligned} A &= \int_C y n_y \mathrm{d}s; \quad S_y^{(2)} = \tfrac{1}{2}\int_C z^2 n_z \mathrm{d}s; \\ S_z^{(2)} &= \tfrac{1}{2}\int_C y^2 n_y \mathrm{d}s; \quad I_y^{(2)} = \tfrac{1}{3}\int_C z^3 n_z \mathrm{d}s; \\ I_z^{(2)} &= \int_C y^3 n_y \mathrm{d}s; \quad I_{yz}^{(2)} = \tfrac{1}{3}\int_C \left(y^3 + yz\right) n_y \mathrm{d}s; \end{aligned} \tag{13.7}$$

where $n_y = \mathrm{d}z/\mathrm{d}s$ and $n_z = -\mathrm{d}y/\mathrm{d}s$. From the quantities obtained in Eq. (13.7), a new position of the CG $y_G = S_y{}^{(2)}/A$ and $z_G = S_z{}^{(2)}/A$ can be computed. This position can be remapped into SYS0 and used to update Eq. (13.1) and all the quantities calculated afterwards. Since the vector **a** is usually almost parallel to $\mathbf{e}_x{}^{(0)}$, this updating procedure converges very quickly to stable values after very few iterations.

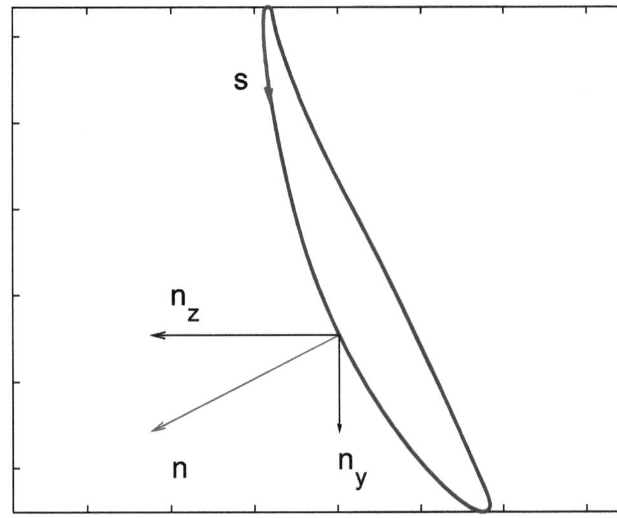

Fig. 13.2 Line $\mathbf{X}(s)$ defining the blade cross-section

Since $y(s)$ and $z(s)$ are parametrized through splines, the integrals in Eq. (13.7) can be computed working directly on the spline coefficients without the need of using any numerical quadrature schema.

Once the integral in Eq. (13.7) are calculated, the third Tait-Bryan angle defining SYS3 can be obtained as

$$\tan \alpha_x^{(3)} = \frac{\theta_z}{\theta_y} \tag{13.8}$$

where θ_y and θ_z are the components along $\mathbf{e}_y^{(2)}$ and $\mathbf{e}_z^{(2)}$, respectively, of the first eigenvector of the matrix

$$\mathbf{I} = \begin{bmatrix} I_z^{(2)} & I_{yz}^{(2)} \\ I_{yz}^{(2)} & I_y^{(2)} \end{bmatrix} \tag{13.9}$$

The eigenvalues of $\mathbf{I}$ provide the principal moments $I_y^{(3)}$ and $I_z^{(3)}$. The position of the cross-section border in SYS3 is given by the equation:

$$\mathbf{x}_3 = \mathbf{R}_x^{(2)}\mathbf{R}_y^{(1)}\mathbf{R}_z^{(0)}\bar{\mathbf{x}}_0 \tag{13.10}$$

where the rotation matrix $\mathbf{R}_x^{(2)}$ is given as (Fig. 13.2)

$$\mathbf{R}_x^{(2)} = \begin{bmatrix} 1 & 0 & 0 \\ 0 & \cos\alpha_z^{(0)} & \sin\alpha_z^{(0)} \\ 0 & -\sin\alpha_z^{(0)} & \cos\alpha_z^{(0)} \end{bmatrix} \tag{13.11}$$

13.4 Beam Model

The cross sections are connected by 2-node FEs formulated on the basis of the beam theory. The axis of the FE coincides with the line connecting the CG of the two cross-sections (Fig. 13.3). This direction, in turns, determines the cross-section normal vector **a** employed in the previous section to identify the plane orthogonal to the blade axis. Al the geometrical parameters are evaluated in SYS3, thus the superscript $^{(3)}$ used to identify coordinates and moments is dropped without producing any ambiguity.

The displacement of each node is represented by 6 kinematic quantities representing displacement and rotation components represented in SYS3. Timoshenko beam model is adopted (e.g. [4, 5]), the element formulation is made consistent to avoid the shear locking phenomena [6, 7]. The stiffness matrix can be expressed in SYS3 as (e.g. [8]).

Fig. 13.3 Beam element

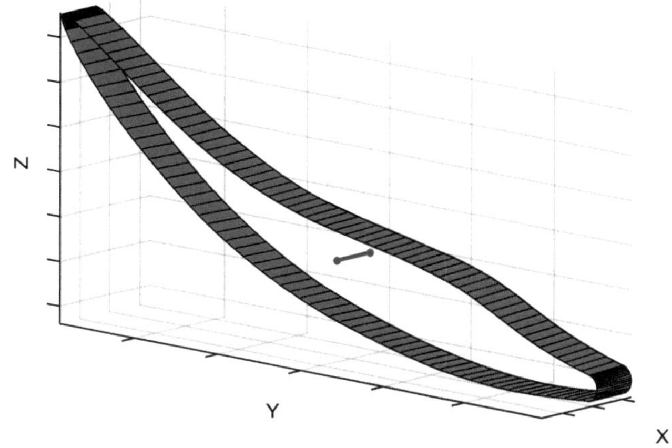

$$\mathbf{K} = \begin{bmatrix} k_{1,1} & & & & & & k_{1,7} & & & & & \\ & k_{2,2} & & & & k_{2,6} & & k_{2,8} & & & & k_{2,12} \\ & & k_{3,3} & & k_{3,5} & & & & k_{3,9} & & k_{3,11} & \\ & & & k_{4,4} & & & & & & k_{4,10} & & \\ & & & & k_{5,5} & & & & k_{5,9} & & k_{5,11} & \\ & & & & & k_{6,6} & & k_{6,8} & & & & k_{6,12} \\ & & & & & & k_{7,7} & & & & & \\ & & & & & & & k_{8,8} & & & & k_{8,12} \\ & & & & & & & & k_{9,9} & & k_{9,11} & \\ & & & & & & & & & k_{10,10} & & \\ & & & & & & & & & & k_{11,11} & \\ & & & & & & & & & & & k_{12,12} \end{bmatrix} \tag{13.12}$$

whose non-zero entries are defined as

$$k_{1,1} = k_{7,7} = -k_{1,7} = \frac{EA}{L} \tag{13.13}$$

$$k_{2,2} = k_{8,8} = -k_{2,8} = \frac{12EI_z}{L^3\left(1+\phi_y\right)} \tag{13.14}$$

$$k_{3,3} = k_{9,9} = -k_{3,9} = \frac{12EI_y}{L^3\left(1+\phi_z\right)} \tag{13.15}$$

$$k_{2,6} = k_{2,12} = -k_{6,8} = -k_{8,12} = \frac{6EI_z}{L^2\left(1+\phi_y\right)} \tag{13.16}$$

$$-k_{3,5} = -k_{3,11} = k_{5,9} = k_{9,11} = \frac{6EI_y}{L^2\left(1+\phi_z\right)} \tag{13.17}$$

$$k_{4,4} = k_{10,10} = -k_{4,10} = \frac{Gk_t}{L} \tag{13.18}$$

$$k_{5,5} = k_{11,11} = \frac{\left(4+\phi_z\right)EI_y}{L\left(1+\phi_z\right)} \tag{13.19}$$

$$k_{5,11} = \frac{(2\text{-}\phi_z)\, E I_y}{L\,(1+\phi_z)} \tag{13.20}$$

$$k_{6,6} = k_{12,12} = \frac{\left(4+\phi_y\right) E I_z}{L\left(1+\phi_y\right)} \tag{13.21}$$

$$k_{6,12} = \frac{\left(2\text{-}\phi_y\right) E I_z}{L\left(1+\phi_y\right)} \tag{13.22}$$

where L is the beam length, E and G are the Young and shear modulus of the material, k_t is the torsional stiffness, ϕ_y and ϕ_z are corrective factors [8] taking into account the shear deformation and defined as:

$$\phi_y = \frac{12 E I_z}{G k_y L^2}; \quad \phi_z = \frac{12 E I_y}{G k_z L^2} \tag{13.23}$$

in which k_y and k_z are the shear areas [9]. The constants k_t, k_y and k_z can be evaluated on the basis of the cross-section geometry through the relationships:

$$\begin{aligned} \tfrac{1}{k_y} &= \tfrac{1}{4{I_z}^2}\int_A \left[{\psi_{y,y}}^2 + {\psi_{y,z}}^2 + \left(y^2 - \overline{\nu} z^2\right)\left(y^2 - \overline{\nu} z^2 - 2\psi_{y,y}\right)\right] \mathrm{d}A \\ \tfrac{1}{k_z} &= \tfrac{1}{4{I_y}^2}\int_A \left[{\psi_{z,y}}^2 + {\psi_{z,z}}^2 + \left(z^2 - \overline{\nu} y^2\right)\left(z^2 - \overline{\nu} y^2 - 2\psi_{z,z}\right)\right] \mathrm{d}A \\ \tfrac{1}{k_t} &= \int_A \left[y^2 + z^2 - {\psi_{t,y}}^2 - {\psi_{t,z}}^2\right] \mathrm{d}A \end{aligned} \tag{13.24}$$

where the functions ψ_y, ψ_z and ψ_t are the solutions of the boundary-value problems

$$\nabla^2 \psi_y = 0; \quad \nabla^2 \psi_z = 0; \quad \nabla^2 \psi_t = 0 \tag{13.25}$$

with the boundary conditions:

$$\begin{aligned} \frac{d\psi_y}{dn} &= \left(y^2 - \overline{\nu} z^2\right) n_y \\ \frac{d\psi_z}{dn} &= \left(z^2 - \overline{\nu} y^2\right) n_z \\ \frac{d\psi_t}{dn} &= z n_y - y n_z \end{aligned} \tag{13.26}$$

The differential problems (13.25)–(13.26) are solved by adopting a FE discretization of the cross-section area based on a triangular mesh (Fig. 13.4).

The mass matrix of the element is expressed as [8]

$$\mathbf{M} = \rho A L \begin{bmatrix} m_{1,1} & & & & & & m_{1,7} & & & & & \\ & m_{2,2} & & & & m_{2,6} & & m_{2,8} & & & & m_{2,12} \\ & & m_{3,3} & & m_{3,5} & & & & m_{3,9} & & m_{3,11} & \\ & & & m_{4,4} & & & & & & m_{4,10} & & \\ & & & & m_{5,5} & & & & m_{5,9} & & m_{5,11} & \\ & & & & & m_{6,6} & & m_{6,8} & & & & m_{6,12} \\ & & & & & & m_{7,7} & & & & & \\ & & & & & & & m_{8,8} & & & & m_{8,12} \\ & & & & & & & & m_{9,9} & & m_{9,11} & \\ & & & & & & & & & m_{10,10} & & \\ & & & & & & & & & & m_{11,11} & \\ & & & & & & & & & & & m_{12,12} \end{bmatrix} \tag{13.27}$$

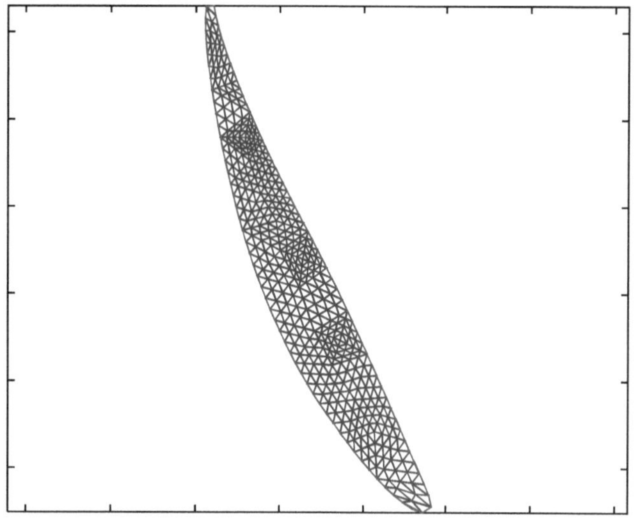

Fig. 13.4 Mesh for the FE solution of Eqs. (13.25)–(13.26)

where

$$m_{1,1} = m_{7,7} = \frac{1}{3} \tag{13.28}$$

$$m_{1,7} = \frac{1}{6} \tag{13.29}$$

$$m_{2,2} = m_{8,8} = \left(\frac{13}{35} + \frac{7}{10}\phi_y + \frac{1}{3}\phi_y^2 + \frac{6I_z}{5AL^2}\right)(1+\phi_y)^{-2} \tag{13.30}$$

$$m_{3,3} = m_{9,9} = \left(\frac{13}{35} + \frac{7}{10}\phi_z + \frac{1}{3}\phi_z^2 + \frac{6I_y}{5AL^2}\right)(1+\phi_z)^{-2} \tag{13.31}$$

$$m_{2,6} = -m_{8,12} = L\left[\frac{11}{210} + \frac{11}{120}\phi_y + \frac{1}{24}\phi_y^2 + \left(\frac{1}{10} - \frac{1}{2}\phi_y\right)\frac{I_z}{AL^2}\right](1+\phi_y)^{-2} \tag{13.32}$$

$$m_{3,5} = m_{9,11} = -L\left[\frac{11}{210} + \frac{11}{120}\phi_z + \frac{1}{24}\phi_z^2 + \left(\frac{1}{10} - \frac{1}{2}\phi_z\right)\frac{I_y}{AL^2}\right](1+\phi_z)^{-2} \tag{13.33}$$

$$m_{2,8} = \left(\frac{9}{70} + \frac{3}{10}\phi_y + \frac{1}{6}\phi_y^2 - \frac{6I_z}{5AL^2}\right)(1+\phi_z)^{-2} \tag{13.34}$$

$$m_{3,9} = \left(\frac{9}{70} + \frac{3}{10}\phi_z + \frac{1}{6}\phi_z^2 - \frac{6I_y}{5AL^2}\right)(1+\phi_y)^{-2} \tag{13.35}$$

$$m_{2,12} = -m_{6,8} = -L\left[\frac{13}{420} + \frac{3}{40}\phi_y + \frac{1}{24}\phi_y^2 - \left(\frac{1}{10} - \frac{1}{2}\phi_y\right)\frac{I_z}{AL^2}\right](1+\phi_y)^{-2} \tag{13.36}$$

$$m_{3,11} = -m_{5,9} = L\left[\frac{13}{420} + \frac{3}{40}\phi_z + \frac{1}{24}\phi_z^2 - \left(\frac{1}{10} - \frac{1}{2}\phi_z\right)\frac{I_y}{AL^2}\right](1+\phi_z)^{-2} \tag{13.37}$$

$$m_{6,6} = m_{12,12} = L^2\left[\frac{1}{105} + \frac{1}{60}\phi_y + \frac{1}{120}\phi_y^2 + \left(\frac{2}{15} + \frac{1}{6}\phi_y + \frac{1}{3}{\phi_y}^2\right)\frac{I_z}{AL^2}\right](1+\phi_y)^{-2} \tag{13.38}$$

$$m_{6,12} = -L^2\left[\frac{1}{140} + \frac{1}{60}\phi_y + \frac{1}{120}\phi_y^2 + \left(\frac{1}{30} + \frac{1}{6}\phi_y - \frac{1}{6}{\phi_y}^2\right)\frac{I_y}{AL^2}\right](1+\phi_y)^{-2} \tag{13.39}$$

$$m_{5,11} = -L^2\left[\frac{1}{140} + \frac{1}{60}\phi_z + \frac{1}{120}\phi_z^2 + \left(\frac{1}{30} + \frac{1}{6}\phi_z - \frac{1}{6}{\phi_z}^2\right)\frac{I_y}{AL^2}\right](1+\phi_z)^{-2} \tag{13.40}$$

The geometric stiffness matrix is obtained using nonlinear strain-displacements relationship pertinent for the longitudinal strain along the direction x. The other non-linear terms are disregarded. It is expressed as [8]

$$\mathbf{G} = \frac{F}{L}\begin{bmatrix}
0 & & & & & & & & & & & \\
 & G_{2,2} & & & & G_{2,6} & & G_{2,8} & & & & G_{2,12} \\
 & & G_{3,3} & & G_{3,5} & & & & G_{3,9} & & G_{3,11} & \\
 & & & G_{4,4} & & & & & & G_{4,10} & & \\
 & & & & G_{5,5} & & & & G_{5,9} & & G_{5,11} & \\
 & & & & & G_{6,6} & & G_{6,8} & & & & G_{6,12} \\
 & & & & & & 0 & & & & & \\
 & & & & & & & G_{8,8} & & & & G_{8,12} \\
 & & & & & & & & G_{9,9} & & G_{9,11} & \\
 & & & & & & & & & G_{10,10} & & \\
 & & & & & & & & & & G_{11,11} & \\
 & & & & & & & & & & & G_{12,12}
\end{bmatrix} \tag{13.41}$$

where F is the axial stress in the beam and

$$G_{2,2} = G_{2,8} = G_{8,8} = \left(\frac{6}{5} + 2\phi_y + \phi_y^2\right)(1+\phi_y)^{-2} \tag{13.42}$$

$$G_{3,3} = G_{3,9} = G_{9,9} = \left(\frac{6}{5} + 2\phi_z + \phi_z^2\right)(1+\phi_z)^{-2} \tag{13.43}$$

$$G_{4,4} = G_{10,10} = -G_{4,10} = \frac{J}{A} \tag{13.44}$$

$$G_{5,5} = G_{11,11} = L^2\left(\frac{2}{15} + \frac{1}{6}\phi_z + \frac{1}{12}\phi_z^2\right)(1+\phi_z)^{-2} \tag{13.45}$$

$$G_{6,6} = G_{12,12} = L^2\left(\frac{2}{15} + \frac{1}{6}\phi_y + \frac{1}{12}\phi_y^2\right)(1+\phi_y)^{-2} \tag{13.46}$$

$$G_{5,9} = G_{11,9} = -G_{3,5} = -G_{9,5} = \frac{L}{10}(1+\phi_z)^{-2} \tag{13.47}$$

$$G_{2,6} = G_{2,12} = -G_{8,6} = -G_{8,12} = \frac{L}{10}(1+\phi_y)^{-2} \tag{13.48}$$

$$G_{5,11} = -L^2\left(\frac{1}{30} + \frac{1}{6}\phi_z + \frac{1}{12}\phi_z^2\right)(1+\phi_z)^{-2} \tag{13.49}$$

$$G_{6,12} = -L^2\left(\frac{1}{30} + \frac{1}{6}\phi_y + \frac{1}{12}\phi_y^2\right)(1+\phi_y)^{-2} \tag{13.50}$$

The stiffness matrix of the beam element can now be calculated as

$$\tilde{\mathbf{K}} = \mathbf{K} + \mathbf{G} - \Omega^2\mathbf{M} \tag{13.51}$$

where Ω is the rotational speed.

13.5 Mesh Assemblage

The blade mesh is built by a standard assemblage procedure, starting from the local stiffness and mass matrices of the FEs. The assemblage procedure provides the global mass and stiffness matrices $\overline{\mathbf{M}}$ and $\overline{\mathbf{K}}$. Figure 13.5 shows a typical blade mesh in which elements with different cross sections are connected one after the other.

13.6 Solution

When the rotational speed of the blade is non-zero, a relevant amount of the blade stiffness derives from the geometric stiffness matrix, which depends of the axial pre-stress of the FEs. This quantities can be evaluated by running a static analysis of the blade loaded with the centrifugal force

$$\mathbf{F}_c = \Omega^2\mathbf{r}\overline{\mathbf{M}} \tag{13.52}$$

where $\mathbf{r}$ is a vector containing, for the translation degrees of freedom, the distance of each node from the rotational axis and, for the rotational degrees of freedom is zero.

The modal analysis consist of the solution of the a classical eigenvalue problem

$$\left(\omega^2\mathbf{I} - \overline{\mathbf{M}}^{-1}\overline{\mathbf{K}}\right)\mathbf{q} = \mathbf{0} \tag{13.53}$$

where ω are the natural circular frequencies and $\mathbf{q}$ the vectors representing the vibration modes.

13.7 Validation

The proposed model is here compared with an accurate 3D FE calculation in terms of modal properties. The reference model is built in ANSYS ® v14.5 using 3D solid elements (SOLID 186). The FE stiffness depends on the rotational speed, thus different simulation are carried out considering different angular velocities. Figure 13.6a shows the first mode shape

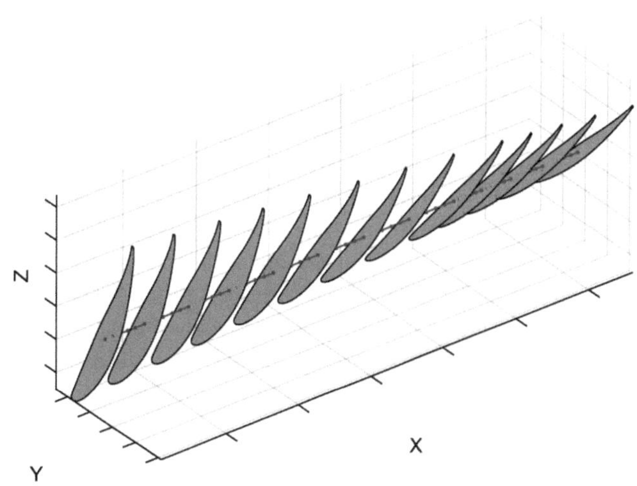

Fig. 13.5 Mesh of the blade model

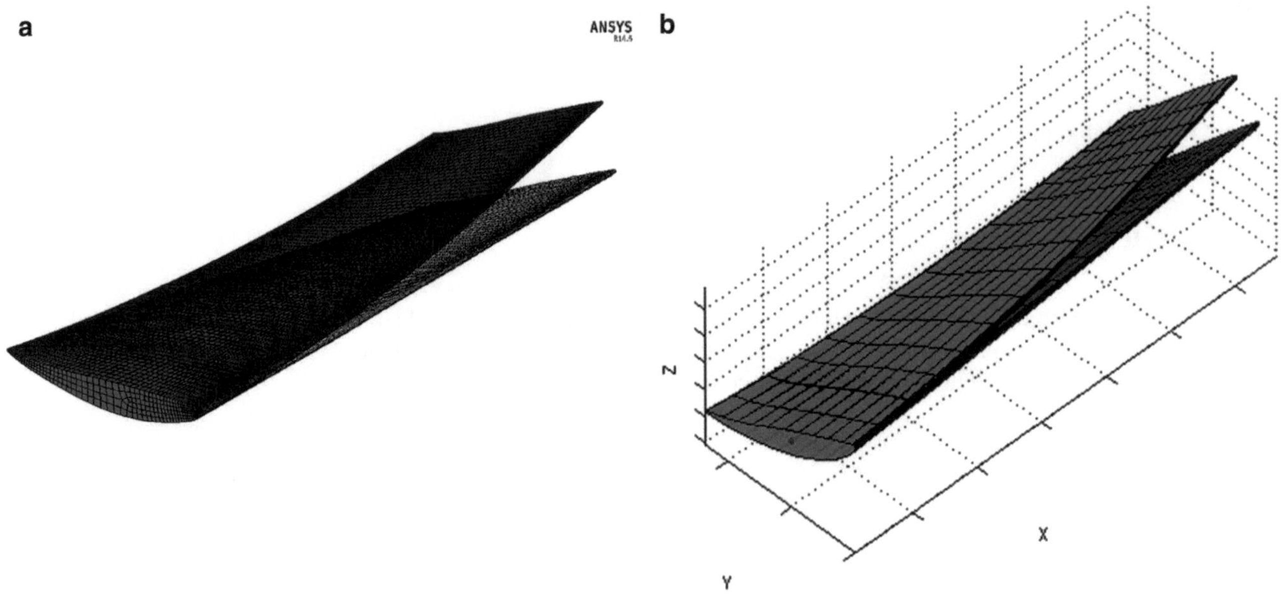

Fig. 13.6 Mode 1: modal shape obtained by the 3D model (a) and 1-D model (b)

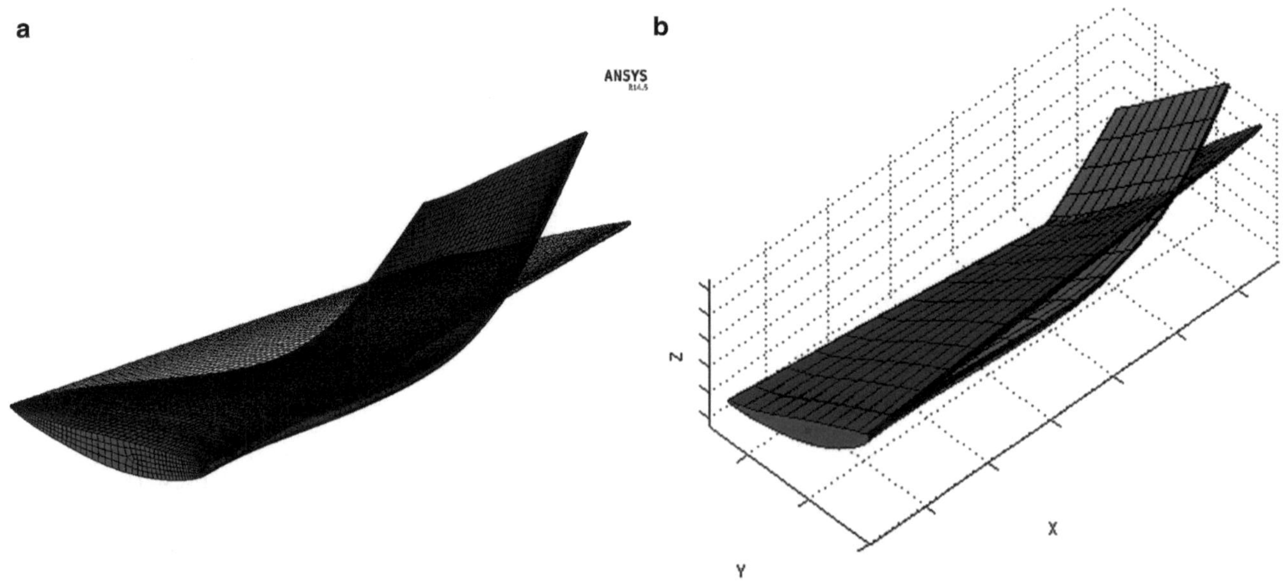

Fig. 13.7 Mode 2: modal shape obtained by the 3D model (a) and 1-D model (b)

of the blade evaluated with the 3D model. Figure 13.6b represents the same mode obtained by the 1D calculation. The two modal shapes appears qualitatively similar. Figures 13.7a and b show the second mode obtained by the 3-D and 1-D model, respectively. Again a good agreement is recognized.

The rotational speed of the rotor is a key element in simulation, since the eigenfrequencies of the blade are highly connected to this parameter due to geometrical stiffness as well as spin softening. Figure 13.8 shows the variation of the first five natural frequencies with the rotational speed. The first natural frequency of the blade is perfectly predicted by the 1-D model, as the difference with respect to the 3-D calculation is below 1 %. The other natural frequencies are predicted with an error below 3 %, with the exception of mode 3.

The third mode is a purely torsional one and, for this case, the 1-D model is inadequate due to the deficiency of the Timoshenko beam in describing torsion.

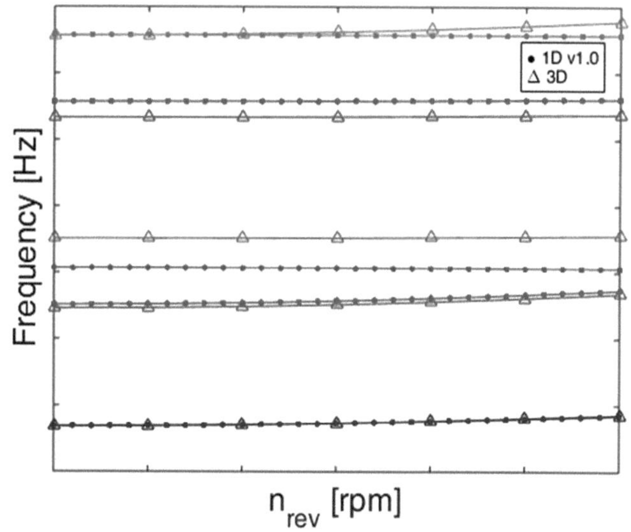

Fig. 13.8 Vatural frequencies as functions of the rotational speed. Comparison 3-D and 1-D modeling

13.8 Closing Remarks

A strategy for the 1-D simulation of turbine blades, that is completely automated and based upon the blade geometry has been presented and validated on a typical compressor blade of a large gas turbine. The prediction of the first few natural frequencies appeared reasonably accurate with the exception of the modes for which the torsional behavior is prevalent. This deficiency suggests improving the description of the torsional behavior implementing a FE formulated on the basis of the thin-walled beam theory for torsion.

References

1. Tsai GC (2004) Rotating vibration behavior of the turbine blades with different groups of blades. J Sound Vib 271(3–5):547–575
2. Castanier MP, Pierre C (2006) Modeling and analysis of mistuned bladed disk vibration: status and emerging directions. J Propuls Power 22(2):384–396
3. Timoshenko S, Stephen P (1970) Theory of elasticity. McGraw-Hill, New York
4. Davis R, Henshell RD, Warburton GB (1972) A Timoshenko beam element. J Sound Vib 22:475–487
5. Friedman Z, Kosmatka JB (1993) Improved two-node Timoshenko beam finite element. Comput Struct 47:473–481
6. Reddy JN (1997) On locking-free shear deformable beam finite elements. Comput Methods Appl Mech Eng 149(1–4):113–132
7. Luo Y (2008) An efficient 3D Timoshenko beam element with consistent shape functions. Adv Theor Appl Mech 1(3):95–106
8. Przemieniecki JS (1985) Theory of matrix structural analysis. Dover, New York
9. Timoshenko S (1921) On the correction for shear of the differential equation for transverse vibration of prismatic bars. Philos Mag 41(6):744–746

Chapter 14
Damping Estimation for Turbine Blades Under Non-stationary Rotation Speed

Luigi Carassale, Michela Marrè-Brunenghi, and Stefano Patrone

Abstract Turbine blades are critical components in thermal power plants and their design process usually includes experimental tests in order to tune or confirm numerical analyses. These tests are generally carried out on full-scale rotors having some blades instrumented with strain gauges and usually involve a run-up and/or a run-down phase. The quantification of damping in these conditions is rather complicated, since the finite sweep velocity produces a distortion of the vibration amplitude with respect to the Frequency-Response Function that would be expected for an infinitely slow crossing of the resonance. In this work, we demonstrate through a numerical simulation that the usual identification procedures procedure lead to a systematic overestimation of damping due both to the finite sweep velocity, as well as to the variation of the blade natural frequency with the rotation speed. An identification procedure based on the time-frequency analysis is proposed and validated through numerical simulations.

Keywords Turbine blades • Damping estimation • Time-frequency analysis • Transient response • Time-variant systems

14.1 Introduction

Turbine blades are critical components in thermal power plants as they are subjected to vibration and high-cycle fatigue. Their dynamic behavior strongly depends on the working conditions, including rotation speed, temperature and flow condition. Regarding the growing importance of renewable energy, the number of run-ups and run-downs in power plants is increasing continuously. The dynamic characteristics of the blades are predicted by numerical analyses that are validated and updated through experimental campaigns.

Rotors are tested first in laboratory inside spin cells. Blades are excited by the injection of an air jet or by a magnet producing a synchronous excitation. Their response is usually measured by strain gauges connected to a telemetry system. Subsequently, in operating condition, several sources of synchronous and non-synchronous excitations are present and the response can be assessed by blade-tip timing (BTT) measurements [1, 2].

The blades natural frequencies can be easily estimated by traditional spectral analyses and direct inspection of the Campbell diagram obtained during run-up or run-down, in which the harmonic content of the vibration is mapped as a function of the rotation speed. On the other hand, the estimation of damping is significantly more complicated and is usually carried out by analyzing the vibration amplitude in the neighborhood of the crossing between a natural frequency and a super-harmonics of the rotation speed. The response amplitude is interpreted as the Frequency-Response Function (FRF) of a s-dof system and damping is estimated by traditional fitting techniques (e.g. [3]).

In this work we investigate the accuracy of damping estimation that can be achieved in realistic test conditions, in which rotation speed transients and/or thermal transients are often unavoidable. In particular, we demonstrate through a numerical simulation that the simple estimation procedure described above leads to a systematic overestimation of damping. A more sophisticated procedure based on time-frequency analysis (e.g. [4]) is proposed and validated through twoclasses

L. Carassale (✉) • M. Marrè-Brunenghi
Department of Civil, Chemical and Environmental Engineering, University of Genova,
Via Montallegro 1, Genoa 16145, Italy
e-mail: luigi.carassale@unige.it; michela.marre.brunenghi@edu.unige.it

S. Patrone
Ansaldo Sviluppo Energia, Via Lorenzi 8, Genoa 16152, Italy
e-mail: stefano.patrone@asen.com

© The Society for Experimental Mechanics, Inc. 2015

R. Allemang (ed.), *Special Topics in Structural Dynamics, Volume 6*, Conference Proceedings of the Society for Experimental Mechanics Series, DOI 10.1007/978-3-319-15048-2_14

of numerical simulations. The first one is extremely simple and is based to the integration of Linear Time-Invariant (LTI) single-degree-of-freedom (s-dof) systems excited by harmonic sweeps. The second class of simulations include multi-modal response of the blade, multiple engine order excitation, random excitation and time-variant mechanical properties.

14.2 Description of the Problem

Modal properties of turbine blades are studied experimentally through vibration measurements in spin cell. The information necessary to the identification is usually gathered from the transient dynamic amplification that appears during run-up or run-down whenever an engine order becomes resonant with a vibration mode. This situation is described in Fig. 14.1 representing the time-frequency energy density of the vibration calculated in the neighborhood of a critical speed. It can be observed that the frequency of the excitation decreases (due to the reduction of the angular velocity) and crosses a resonance producing an increment of the vibration amplitude. The trace of the resonance, on its turn, leans downward since the blade natural frequency decreases as the angular frequency decreases (due to the reduction of the centrifugal force). Once the rotor passes the critical speed, the vibration energy remains concentrated into two frequency bands: the first one is the band with the higher slope in Fig. 14.1, it is determined by the excitation and contains the synchronous vibration; the second one is governed by the natural frequency and contains the free decaying vibration (represented in Fig. 14.1 by the trace with minor slope). The described scenario has been deeply investigated through analytical and numerical methods (e.g. [5–8]) and it is very much determined by the sweep rate and by the damping of the system [9].

14.3 Effects of the Sweep Rate

This session discusses the effects of the sweep rate on the response of a very simplified model of blade. It is assumed that the blades of the disc acts as independent Linear Time-Invariant (LTI) single-degree-of-freedom (s-dof) systems and that are excited by a single harmonic sweep corresponding to some engine order N. Under these hypotheses, the motion some characteristic point of the blade y is governed by the differential equation:

$$\ddot{y}(t) + 2\xi\overline{\omega}\dot{y}(t) + \overline{\omega}^2 y(t) = \frac{1}{m} f(t) \tag{14.1}$$

where the dot represents the derivation with respect to the time t, $\overline{\omega}$ is the natural circular frequency, ξ the damping ratio and m the mass. The external dynamic force f is assumed in the form:

$$f(t) = F \sin\left[N\left(\omega_0 t + \dot{\omega}\frac{t^2}{2}\right)\right] \tag{14.2}$$

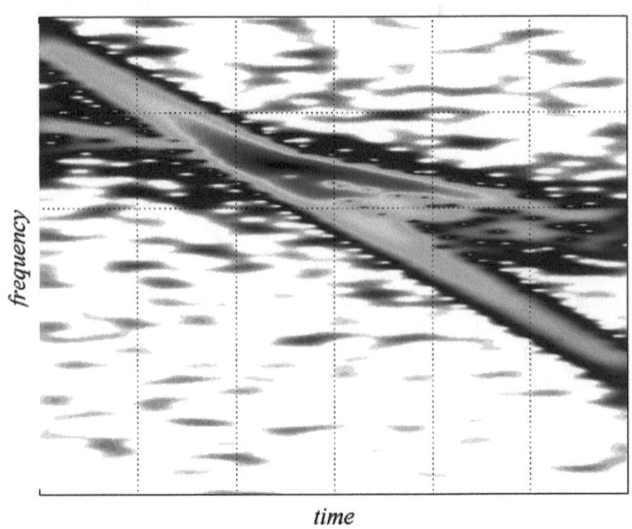

Fig. 14.1 Time-frequency energy density of the vibration in the neighborhood of a critical speed (experimental)

where F is a constant, ω_0 the rotation velocity at $t=0$ and $\dot{\omega}$ its time derivative, referred below as sweep rate. Defining the instantaneous phase angle as $\phi(t) = \omega_0 t + \dot{\omega}t^2/2$, it can be readily observed that its time derivative can be interpreted as an instantaneous frequency $\omega(t) = \dot{\phi}(t) = \omega_0 + \dot{\omega}t$ whose variation is linear in time. On the other hand, time and angular velocity are related by a linear map and, given one of these two quantities, the other is defined univocally. In particular, the force is perfectly resonant for $t = t_{res} = (\overline{\omega}/N - \omega_0)/\dot{\omega}$.

In order to perform a systematic analysis of the system response, Eq. (14.1) must be scaled as:

$$\tilde{y}''(\tilde{t}) + 2\xi\tilde{y}'(\tilde{t}) + \tilde{y}(\tilde{t}) = \tilde{f}(\tilde{t}) \tag{14.3}$$

introducing the following non-dimensional quantities

$$\begin{aligned} y &= Y\tilde{y}; \quad f = \frac{F}{m\overline{\omega}^2 Y}\tilde{f}; \quad t = \frac{\tilde{t}}{\overline{\omega}}; \\ \omega &= \overline{\omega}\tilde{\omega}; \quad \omega_0 = \overline{\omega}\tilde{\omega}_0; \quad \dot{\omega} = \overline{\omega}^2\tilde{\omega}' \end{aligned} \tag{14.4}$$

and denoting the derivation with respect to the non-dimensional time $\tilde{t}$ with the symbol $'$. Time and frequency are scaled with the system natural frequency $\overline{\omega}$ that, in practical cases, is unknown; Y is a scaling factor. In these cases, any guess value for the natural frequency can be used.

Adopting this scaling, the non-dimensional force is defined (up to an inessential constant factor) as:

$$\tilde{f}(\tilde{t}) = \sin\left(\tilde{\omega}_0\tilde{t} + \tilde{\omega}'\frac{\tilde{t}^2}{2}\right) \tag{14.5}$$

and the solutions of Eqs. (14.3)–(14.5) can be studied into the parameter space constituted by $\tilde{\omega}'$ and ξ. Without loss of generality it is assumed $N=1$ and $\tilde{\omega}_0$ far enough from unity in such a way to avoid effects due to initial conditions. From now on, the sign tilde used to denote non-dimensional quantities is dropped to simplify the notation.

The dynamic response in the neighborhood of the resonance ($t \simeq t_{res}$) can be conveniently represented as a harmonics modulated in amplitude and phase as:

$$y(t) = A(t)\sin\phi_y(t) \tag{14.6}$$

The pair of functions $A(t)$ and $\phi_y(t)$ providing $y(t)$ is not unique, however a classical approach to estimate them is to consider the analytic function $y_a = y + i\,\mathrm{H}[y]$, i being the imaginary unit and $\mathrm{H}[\bullet]$ the Hilbert transform. Using this tool, the amplitude modulating function is readily obtained as $A(t) = |y_a(t)|$ and is called Analytic Amplitude (AA).

It can be verified that if $\omega' \to 0$, $A(t)$ can be related with the Frequency-Response Function (FRF) of the system $H(\omega)$ as:

$$|H(\omega)| = A(t(\omega)) \tag{14.7}$$

where the function $t(\omega)$ maps the angular velocity into the time and

$$H(\omega) = \frac{S}{\overline{\omega}^2 - \omega^2 + 2i\xi\overline{\omega}\omega} \tag{14.8}$$

in which $\overline{\omega}$ and S are, respectively, the non-dimensional natural frequency and a scale factor. These two parameters are unitary for the present simulated case, but in general are unknown and must be estimated. The scale parameter is usually irrelevant, while $\overline{\omega}$ and ξ can be evaluated through the well-known half-power band method as:

$$\begin{aligned} \overline{\omega} &= \arg\max[A(t(\omega))] \\ \xi &= \frac{\omega^+ - \omega^-}{2\overline{\omega}} \end{aligned} \tag{14.9}$$

where ω^+ and ω^- are, respectively, the frequency values to the right and to the left of the peak corresponding to the level $A(\omega^+) = A(\omega^-) = 2^{-0.5}\max(A(\omega))$. A more robust approach abstains an estimation of $\overline{\omega}$ and ξ from the minimization of the error measure:

$$D = \int_{1-\alpha}^{1+\alpha} \left|A(\omega(t)) - |H(\omega)|\right|^2 d\omega \tag{14.10}$$

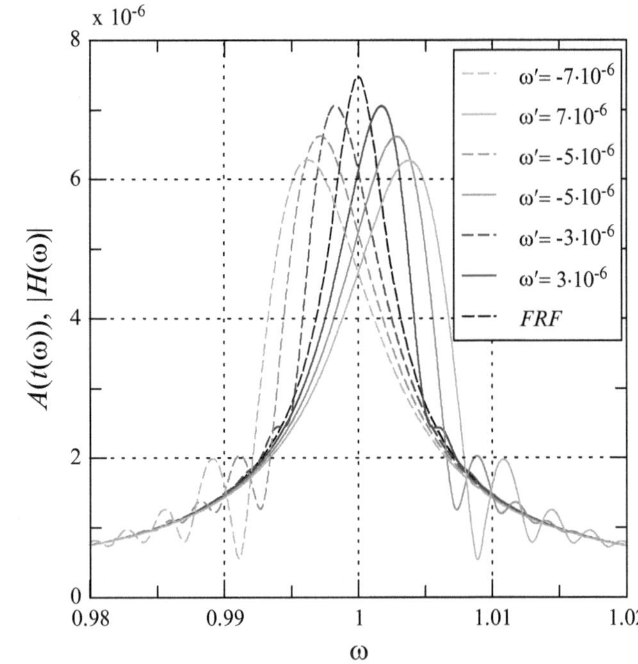

Fig. 14.2 AA for different sweep rates compared with FRF

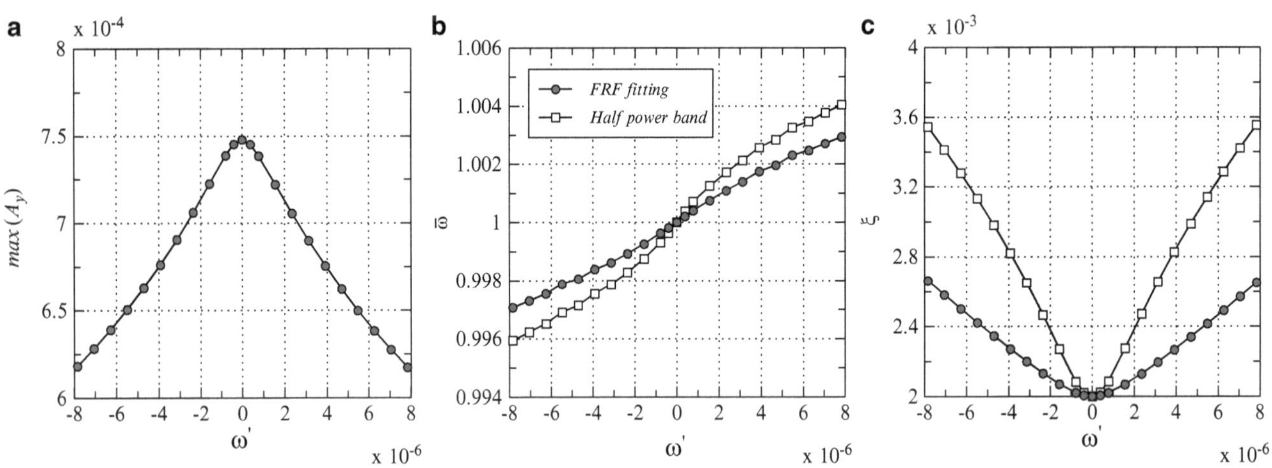

Fig. 14.3 Maximum amplitude of vibration (**a**), estimated natural frequency (**b**), estimated damping ratio (**c**) as a function of the sweep rate

where α defines an user-selected bandwidth adopted for the estimation. In the following, this latter approach is denoted as FRF fitting.

Figure 14.2 shows the effects of the sweep rate on the response of a system modelled according to Eqs. (14.3)–(14.5) with $\xi = 0.2$ %. It can be observed that the transient response amplitude A is a good approximation of $|H|$ for small $|\omega'|$, while they become quite far apart when $|\omega'|$ increases. In particular, the increment of $|\omega'|$ produces four effects: (1) the maximum of A is achieved for $t > t_{res}$, which means that it appears for $\omega > 1$ during run-up and for $\omega < 1$ during run-down; (2) the maximum amplitude decreases; (3) the beating phenomenon due to the interaction between the synchronous vibration and the free decaying vibration becomes more relevant; (4) the peak of the response in the neighborhood of the resonance becomes larger.

Figure 14.3 shows the maximum vibration amplitude (a), as well as the estimated natural frequency (b) and damping ratio (c) as a function of the sweep rate. The considered range for the sweep rate is realistic for tests on large turbine rotors carried out in spin cell. It can be observed that, while the error in the estimation of $\overline{\omega}$ is always very small, as $|\omega'|$ increases, the estimation error for ξ becomes very large, in particular when the half-power band method is used. This behavior is practically symmetric with respect to the sign of ω', making no difference for run-up and run-down condition, at least within the considered range of the parameters.

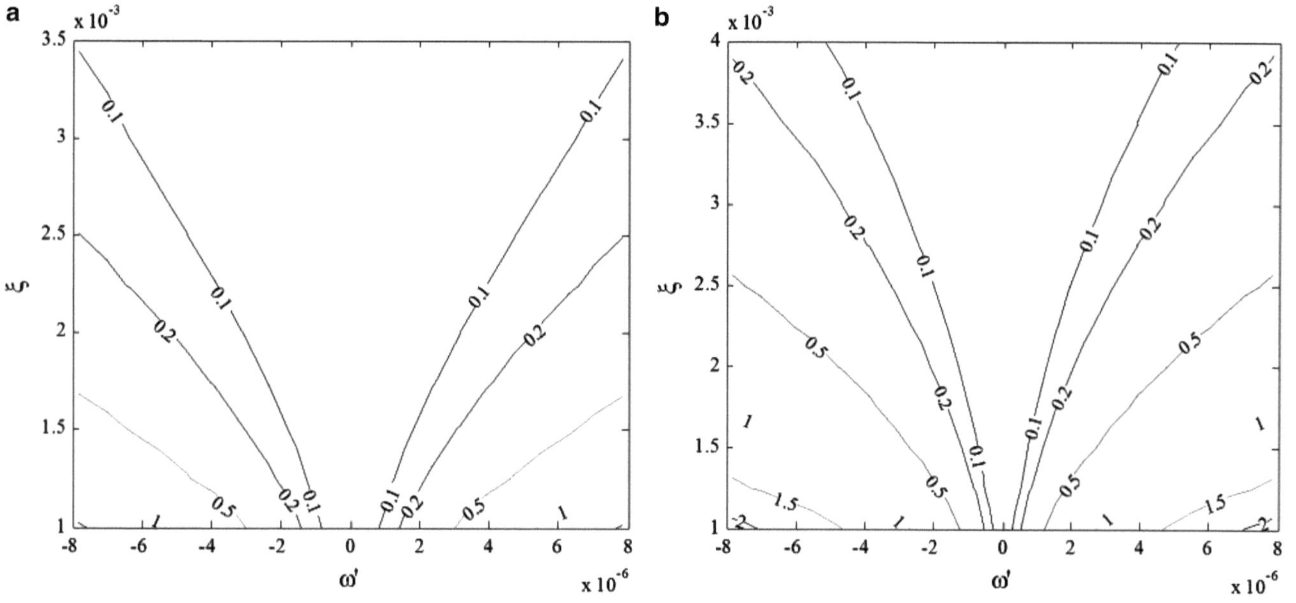

Fig. 14.4 Relative error in the estimation damping through the half-power band method (**a**) and the FRF fitting (**b**)

Figure 14.4 shows the relative error for the estimation of damping using the half-power band technique (a) and the FRF fitting (b). With both these techniques, the estimation error increases as ξ decreases and can be very large (more than 100 %) for realistic values of sweep rate and damping.

14.4 Time-Frequency Analysis

The beating phenomenon observed in Fig. 14.2 is related to the interaction of the synchronous response and the free decaying vibration that, immediately after the crossing of the critical speed, are closely spaced in the frequency domain. In order to isolate the synchronous component of the response, the continuous Wavelet Transform (WT) of the simulated signal is considered [4, 10]:

$$W_{u,s}[y] = \int_{-\infty}^{\infty} y(t)\psi_{u,s}^{*}(t)\,dt \tag{14.11}$$

representing the correlation of the signal y with the time-frequency atom $\psi_{u,s}$; the superscript * identifies the complex conjugate. The time-frequency atom $\psi_{u,s}$ is defined by translating, of the quantity u, and dilating, of the quantity s, a mother wavelet ψ, i.e.

$$\psi_{u,s}(t) = \frac{1}{\sqrt{s}}\psi\left(\frac{t-u}{s}\right) \tag{14.12}$$

The mother wavelet adopted in the present study is the complex Morlet wavelet, that is defined as:

$$\psi(t) = \frac{1}{\sqrt[4]{\pi f_b}}\exp\left(2\pi i\, f_c t\right)\exp\left(-\frac{t^2}{2 f_b}\right) \tag{14.13}$$

where f_b and f_c have, respectively, the meaning of bandwidth parameter and central frequency. The scale is related to the frequency by the relationship $\omega = 2\pi f_c/s$.

The wavelet transform can be used to estimate the amplitude of the signal y in the neighborhood of a particular time t and frequency ω. Accordingly, the amplitude of the synchronous vibration at the frequency ω during the harmonic sweep can be obtained from Eq. (14.11) letting.

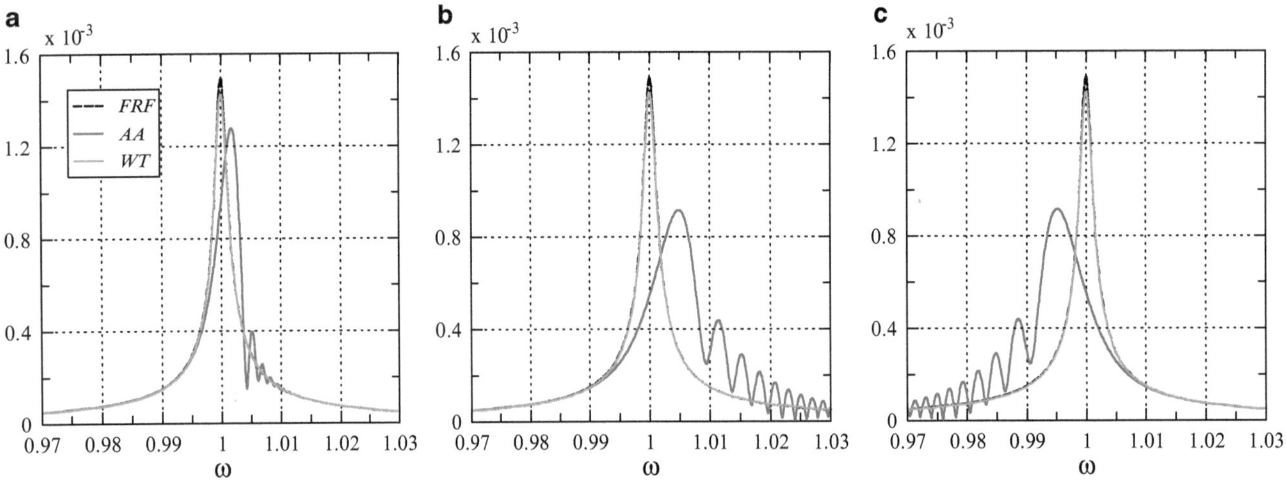

Fig. 14.5 Amplitude of wavelet ridge and response envelope compared to the FRF; $\omega' = 2 \cdot 10^{-6}$ (**a**), $7 \cdot 10^{-6}$ (**b**), $-7 \cdot 10^{-6}$ (**c**); $\xi = 0.1$ %

$$u = \frac{\omega - \omega_0}{\omega'}; \quad s = \frac{2\pi f_c}{\omega} \tag{14.14}$$

Figure 14.5 shows the comparison of the AA of y, the WT of y along the line defined by Eq. (14.14) and the FRF of the system for the sweep ratio $\omega' = 2{\cdot}10^6$ (a), $\omega' = 7{\cdot}10^6$ (b) and $\omega' = -7{\cdot}10^6$ (c). It can be observed that, while the analytic amplitude suffers of the problems described in the previous section, the wavelet transform is practically insensitive to the sweep ratio and reproduces the FRF almost perfectly. This excellent matching obviously reflects into a perfect estimation of the damping ratio even for very high sweep rates.

14.5 Identification of Time-Variant Systems

In the previous section the scenario described in Fig. 14.1 was drastically simplified. A more realistic simulation should include (1) the multi-modal response of the blades, (2) the excitation with several engine orders, (3) the random excitation due to buffeting and (4) the variation of the natural frequencies during the sweep due to the change of the centrifugal force. This result is achieved through an explicit time-domain simulation of a rotating disk represented through a reduced-order model. As a case study, it is considered the first compressor stage of a large gas turbine.

The blades are considered mutually uncoupled and their dynamics is represented through their first five vibration modes. The variation of the natural frequencies with respect to the angular velocity is approximated through polynomial laws experimentally identified. The external excitation is produced by a fixed force hitting the blades once per revolution to simulate the effect of the air jet used during tests in spin cell. Besides, a random excitation modeled as a Gaussian white noise is added.

Figure 14.6 shows the result of the simulation of a run-down sweep across a critical velocity. The random components of the response is well visible, as well as several amplifications due to resonances of different modes at various engine orders.

In the present case, the identification process based on the AA of the signal cannot be directly applied since the dynamic response is multi-component. This problem can be circumvented to some extent by pass-band filtering the signal to retain only the information in a narrow frequency band containing the natural frequency. In the present application a filter with pass band [0.95, 1.05] is used to pre-process data. It is just the case to note that the mentioned filtering must be designed in such a way to have zero phase to avoid the alteration of the relationship between ω and t.

On the contrary, the identification based of the WT does not require pre-filtering since it is essentially designed to select a desiderated time-frequency neighborhood of each investigated point of the time-frequency plane.

Figure 14.7 shows the AA of the filtered signal, the WT along the line defined by Eq. (14.14) and the FRF of the system for the sweep rate $\omega' = 7{\cdot}10^6$ (a) and $\omega' = -7{\cdot}10^6$ (b) with $\xi = 0.15$ %. It can be clearly observed that also in this case the WT has a much better performance than the analytic amplitude, being practically insensitive to beating and more robust with respect to the random component of the response.

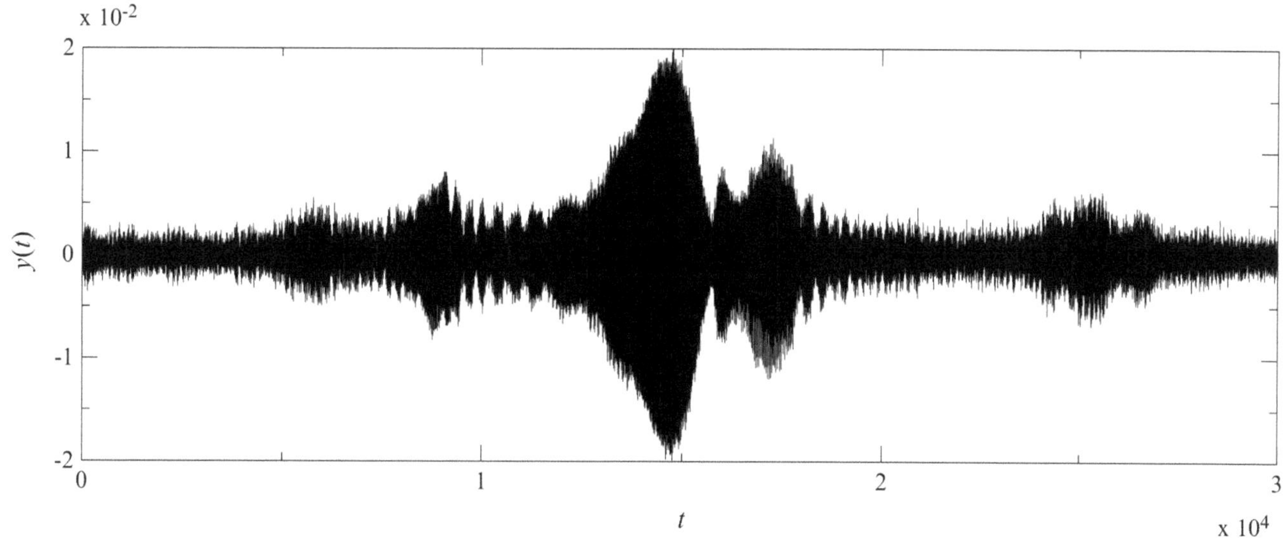

Fig. 14.6 Simulation of a critical velocity crossing. Sweep rate $\omega' = 6{\cdot}10^{-6}$

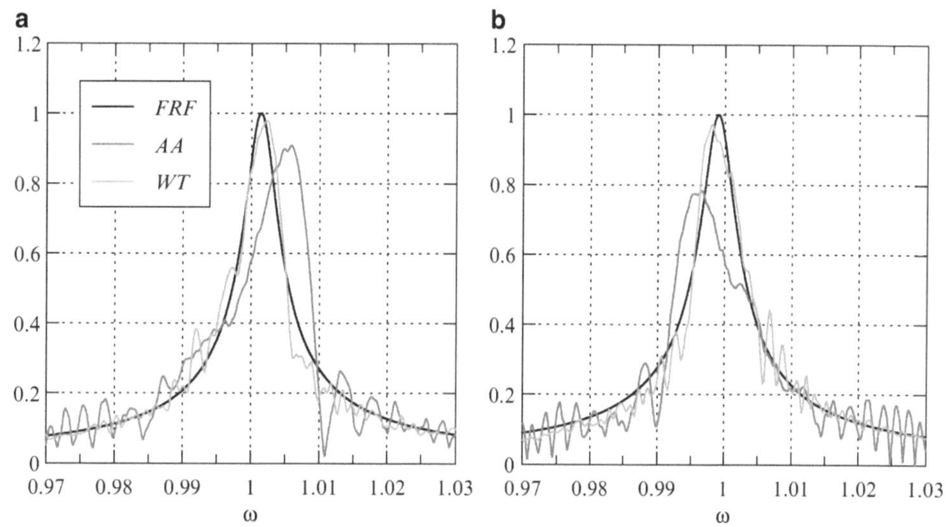

Fig. 14.7 Analytic amplitude and wavelet transform compared to FRF for run-down (a) and run-up (b)

Figure 14.8 shows the damping value estimated through the FRF fitting of the AA and WT (solid markers). Beside the larger spread of the estimation mostly due to the presence of the random component of the response, it can be clearly observed that the estimation does not converge to the exact damping value (0.15 %) even for very low sweep rate. The reason of this bias is the variation of the natural frequency with the angular velocity, which makes the system under observation a Linear Time-Variant (LTV) system.

To proceed with the identification, we assume that the natural frequency of the mode under investigation is known with a reasonable accuracy and that its variation in the neighborhood of the critical velocity crossing is smooth. In this case, the time-variant natural frequency can be expressed as

$$\overline{\omega}(\omega) = 1 + r_{\omega}\omega \tag{14.15}$$

where the constant r_{ω} represents the local slope of the natural frequency observed in a Campbell diagram and can be easily estimated a linear fitting. This further information can be included into the FRF model generalizing Eq. (14.8) as:

$$H(\omega) = \frac{S}{(\overline{\omega} + r_{\omega}\omega)^2 - \omega^2 + 2i\xi(\overline{\omega} + r_{\omega}\omega)\omega} \tag{14.16}$$

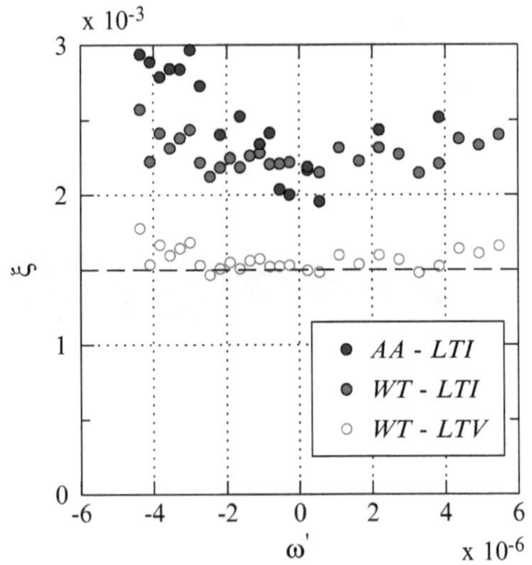

Fig. 14.8 Damping ratio estimated by fitting AA and WT with the LTI FRF, as well as fitting the WT with the LTV FRF. Exact value $\xi = 0.15\ \%$

The empty markers in Fig. 14.8 represent the estimation of damping obtained by fitting the WT with the LTV FRF given by Eq. (14.16). It can be observed that the estimation is very accurate no matter what the sweep rate is.

14.6 Concluding Remarks

An extensive numerical simulation demonstrated that the identification of damping in rotating blades crossing a resonance may be affected by a systematic overestimation due to the finite sweep velocity, as well as the dependence of the natural frequency with the rotation velocity. In contrast with the traditional method based on the analytic amplitude of the measured signal, the use of the wavelet transform leads to estimations that are insensitive to the sweep rate and are particularly robust towards random noise. Besides, if the amplitude extracted by WT is fitted with a time-variant FRF, also the error due to the variation of the natural frequency can be removed.

References

1. Heath S, Imregun M (1996) An improved single-parameter tip-timing method for turbomachinery blade vibration measurements using optical laser probes. Int J Mech Sci 38:1047–1058
2. Dimitriadis G, Carrington IB, Wright JR, Cooper JE (2002) Blade-tip timing measurement of synchronous vibrations of rotating bladed assemblies. Mech Syst Signal Process 16(4):599–622
3. Ewins DJ (2000) Modal testing: theory, practice and application. Research Studies Press, Philadelphia, Baldock
4. Mallat S (1998) A wavelet tour of signal processing. Academic Press, San Diego
5. Markert R, Seidler M (2001) Analytically based estimation of the maximum amplitude during passage through resonance. Int J Solids Struct 38:1975–1992
6. Vyas NS, Rao JS (1984) Equations of motion of a blade rotating with variable angular velocity. J Sound Vib 156:327–336
7. Omprakash V, Ramaturti V (1989) Spectral analysis of the transient characteristics of a bladed disk during run-up. Comput Struct 37:983–992
8. Ayers JP, Feiner DM, Griffin JH (2006) A reduced-order model for transient analysis of bladed disk forced response. J Turbomach 128:466–473
9. Bonhage M (2014) Transient amplitude amplification of mistuned structures. In: Proceedings of the ASME 19th blade mechanics seminar, Zurich
10. Chancey VC, Flowers GT, Howard CL (2003) A harmonic wavelets approach for extracting transient patterns from measured rotor vibration data. J Eng Gas Turbines Power 125:81–89

Chapter 15
Finite Element Modeling of a 40 m Space Frame Wind Turbine Tower

S.A. Smith, W.D. Zhu, and Y.F. Xu

Abstract The "20 % Wind Energy by 2030" initiative by the U.S. Department of Energy initiated the investigation of taller wind turbines. The highway infrastructure in the U.S. is causing the wind energy industry to investigate alternative designs; the lattice tower design is one such solution. One company designed a lattice tower that utilizes interference between the bolt and the clamped components. To study the dynamics of this tower an intensive model was created using beam, shell, and solid finite elements. Due to experimental results the model was originally modeled using a fixed boundary; however, the resulting frequencies for the first two modes were high. To investigate if the boundary was the cause of this error, the soil surrounding the tower was modeled using a relatively large linear solid continuum. The addition of the solid continuum greatly improved the results; resulting in a maximum difference of −7 % and mode shapes for the experiment and model follow the same trend.

Keywords Finite element modeling • Lattice tower • Wind turbine • Modal testing • Model simplification

15.1 Introduction

The increasing cost of fossil fuels has caused renewable energy to become a major focus of the energy industry in recent years [1]. In July 2008, the focus was cast onto wind energy with the release of the "20 % Wind Energy by 2030" report by the U.S. Department of Energy; to meet this goal an additional 300 GW of wind energy would need to be harvested [2]. To meet the additional requirements the turbines have to grow in height since more energy can be harvested at higher elevations [1, 3].

Currently, there are three designs for wind turbine towers available: tubular (steel or concrete), hybrid steel-concrete, and steel lattice [1, 3]; examples of the designs are shown in Fig. 15.1. The utilization of the designs depend on the height of the tower, the transportation limitations to the wind farm location, as well as location for aesthetic reasons. The limitation on the height of steel tubular towers are: the thickness of the walls increasing with height, increasing the manufacturing cost; and larger base diameters with greater heights, causing transportation issues due to the highway infrastructure [1, 3]. This design however has high fatigue strength and is aesthetically pleasing. Concrete tubular structures have an issue with wall thickness; however, since concrete has higher density than steel the increased weight causes the foundation to be larger, increasing cost, and larger weight leads to greater stress [1]. These structures have the same advantages as their steel counterparts, and the tower structure can be modular alleviating the transportation issue. A hybrid design includes a steel tubular structure on a concrete pedestal. This design capitalizes on the advantages of both steel and concrete towers; the combination allows the towers to be higher with lower weight than pure concrete and lower transportation costs due to the smaller diameter of the steel section [1]. Steel lattice structures have the greatest efficiency in supporting all loads, wind loads are reduced due to the lattice topology, and have lower manufacturing and transportation costs [3]. The lattice structure though has low fatigue strength and is less aesthetically pleasing.

The tower presented in this paper is a 40 m section of a 100 m lattice tower. To overcome the low fatigue strength of latice towers, a design that utilizes an interference-bolts was implemented; interference-bolts eliminate the gaps between the bolts and the clamped components increasing the fatigue strength of the joint. With the increased fatigue strength the height of the lattice tower design can grow while maintaining structural integrity.

S.A. Smith • W.D. Zhu (✉) • Y.F. Xu
Department of Mechanical Engineering, University of Maryland, Baltimore County, Baltimore, MD 21250, USA
e-mail: wzhu@umbc.edu

© The Society for Experimental Mechanics, Inc. 2015

R. Allemang (ed.), *Special Topics in Structural Dynamics, Volume 6*, Conference Proceedings of the Society for Experimental Mechanics Series, DOI 10.1007/978-3-319-15048-2_15

Fig. 15.1 Various designs of wind turbine towers; (*left to right*) Tubular [4], Steel-Concrete Hybrid [5], and Lattice [6]

The objective of this work is to experimentally measure and model a 40 m lattice tower. Modal testing is performed on the tower using the multiple impact method outlined in Sect. 15.2. The tower is then modeled using the finite element program ABAQUS, outlined in Sect. 15.3. The individual components are modeled using simplification techniques are outlined in Sects. 15.3.1 and 15.3.2. Modal analysis was performed on the tower using a fixed boundary in Sect. 15.3.4. Soil was then added to the model using a solid continuum model, outlined in Sect. 15.3.5.

15.2 Experimental Testing of the Tower

Modal field testing was performed on a 40 m space frame tower. The tower is divided into three sections: Section 5 (bottom), Section 6 (middle), and Section 8 (top); each section has five legs and faces, Leg 1–5 and Face 1–5; the nomenclature and numbering of the towers components can be seen in Fig. 15.2. The tower was measured at ten levels on each leg and nine levels on each face, levels 2 through 10, the levels can be seen in Fig. 15.2. The points of measurement are located on the line of symmetry of each leg and the intersections of the cross-braces of each face; the top of the tower (Level 10) is a mounting ring where the measurement points were placed as close as possible on the same vertical line as those of lower levels. The tower was excited using a PCB 086D50 large-sledge impulse hammer; acceleration was measured in the radial, tangential and vertical directions using PCB 356A17 tri-axial accelerometers. A LMS 36-channel spectrum analyzer was used to collect the data; and the data was analyzed using the program LMS TEST.LAB to obtain the frequency response functions (FRFs) and power spectral densities (PSD). The tower is geometrically symmetric to a certain level indicating that orthogonally closely spaced modes were expected. To distinguish these modes the multiple impact method, proposed by Zhu et al. [7], was used with a frequency resolution of 0.0078125 Hz, resulting in a 128 s acquisition time.

The rigidity of the ground was tested to determine the boundary condition of the tower. Accelerometers were placed at Level 1 on Leg 4, the footpad of the same leg, and on the ground at the center of the tower; the tower was impacted on Leg 3 at Level 7. Comparing the measured FRFs, Fig. 15.3, it can be seen that the footpad and ground are much smaller than the FRF of Level 1, meaning the boundary of the tower may be considered rigid. After determining the boundary condition, modal testing using the roving sensor method was performed. The tower was excited in the radial and tangential directions on Legs 3 and 4, and the accelerations were measured at each level on all five faces and legs. The natural frequencies and descriptions of the first seven modes are listed in Table 15.1, and the mode shapes of Modes 1, 3, 4, and 6 can be seen in Fig. 15.4.

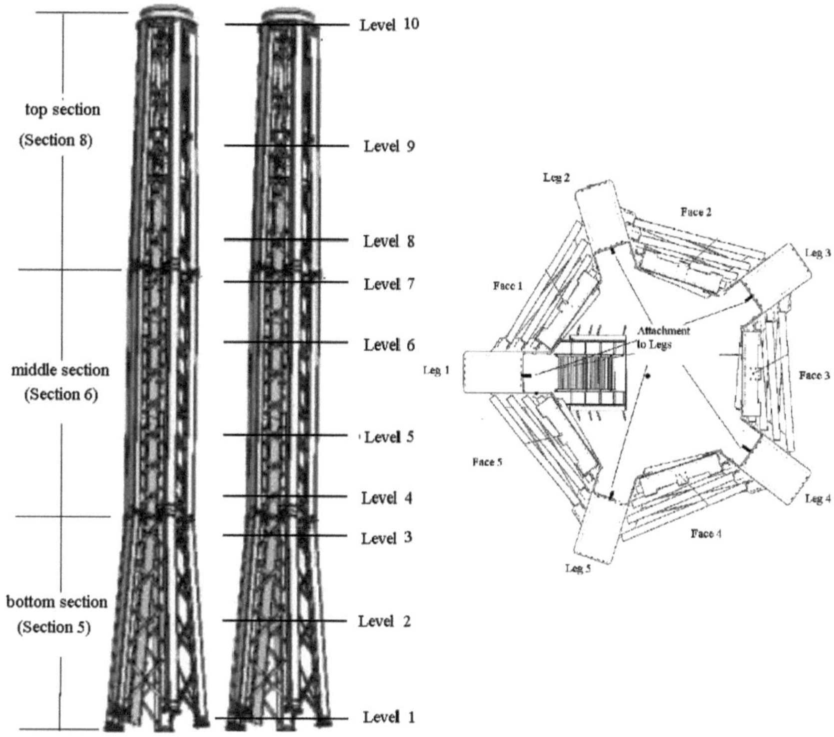

Fig. 15.2 Nomenclature and numbering of tower components

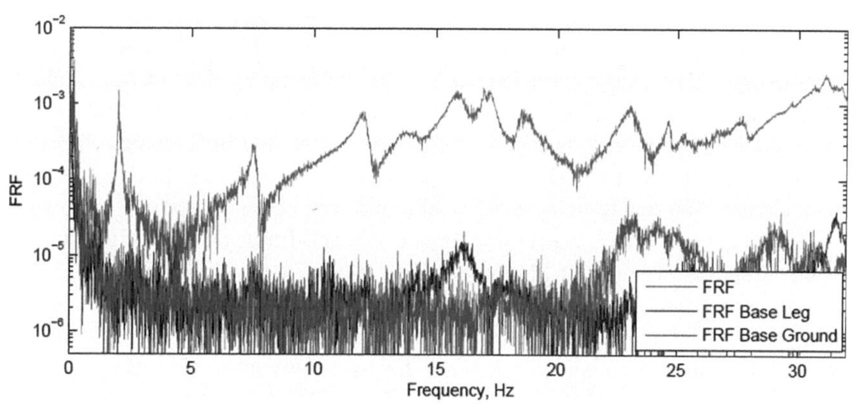

Fig. 15.3 The measured FRFs of the leg (*black*), footpad (*red*), and ground (*blue*) to determine the boundary condition

Table 15.1 Measured natural frequencies and descriptions of the first seven modes

Mode	Natural frequency (Hz)	Description
1	2.006	First bending
2	2.072	First bending (orthogonal)
3	7.056	First torsional
4	7.531	Second bending
5	7.629	Second bending (orthogonal)
6	11.981	Third bending
7	11.984	Third bending (orthogonal)

15.3 Modeling of the Tower

Studies have been performed on space-frame transmission line towers that model the components as beam/truss elements [1, 8–14]; Ahmed et al. added non-linear springs at the joints to simulate the bolted joint [8], a study on an airport lightmast was modeled using beam elements for the cross-members and legs and shell elements for redundant splice plates [15]. There have been studies on tubular wind turbine towers; some studies used beam elements to model the tower of the turbine and shell elements for the blades [16–20]; some using shell elements for the tower and solid elements for the nacelle [21–23];

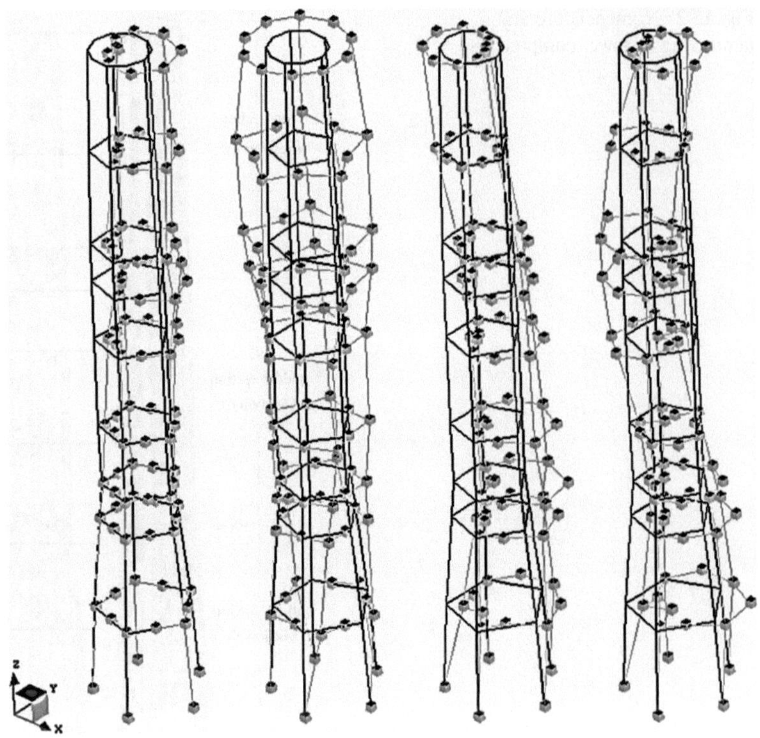

Fig. 15.4 Measured mode shapes of *(left to right)* Mode 1, Mode 3, Mode 4, and Mode 6

and some using solid elements for all parts of the tower [24]. To help reduce the computation time a bridge with a truss core in the cross-members was modeled using beam elements for the truss and solid elements for the casing [25]. The use of just beam elements cannot be used due to the lack of proper joint stiffness and the non-linear springs are for regular bolted joints. The model cannot be completely solid due to the computational demand for said model. A model that implements the three different elements would reduce the negatives of using just one; this would allow sections that need higher accuracy to be modeled as solids and those that can be simplified can be modeled using shell and beam elements, or combination of both.

The 40 m wind turbine tower implements a space-frame design to increase the stiffness-mass ratio. To achieve this ratio complex shapes were used: filleted L-beams, bolted joint connections, and interference bolted joints. These components are easily modeled as a solid element; however, to reduce the computational time the models where simplified as much as plausible to shell and beam elements. The filleted L-beam was modeled using the methods developed by He and Zhu [26]; the bolted joints were modeled following the procedure outlined by He and Zhu [27]; and the interference bolted joint were modeled using the method outlined, to be published later. All of the tower's components are composed of Grade 50 steel with Modulus of Elasticity $E = 200\,\text{GPa}$, Poisson's Ratio $\nu = 0.3$, and Density $\rho = 7{,}800\,\text{kg/m}^3$. The following sections outline the methods used for calculating the parameters for the various components encompassing the tower. These properties only vary with the simplification of the model outlined in the following sections.

15.3.1 Modeling of the Filleted L-Beam

The fillet in an L-beam cannot always be neglected due to the effects it has on the frequency and mode shapes of a structure [26]. Different methods for modeling a simplified fillet were reviewed by He and Zhu and the methods have issues: large element count, incompatible shape functions between solid and shell elements, and not considering out-of-plane effects of the fillet [26]. He and Zhu proposed a method that models the beam using a combination of shell (S4R) and beam (B31) linear elements to simplify the L-beam.

The L-beam can be broken down into three components: two flanges and the fillet region. The flanges are modeled as shell elements along the mid-plane using the base material and geometric properties of the flanges. The fillet region is modeled as a pair of shell elements, located at the midplane of the flanges, and a beam element, located at the centroid of the region; the shell elements of the region are then connected to the beam element via rigid links. The stiffness of the shell elements is defined by the thickness and length of the shell which are obtained by:

$$C_{m\theta} = \frac{12L}{s^3}, C_{p\theta} = \frac{6L^2}{s^3} \tag{15.1}$$

where $C_{m\theta}$ and $C_{p\theta}$ are given by:

$$\begin{gathered} C_{m\theta} = \sum_{i=1}^{n} \frac{1}{A_i R_i h_i} ds_i \\ C_{p\theta} = \sum_{i=1}^{n} \frac{r_i \sin \alpha_i - h_i \sin \alpha_i}{A_i R_i h_i} ds_i \end{gathered} \tag{15.2}$$

The fillet region is divided into i small components that have angular coordinate α_i, that the radius of curvature R_i is calculated at; is given as:

$$R_i = \frac{\left[r_i^2 + \left(\frac{dr}{d\alpha}\right)_i^2\right]^{3/2}}{r_i^2 + 2\left(\frac{dr}{d\alpha}\right)_i^2 - r_i \left(\frac{d^2 r}{d\alpha^2}\right)_i} \tag{15.3}$$

where

$$\begin{aligned} r_i &= \frac{t+r_o}{2\cos\alpha_i} + \frac{r_o}{2} \\ \left(\frac{dr}{d\alpha}\right)_i &= \frac{t+r_o}{2} (\sec\alpha_i)(\tan\alpha_i) \\ \left(\frac{d^2 r}{d\alpha^2}\right)_i &= \frac{t+r_o}{2}\left[(\sec\alpha_i)^3 + \sec\alpha_i\,(\tan\alpha_i)^2\right] \end{aligned} \tag{15.4}$$

where t is the thickness of the flanges and r_o is the radius of the fillet. The last variables needed to calculate $C_{m\theta}$ and $C_{p\theta}$ are the cross-sectional area A_i, Eq. (15.5a); the distance between the centroidal and neutral axes h_i, Eq. (15.5b); and the length of the centroidal axis of the small section ds_i, Eq. (15.5c).

$$\begin{aligned} A_i &= 2\frac{t+r_o}{2\cos\alpha_i} - \frac{r_o}{2} && \text{(a)} \\ h_i &= R_i - \frac{A_i}{\ln\left(\frac{2R_i + A_i}{2R_i - A_i}\right)} && \text{(b)} \\ ds_i &= d\alpha_i \sqrt{r_i^2 + \left(\frac{dr}{d\alpha}\right)_i} && \text{(c)} \end{aligned} \tag{15.5}$$

The out-of-plane stiffness of the half-fillet is modeled as a beam with an irregular cross-section located at the centroid of the half-fillet [26]. The irregular cross-section beam is modeled using the Generalized Shape profile in ABAQUS. The bending stiffness of the fillet is calculated by finding the area moment of inertia, AMI, about the centroid of the half-fillet. Due to the symmetry of the fillet the area moment of inertia about the x-axis and y-axis are equal. The total AMI of the fillet region using the parallel-axis theory is:

$$\begin{aligned} I_{x_1} &= I_{x_1}^{beam} + I_{x_1}^{shell-1} + A_{shell} dy_1^2 + I_{x_1}^{shell-2} + A_{shell} dy_2^2 \\ I_{x_1 y_1} &= I_{x_1 y_1}^{beam} + 2A_{shell} dy_i dy_2 \end{aligned} \tag{15.6}$$

where $I_{x_1}^{beam}$ and $I_{x_1 y_1}^{beam}$ are to be determined, A_{shell} is area of the shell (given by $A_{shell} = sL$), $I_{x_1}^{shell-1}$ is the AMI of the vertical shell (given by $I_{x_1}^{shell-1} = s^3 L/12$), $I_{x_1}^{shell-2}$ is the AMI of the horizontal shell (given by $I_{x_1}^{shell-2} = L^3 s/12$), d_{y_i} are the distances between shell elements and the centroid of the fillet, and I_{x_1} and $I_{x_1 y_1}$ are the AMI of the fillet regions found using integration. For the half-fillet cross-section the polar AMI is not the torsional stiffness factor due to the cross-section not being a circle, the stiffness is given by [26]:

Fig. 15.5 Cross-sectional view of the simplified filleted L-beam

$$
\begin{aligned}
K_b &= \tfrac{d-L}{d} K \\
K &= K_1 + K_2 + \alpha D^4 \\
K_1 &= (r_o + t)t^3 \left[\tfrac{1}{3} - 0.21 \tfrac{t}{r_o+t} \left(1 - \tfrac{t^4}{12(r_o+t)^4} \right) \right] \\
K_2 &= r_o t^3 \left[\tfrac{1}{3} - 0.105 \tfrac{t}{r_o} \left(1 - \tfrac{t^4}{192 r_o^4} \right) \right] \\
\alpha &= 0.07 + 0.076 \tfrac{r_o}{t} \\
D &= 2 \left[2t + 3r_o - \sqrt{2 (2r_o + t)^2} \right]
\end{aligned}
\qquad (15.7)
$$

To maintain the same total mass and density the cross-sectional area of the beam A_b is that of the whole fillet minus the cross-sectional area of the fillet's shell elements.

The methodology described is applied to a $0.1524 \times 0.1016 \times 0.01588$ m thick L-beam with a 0.01588 m fillet. The out-of-plane parameters were calculated using Eq. (15.1), the length of the shell was calculated to be 0.013872 m and the thickness was calculated to be 0.016958 m. The in-plane parameters are calculated by using Eqs. (15.6) and (15.7); the AMI are $I_x = 1.49218 * 10^{-8}\,\mathrm{m}^4$ and $I_{xy} = 8.49594 * 10^{-9}\,\mathrm{m}^4$, the area of the beam is calculated to be $0.000689854\,\mathrm{m}^2$, and the torsional stiffness factor of $2.34587 * 10^{-8}\,\mathrm{m}^4$. The final cross-section of the filleted L-beam can be seen in Fig. 15.5.

15.3.2 Modeling of the Bolted Joint Connections

15.3.2.1 Modeling of the Regular Bolted Joints

The modeling of the bolted joints is important due to the dynamic behavior of a structure being affected by the mass, stiffness, and damping of the joints; when the joints are modeled as fixed point, large errors occur in the results [27]. While the mass of the bolted joint is relatively simple; the stiffness needs to take into account the normal, shear, and rotational motions. Stiffness of the joint depends on the contact area and the pressure applied during the tightening processes; the contact area is defined as the area in which pressure exists in the components' interface [27]. The parameters There are ten different connections that use the regular bolted joint in the tower studied; the dimensions can be found in Table 15.2; where T is the thickness of the top clamped component, B is the thickness of the bottom component, and M is the thickness of the middle component for three clamped components; the axis-symmetric model with dimensions is shown in Fig. 15.6. The contact areas of the first and third set of joints listed in Table 15.2 were found to have an outside radius of 0.0276 m, and

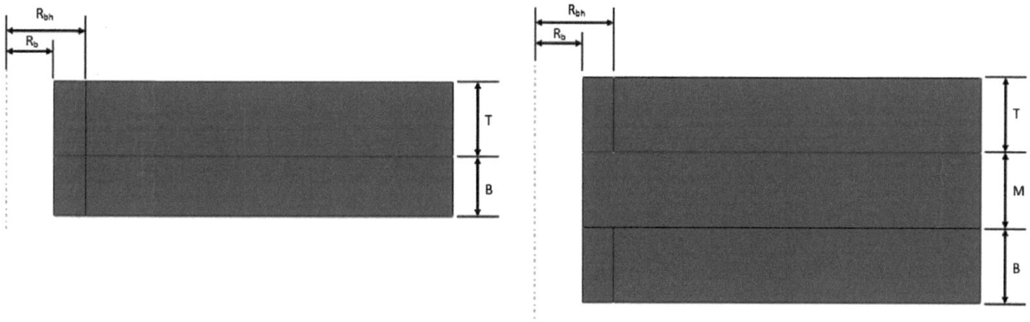

Fig. 15.6 Axis-symmetric model of the bolted joint connection with dimensions for the two clamped components (*left*) and three clamped components (*right*)

Table 15.2 Contact radius and dimensionless Young's modulus for the various regular bolts

Joint number	1	2	3	4	5	6	7	8	9	10
Component thickness (cm)	1.5875 1.746	1.5875 1.27	1.5875 1.746 1.5875	1.5875 1.27 1.5875	1.113 1.746 1.43	1.43 1.746 1.113	1.27 1.746 1.27	1.746 3.175	4.14 1.27	1.27 1.27 1.27
Contact radii (cm)	2.760	2.390	2.750	2.700	2.275 2.216	2.426 2.524	2.37 2.376	3.230	2.890	2.323
Dimensionless Young's modulus	0.599	0.538	0.761	0.508	0.659 0.815	0.665 0.702	0.624 0.719	0.514	0.752	0.532

Note: Contact radius and Dimensionless Young's Modulus are ordered such that the values are with the associated two components, e.g., for Joint Number 5 2.275 cm radius and 0.659 dimensionless modulus are associated with the cylinder for 1.113 cm and 1.746 cm component order, while 2.2216 cm radius and 0.815 dimensionless modulus are associated with the latter

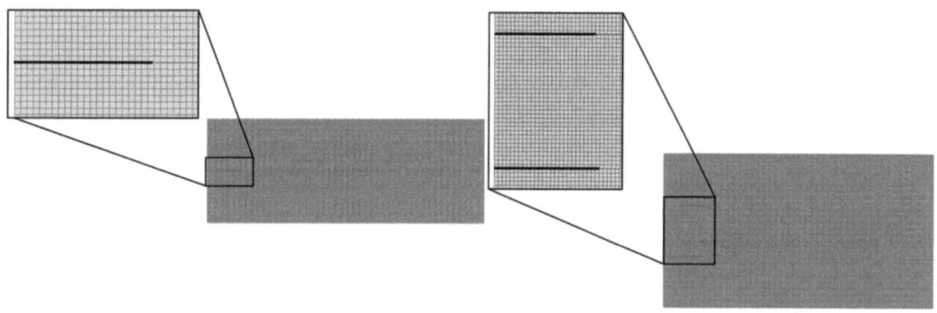

Fig. 15.7 Results for the contact area for the two clamped components (*left*) and three clamped components (*right*)

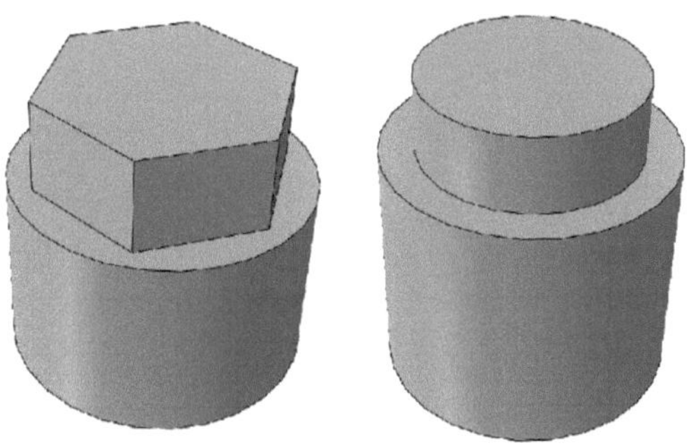

Fig. 15.8 Examples of the bolted joint cylinders, (*left*) a two component bolted connection and (*right*) a three component bolted connection

0.0275 m, respectively, as shown in Fig. 15.7. The connection is modeled as a solid cylinder, using linear tetrahedral solid (C3D4) elements, with the outside radius of the contact area and the length of the distance between the two mid-planes of the outside clamped components; the components are modeled using shell (S4R) elements. The head of the bolt plus extra length is added to ensure that all of the mass of the bolt and nuts are included. The length of the cylinder of the two component joints are 0.03122 and 0.0234 m; and the length for the three component joints are 0.03712 and 0.033695 m, see Fig. 15.8 for the modeled cylinders.

The final property of the bolted joint is stiffness. From tests conducted He and Zhu found that the natural frequencies are more susceptible to the change in the contact area than the material properties [27]; however, the study indicates that when the non-dimensional stiffness is less than 0.3 the least-squares relative error is almost double that when the stiffness is greater than 0.3. The stiffness needs to be calculated to ensure that the error is minimized. Using Rotsher's pressure-cone method and springs in parallel and/or series the stiffness can be calculated [27]. From the simulation to find the contact area it can be seen that there are four pressure cones in series and the spring stiffness of one clamped component is given by:

$$k_m^j = \frac{4\pi}{\sum_{i=1}^{2} \frac{1}{E^j r \tan\alpha_i^j} \ln\left(\frac{(h^j \tan\alpha_i^j + 2b)(2b+4r)}{2b(h^j \tan\alpha_i^j + 2b + 4r)}\right)} \tag{15.8}$$

where k_m^j is the stiffness of the jth component ($j = 1, 2, (3)$), E^j is the Young's modulus of the jth component, α_i^j is the apex angle of the ith ($i = 1, 2$) cone in the jth component, h^j is the thickness of the jth component, b is the radius of the applied pressure minus the radius of the bolt, and r is the radius of the contact area minus the radius of the bolt [27]. The stiffness coefficient for the clamped component is calculated as the equivalent stiffness of the springs in series. The Young's modulus of the cylinder is then calculated using the spring stiffness of an axially loaded rod [29]:

$$E_c = \frac{k_c l_b}{A_c} = \frac{(k_b + k_m)(h^1 + h^2)}{2\pi a^2} \tag{15.9}$$

where k_b is the stiffness of the bolt given by $k_b = A_b E_b / l_b$, where A_b is the cross-sectional area of the bolt, E_b is the Young's modulus, and l_b is the length of the bolt; $h^{1,2}$ is the thickness of the components; and a is the radius of the contact area [27]. To calculate the stiffness the angles of the pressure cones were measured using the angle queery tool in ABAQUS. The angles were then plugged into Eqs. (15.8) and (15.9), and the resulting dimensionless Young's Modulus can be found in Table 15.2.

15.3.2.2 Modeling of the Interference-Pin Bolted Joint Connection

The wind turbine tower contains two bolted joints: regular bolts, described in the previous section, and interference bolts. Interference bolts have a ribbed neck that is slightly larger than holes in the clamped components, seen in Fig. 15.9. The bolt is installed by driving the bolt into the clamped components, which causes the ribbed neck to deform and wear-out the clamped components. The deformation and wear-out cause the bolt and clamped components to have interference with each other, the interference can be seen in Fig. 15.10. The stiffness of the clamped components can be calculated using:

$$E_p = \frac{FL}{A\Delta L} \tag{15.10}$$

Fig. 15.9 Image of an interference pin

Fig. 15.10 Images (*left*) of an interference joint with one component removed, and (*right*) the hole of the removed component

Table 15.3 Contact radius and dimensionless Young's modulus for the various interference-pin bolted joints

Joint number	1	2	3	4
Component thickness (cm)	1.5875 1.746	1.5875 1.27	1.27 1.746 1.27	1.27 0.48 1.27 1.27
Contact radii (cm)	3.27	2.95	2.75 2.81	2.76 2.77 2.8
Dimensionless Young's modulus	0.4522	0.4831	0.1798 0.2846	0.4286 0.2411 0.3355

where E_p is the stiffness of the clamped components, F is the clamping force, L is the distance between the top and bottom of the clamped components before the force is applied, A is the area under compression, and ΔL is the change in the distance between the top and bottom of the clamped components after the force is applied. The stiffness of the joint cylinder includes the stiffness of the clamped components and the stiffness of the bolt, is given as:

$$E_c = \frac{\left(E_p A + E A'\right)}{(A + A')} \tag{15.11}$$

where E_c is the stiffness of the cylinder, E is the material's Young's modulus, and A' is the area of the bolt neck. The interference joint is modeled the same as the regular bolt joint, with a solid cylinder that has the same radius as the contact area in the joint and is the length of the distance between the mid-planes of the clamped components; the clamped components are modeled as shell elements at the mid-plane of the components.

The tower contains eight different interference bolt joints that need to be modeled. There are two different two-clamped components, three three-clamped components, and a four-clamped component, that were calculated. The different joints with the contact radius and dimensionless Young's Modulus are listed in Table 15.3.

15.3.3 *Simulation of L-Beams and Joints*

To verify if the simplification in the previous sections are needed, a 10 m section of the 40 m tower was studied. The section includes forty L-beams of varying lengths, 144 joint Number 1 and 40 Number 8 regular bolts, 20 joint Number 3 regular bolts, 96 joint Number 1 interference-pin bolts, 10 footpads and 5 legs, the section can be seen in Fig. 15.11. The footpads are modeled using tetrahedral solid elements due to the importance of boundary conditions in modal analysis. The legs are modeled using shell elements along the mid-plane of the thickness. The components of the joints are constrained using the "tie" constraint in ABAQUS. These studies were performed due to the volume of the simplified regions being much smaller than the whole structure. The first test simulates if the filleted L-beam can be modeled as just an L-beam. The next simulations are done to verify if the bolts are needed and if all regular bolts could be used instead of the combination.

15.3.3.1 Simulation of Filleted L-Beam Versus Non-filleted L-Beam

To simulate if the fillet L-beam is required, a L-beam with the same dimensions minus the fillet is modeled; the simplified and L-beam can be seen in Figs. 15.5 and 15.12. The first 15 modes were calculated using the Linear Perturbation Frequency step in ABAQUS 6.9EF. The resulting frequencies and percent difference to the original setup and Modal Assurance Criterion (MAC), calculated using FEMtools, number can be found in Tables 15.4 and 15.5, respectively; and the mode shapes of Mode 1, 3, and 9 for the (a) simplified filleted L-beam and (b) L-beam can be seen in Fig. 15.13. From the simulation the frequencies change less than 2 %; however the MAC numbers for five of the 15 modes are below 80, indicating that the mode shapes have changed. The change in the shapes indicate that the non-filleted L-beam cannot replace the simplified filleted L-beam.

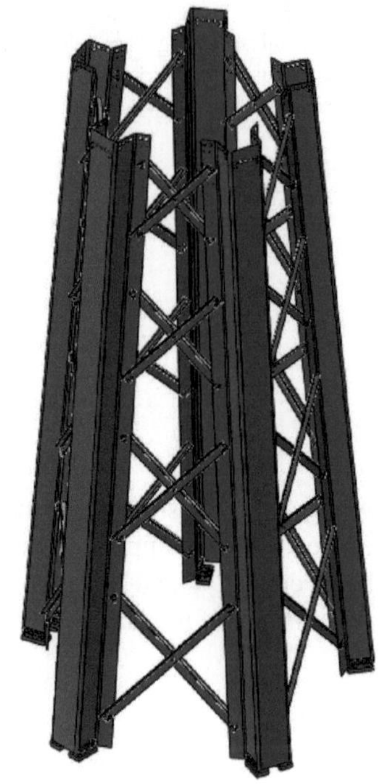

Fig. 15.11 FE model of the section used in the various component simulations

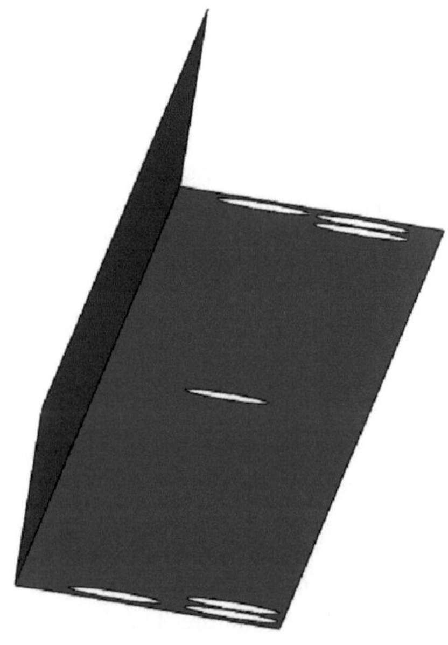

Fig. 15.12 FE model of the non-filleted L-beam

15.3.3.2 Simulation of Actual Tower Lay-Out Bolted Joints Versus No Bolts and All Regular Bolts

After determining that the simplified filleted L-beam is needed the bolts where then tested. All of the bolts were removed from the model and the degrees of freedom at the joint locations were tied using the "tie" constraint in ABAQUS. The first 15 modes were calculated; the resulting frequencies and percent difference to the original setup can be found in Table 15.6 and the MAC in Table 15.7. The mode shapes of Mode 1, 3 and 9 of the original simulation can be seen in Fig. 15.14a and the shapes for the no bolt mode can be seen in Fig. 15.14b. From the simulation there is a 10.45 % frequency change and the MAC number of all but three of modes are below 80. The change in both frequency and mode shape indicates that the bolts are required for an accurate model.

Once the need for bolts was determined, the final simulation is to determine if the interference bolts are needed. The 96 interference-pin joints were replaced with the corresponding regular bolted joint. The resulting frequencies and percent difference to the original setup can be found in Table 15.6 and the MAC in Table 15.8 for the first 15 modes; the mode shapes

Table 15.4 Simulation frequency results of simplified filleted L-beam vs. L-beam

Mode number	Frequency simplified fillet (Hz)	Frequency L-beam (Hz)	Percent difference (%)
1	9.0858	9.1412	0.61
2	9.1094	9.1668	0.63
3	9.2962	9.335	0.42
4	9.3173	9.3606	0.46
5	11.458	11.606	1.29
6	14.961	14.893	−0.45
7	15.025	14.954	−0.47
8	15.652	15.519	−0.85
9	19.288	19.287	−0.01
10	19.398	19.384	−0.07
11	19.53	19.853	1.65
12	19.96	20.291	1.66
13	19.998	20.322	1.62
14	22.545	22.522	−0.10
15	23.265	23.328	0.27

Table 15.5 MAC for the filleted L-beam vs L-beam simulation

	Simplified filleted L-beam															
	Mode	1	2	3	4	5	6	7	8	9	10	11	12	13	14	15
Non-filleted L-beam	1	99.6	0.7	11.2	0.2	0	2.1	0.9	0	0	0	0.1	2.9	0.9	0.1	0.1
	2	0.1	99.3	0.2	2.8	8.1	0.8	2.1	0.1	0.2	2.5	0.1	0.3	0.3	0	0
	3	11.2	0.4	100	0.4	0.1	2.1	0.1	0	0.4	0.2	0	3.5	0.4	0.1	0
	4	0.2	7	0.2	99.8	14.6	0.6	3.8	0.5	0.2	0.1	0	0.1	0.5	0.2	1.3
	5	0	9.5	0.1	14.6	100	0.4	4.2	0.1	0	0.1	1.6	0.2	0.7	0	0.2
	6	1.8	1.2	1.7	0.8	0.8	99.1	0.3	2.4	1	3.6	1.4	1.3	0.4	0	1.6
	7	1.3	1.8	0.2	2.8	3.8	1.7	99.1	1.7	0.1	3.6	0.5	0.5	0.6	0	0.5
	8	0	0.2	0	0.5	0	3.1	1.6	99.9	5.1	0.3	0.9	0.3	0.2	1.7	0.2
	9	0.4	0.8	0.7	0.2	1.3	0.3	1.6	0	41.1	21.8	24.8	0	9	0	1.7
	10	0.1	0.7	0.1	0.2	0.3	0	3.1	6.3	43.6	27.3	12.5	3.9	2	0.2	0.3
	11	0.1	3	0.9	0	0.7	2.9	0.6	0.4	5.8	53.2	49.6	5.6	0.1	0	0.1
	12	1.6	0.1	0.8	0.3	0.5	1	0.6	1.6	1.9	1.7	2	12.6	88.3	0.1	1.6
	13	1.4	0	1.3	0	0.2	3.2	2.5	0.1	1.8	37.8	1.9	71.2	2.4	0.1	0
	14	0.1	0	0.1	0.2	0	0	0	2	0	0.1	0.2	0.1	0	99.8	1.6
	15	0.6	0.1	3	0.4	0.1	0.6	0.2	0	0.1	0.2	0.5	0	0.1	0.6	31.4

of Mode 1, 3, and 9 can be seen in Fig. 15.14c. The maximum frequency shift is 0.36 % and the MAC number of two of the modes are 80 and the order of two of the modes changed. The change in the first two mode shapes is great indicating that the combination of regular bolts and interference-pin joints is required.

15.3.4 Assembling the Tower

Once the proper simplified models are determined, the tower is assembled, shown in Fig. 15.15. As can be seen there is a large ring at the top of the tower, the ring is modeled using linear tetrahedral solid (C3D4) elements; modeled the ring as a solid element is so the boundary condition is modeled with greater accuracy. The tower has a number of aluminum ladders and a Mylar wrapping that span the whole height of the tower, these elements do not provide any form of structural stiffness due to their location and lower stiffness. These components are modeled using ABAQUS's Engineering Feature "Nonstructural Mass" at the appropriate surfaces and regions. The use of "Nonstructural Mass" alleviates the much larger vibration amplitudes of the components, which tend to "hide" the elastic shapes of the tower model. Figure 15.2 indicates that there are four platforms spaced throughout the tower; the platforms are modeled using shell elements and are constrained

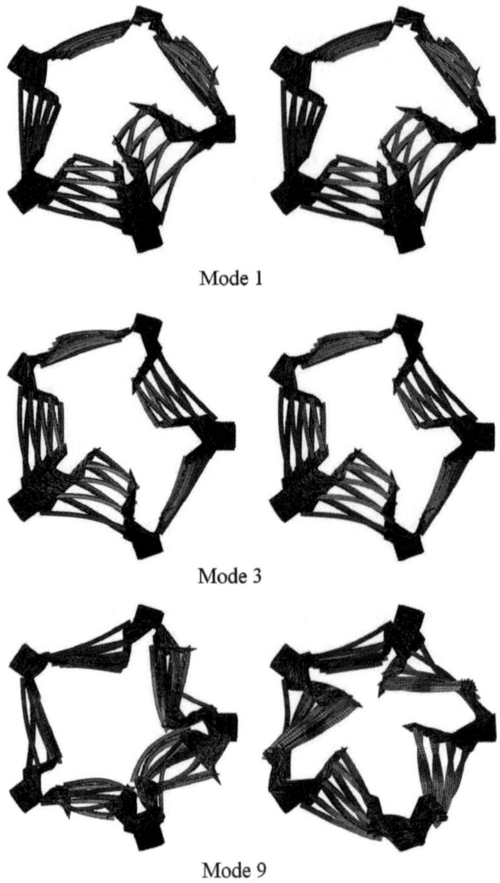

Fig. 15.13 Mode shapes of Mode 1, 3 and 9 for the (*left*) simplified filleted L-beam and (*right*) L-beam

Table 15.6 Frequency results of the bolted joint simulations

Mode number	Original bolt model frequency (Hz)	No bolt model frequency (Hz)	Percent difference to original model (%)	All regular bolt model frequency (Hz)	Percent difference to original model (%)
1	9.0858	9.6775	6.51	9.0999	0.16
2	9.1094	9.691	6.38	9.1103	0.01
3	9.2962	9.9248	6.76	9.2846	−0.12
4	9.3173	9.986	7.18	9.3191	0.02
5	11.458	11.978	4.54	11.464	0.05
6	14.961	15.933	6.50	14.995	0.23
7	15.025	16.052	6.84	15.057	0.21
8	15.652	17.287	10.45	15.624	−0.18
9	19.288	19.598	1.61	19.295	0.04
10	19.398	20.282	4.56	19.400	0.01
11	19.530	20.356	4.23	19.506	−0.12
12	19.960	21.341	6.92	19.978	0.09
13	19.998	21.611	8.07	19.999	0.01
14	22.545	23.87	5.88	22.546	0.00
15	23.265	24.173	3.90	23.348	0.36

to the legs at the appropriate locations. The pretesting of the foundation in Sect. 15.2 indicates that the boundary condition on footpads can be modeled as a fixed boundary. The first seven modes were calculated and the frequencies are compared to the experimental results in Table 15.9; the mode shapes of Mode 1, 3, 4, and 6 can be seen in Fig. 15.16. The frequency difference is large, over 11 %, for Modes 1 and 2, while the rest are within ±5.20 %; this error indicates that there could be error in the simplification or the boundary condition.

Table 15.7 MAC for the no bolt model vs original model

	Original bold model															
	Mode	1	2	3	4	5	6	7	8	9	10	11	12	13	14	15
No bold model	2	44.9	52.6	0.2	0	0	0	0.1	0.1	0	0.5	2.8	0.2	0.1	0.1	0.5
	2	54.8	46	0	0.2	0.3	0	0	0	0.1	4.9	0.7	0	0	0	0.1
	3	0.1	0	70.7	27.7	0	0	0.2	0	0	0	1.9	0	0	0	0.2
	4	0.8	1.1	27.8	71.7	0	0.4	0.6	0	0	0.5	0.1	0	0	0	0
	5	0.1	0	0	0	99.2	0	0.1	0.1	6.2	0.7	0.1	0	0.1	0.1	0
	6	0	0.1	0.4	0	0.2	55.6	42.7	0	0	0	0	0	0	0.1	0
	7	0	0.3	0	1.5	0	42.8	56.8	0	0	0	0	0.1	0	0.1	0
	8	0.1	0.1	0.1	0	1.5	0.1	0	99.1	0	0	0	0.1	0	3.7	0
	9	0.5	0	0.2	0	2.5	0.1	0	0	57.1	9.4	9.5	14	4.1	0.6	0
	10	0.4	0.7	0.2	0.1	0.5	0.3	0	0.1	16.9	33.3	30	42.5	3.8	0.5	1.6
	11	1.6	0	0.1	0	2	0	0	0.1	23.8	45.6	0.2	0.1	27.7	0.7	0.4
	12	0	3.9	0.6	0.5	0.3	0.1	0	0	0.3	4.3	50	41.7	4.2	0	1.3
	13	2.7	0	0.6	0	0.5	0	0	0	0.4	33.2	2.9	4.3	58.4	0.1	0.3
	14	0.1	0	0	0	0.3	0.1	0	4.1	0.4	0	0	0.3	0	91.1	0.5
	15	0.3	0	0	0	0	0	0	0	0.1	3.4	0.6	0	0.1	2.7	3.9

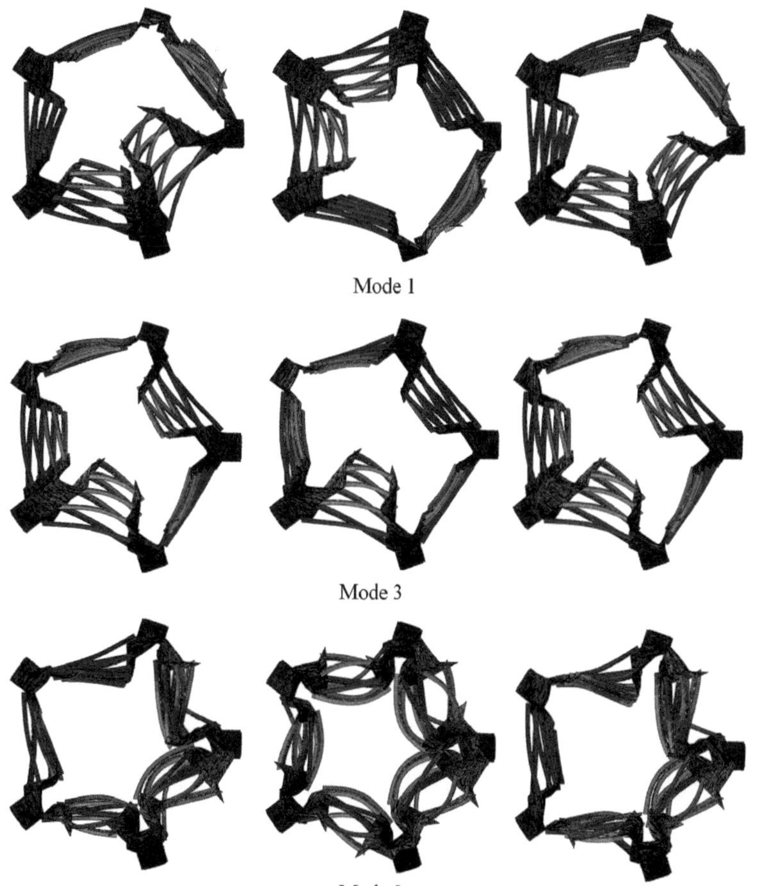

Fig. 15.14 Mode shapes of Mode 1, 3, and 9 for the *(left to right)* original model, no-bolt model, and all regular bolt model

15.3.5 Modeling of the Ground Surrounding the Tower

Since the simplifications have been proven on complex structures the boundary is tested. In modal analysis frequencies are dependent on the mass and stiffness of the structure and boundaries. Modeling the boundary as fixed or flexible will result in different dynamic responses [28]. The tower is located in an area that the soil is largely "hard" clay which has a much lower stiffness than that of the tower; this results in the boundary being flexible. Many researchers have studied the effects of soil

Table 15.8 Mac for all regular bolt model vs original model

	Original bold model															
	Mode	1	2	3	4	5	6	7	8	9	10	11	12	13	14	15
All regular bold model	1	80.7	18.7	0.6	0	0	0.2	0.1	0	0.2	0.2	0.1	0.9	0.2	0	0.1
	2	18.6	80.5	0	1.3	0	0.1	0	0	0.6	0.4	0.2	0.2	1.1	0	0.1
	3	0.3	0	99.5	0.1	0	0.5	0.2	0	0.1	0.3	0	0.2	0.1	0	0.1
	4	0.1	0.5	0.1	99	0	0.4	0.8	0	0	0	0	0	0	0	0
	5	0	0	0	0	100	0	0	0.1	0.8	0	0.8	0.1	0.1	0	0
	6	0.3	0	0.8	0.2	0	98	1.7	0.2	0	0	0	0	0	0	0
	7	0	0.1	0.1	1.2	0	2.1	97.7	0.5	0	0	0	0	0	0.1	0
	8	0	0	0.1	0	0.1	0	0.2	99.6	0.1	0	0	0	0	1.3	0
	9	0.1	0.5	0.1	0	0.9	0	0	0.1	90.8	7.9	1.6	0	0	0	0.2
	10	0.6	0.2	0.2	0	0.2	0	0	0	4.8	91.3	3	0.1	0	0	0
	11	0	0.7	0	0.1	0.6	0	0	0.1	2.6	1.9	94	1.1	0	0	0.2
	12	1	0.2	0.2	0.1	0.1	0	0	0	0.1	0.8	0.1	6.6	92.8	0	0
	13	0.2	0.6	0.1	0	0.1	0	0	0	1.1	0.2	2.8	90.8	7.3	0	0.3
	14	0	0	0	0	0	0	0	1.4	0.1	0	0	0	0	99.9	0.1
	15	0.2	0.1	0.1	0	0	0	0	0	0.1	0.2	0	0.3	0.4	0.1	57

Fig. 15.15 FE model of the assembled tower

as flexible boundaries by modeling it as sets of springs [17, 29]; however this simplification causes errors in the structure at resonance [17]. Modeling the soil as a continuum using linear elastic elements could be used to avoid the source of the resonance error [17, 29]. Due to soil having a porous structure the mass and stiffness must take into account, the mass of the soil and water, the stiffness of the soil skeletal structure, as well as the forces associated with the fluid; these parameters result in a complex, damped, coupled equations of motion [30]. The soil substructure is assumed to be "Drained, Saturated Porous without Phase Change" which assumes that the liquid pressure does not change with time resulting in linear equations of motion.

The boundary is modeled with the concrete pad the tower is built on and the soil surrounding the pad; the model containing the solid continuum can be seen in Fig. 15.17. The soil is modeled using a 200 m cubic continuum with fixed boundaries;

Table 15.9 Frequency comparison of the FE model to the experimental results

Mode	FE results (Hz)	Experimental results (Hz)	Percent difference (%)
1	2.3042	2.006	14.87
2	2.3049	2.072	11.24
3	6.6893	7.056	−5.20
4	7.5424	7.531	0.15
5	7.5507	7.629	−1.03
6	11.558	11.981	−3.53
7	11.973	11.984	−0.09

Fig. 15.16 Tower model mode shapes for (*left to right*) Mode 1, 3, 4, and 6

Table 15.10 Frequency comparison of FE model with soil to experimental results

Mode	FE results (Hz)	Experimental results (Hz)	Percent difference (%)
1	2.0019	2.006	−0.20
2	2.0720	2.072	−0.97
3	6.5672	7.056	−6.93
4	7.6403	7.531	1.45
5	7.6562	7.629	0.36
6	11.843	11.981	−1.15
7	11.862	11.984	−1.02

this size is selected to reduce the effects of the fixed boundary condition. Prowell et al. used a 400 m square by 200 m deep continuum to avoid boundary conditions [17]. The model of the wind turbine tower has a large number of elements which does not allow for a large continuum, which is in agreement with Králik and Králik [31]. Using the assumption from Anandaraja [30], the concrete pad and soil are assumed to be linear with the material properties of; Young's Modulus of 24.67835 GPa, density of 2,827.55 kg/m^3, and Poisson's ratio of 0.2 for concrete; and Young's Modulus of 130 MPa, density of 1,714 kg/m^3, and Poisson's ratio of 0.4 for hard clay. The first seven modes were calculated and the frequencies are compared to the experimental values in Table 15.10; Mode 1, 3, 4, and 6 mode shapes can be seen in Fig. 15.18. The frequency difference for Mode 3 is still large at −6.93 %, while the rest are within ±1.5 %.

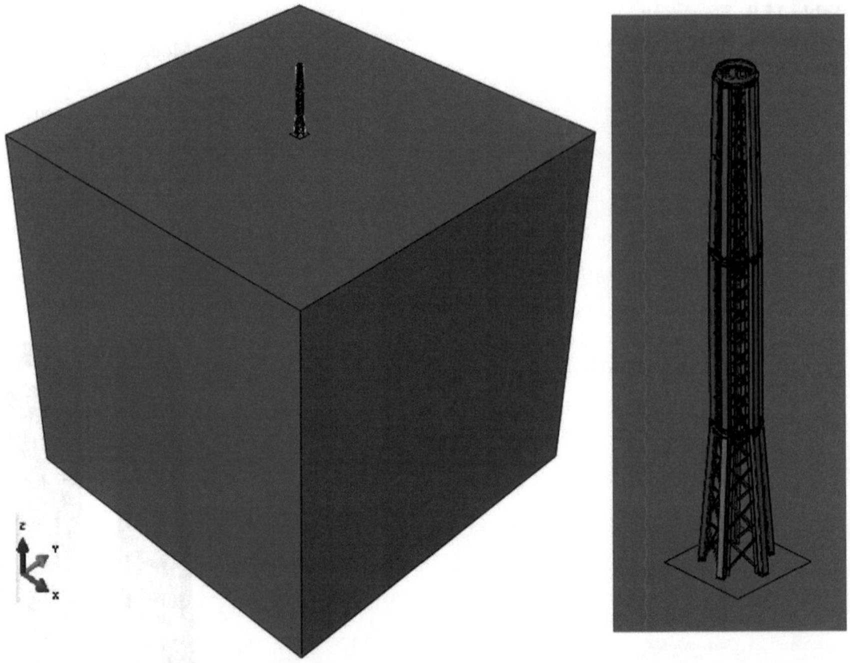

Fig. 15.17 FE model of the tower with the concrete pad and soil; (*left*) whole model, (*right*) zoomed in view of tower and pad

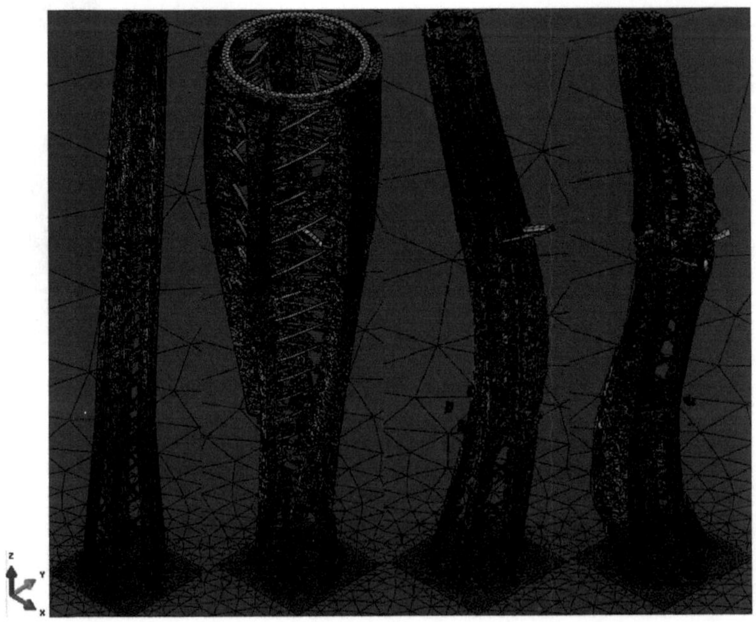

Fig. 15.18 Tower model mode shapes for (*left to right*) Mode 1, 3, 4, and 6

15.4 Conclusion

With the increasing wind turbine height, accurate FE models are needed to test the viabilities of designs. A technique to model a space frame tower was outlined using a combination of solid, shell and beam elements; and simplifies non-structural components into non-structural mass applied to the model. The simplification techniques of the various components were shown to have great influence on the response of the tower and are required for an accurate model. The experimental evaluation of the tower indicated that the foundation could be modeled as a fixed boundary; however, during the modeling of the tower it was noticed that the foundation and the surrounding soil needs to be modeled. The results of the updated model is accurate for the seven modes measured during the experiment.

Acknowledgements This research was funded by the NSF Grant No. CMMI-1335024. The authors would also like to thank Kai Wu and Chao Hu for helping with the experimental testing of the structure; as well as thank Rong Cong for assistance in modeling of the tower.

References

1. Lewin TJ (2010) An investigation of design alternatives for 328-ft (100-m) tall wind turbine towers. Graduate Theses and Dissertations, Iowa State University
2. U.S. Department of Energy (2008) 20% Wind power by 2030. http://www.20percentwind.org
3. Gencturk B, Attar A, Tort C (2012) Optimal design of lattice wind turbine towers. In: Proceedings of the 15th world conference on earthquake engineering, Lisbon, Portugal
4. Dvorak P (2010) GE introduces 2.75 MW wind turbines. Windpower Eng Dev. http://www.windpowerengineering.com, visited 2014
5. Siemens wind turbine with ATS hybrid tower completed. Sun Wind Energy, 2009. http://www.sunwindenergy.com, visited 2014
6. Fortis 10 kW wind turbine college installation. Fortis Wind Energy, 2008. http://www.fortiswindenergy.com, visited 2014
7. Zhu WD, Zheng NA, Wong CN (2007) A stochastic model for the random impact series method in modal testing. ASME J Vib Acoust 129:265–275
8. Ahmed KIE, Rajapakse RKND, Gadala MS (2009) Influence of bolted-joint slippage on the response of transmission towers subjected to frost-heave. Adv Struct Eng 12:1
9. Gani F, Légeron F (2010) Dynamic resonse of transmission lines guyed towers under wind loading. Can J Civ Eng 37:3
10. Saudi G (2014) Structural assessment of a guyed mast through measurement of natural frequencies. Eng Struct 59:1
11. Xie L, Tang J, Tang H, Xie Q, Xue S (2012) Seismic fragility assesment of transmission towers via performance-based analysis. In: Proceedings of the 15th world conference on earthquake engineering, Lisbon, Portugal
12. Mohammed AF, Özakça M, Tayşi N (2012) Optimal design of transmission towers using genetic algorithm. SDU Int J Technol Sci 4: 2
13. Rao NP, Knight GMS, Seetharaman S, Lakshmana N, Iyer NR (2011) Failure analysis of transmission line towers. J Perform Constr Facil 25: 3
14. Sato Y, Ishikawa T (2012) Earthquake resistance of transmission steel towers. In: Proceedings of the 15th world conference on earthquake engineering, Lisbon, Portugal
15. Zimcik DG, Ensan MN, Jenq ST, Chao MC (2004) Finite element analysis simulation of airport approach lighting towers. J Struct Eng 130:5
16. Prowell I, Veletzos M, Elgamal A, Restrepo J (2009) Experimental and numerical seismic response of a 65 kW wind turbine. J Earthq Eng 13:8
17. Prowell I, Elgamal A, Lu J (2010) Modeling the influence of soil structure interaction on the seismic response of a 5MW wind turbine. In: Proceedings of the 5th international conference on recent advances in geotechnical earthquake engineering and soil dynamics and symposium in honor of Professor I.M. Idriss, San Diego, CA, 24–29 May 2010
18. Saravia CM, Machado SP, Cortínez VH (2013) A composite beam finite element for multibody dynamics: application to large wind turbine modeling. Eng Struct 56: 1164–1176
19. Olariu CP (2013) Soil-structure interation in case of a wind turbine. Bull Polytechnic Inst Iasi Construction Architecture Section 63:1
20. Graczykowski C, holnicki-Szulc J (2009) Protecting offshore wind turbines against ship impacts by means of adaptive inflatable structures. Shock Vib 16:4
21. Nuta E, Christopoulos C, Packer JA (2011) Methodology for seismic risk assessment for tubular steel wind turbine towers: application to Canadian seismic environment. Can J Civ Eng 38:3
22. Wang Z, Zhao Y, Li F, Jiang J (2013) Extreme dynamic responses of MW-level wind trubine tower in the strong typhoon considering wind-rain loads. Math Prob Eng 1–13, 2013
23. Al Satari M, Hussain S (2008) Vibration based wind turbine tower foundation design utilizing soil-foundation- structure interaction. In: AIP conference proceedings
24. Ghaemmaghami A, Kianoush R, Yuan XX (2013) Numerical modeling of dynamic behaviour of annular tuned liquid dampers for applications in wind turbines. Comput Aided Civ Infrastruct Eng 28:38–51
25. Fei QG, Xu YL, Ng CL (2007) Structural health monitoring oriented finite element model of tsing Ma bridge tower. Int J Struct Stability Dyn 7:4
26. He K, Zhu WD (2009) Modeling of fillets in thin-walled beams using shell/ plate and beam finite elements. ASME J Vib Acoust 131:5, 051002-1–051002-16
27. He K, Zhu WD (2011) Finite element modeling of structures with L-shaped beams and bolted joints. ASME J Vib Acoust 133:1, 011011-1–011011-12
28. Dai KS, Chen SE, Qi WY, Conner E, Erdle J, Galloway C (2011) Field testing of directly embedded poles. Exp Tech 35:2, 14–23
29. Paullo MLF, Goncalves PB, Silveira RAM, Silva ARD (2013) Nonlinear dynamics analysis of frame structures under seismic excitation considering soil-structure interaction and elasto-plastic soil behavior. In: Proceedings of the 11th international conference on vibration problem, Lisbon, Portugal, 9–12 Sept 2013
30. Anandarajah A (2010) Computational methods in elasticity and plasticity: solids and porous media. Springer, New York
31. Králik J, Králik J Jr (2013) Sensitivity analysis of soil-foundation-machine interaction on layered soil. In: Proceedings of the 11th international conference on vibration problem, Lisbon, Portugal, 9–12 Sept 2013

Chapter 16
Experimental Validation of Modal Parameters in Rotating Machinery

Bram Vervisch, Kurt Stockman, and Mia Loccufier

Abstract In this paper, instability of rotating machinery systems due to rotating damping is investigated through experimental modal analysis. Generally, advanced methods are needed for these kind of systems because of the asymmetry of the matrices. However, it is shown by means of cyclic energy dissipation that the rotating damping can be handled as damping which changes due to the rotating speed. The use of damping matrix estimation techniques is discussed and a method is proposed to estimate and model the influence of the rotating damping matrix. A dedicated experimental setup, with negligible gyroscopic effect, is presented for validation purposes. It is shown experimentally that the estimation of the rotating damping matrix is able to predict the decay rate of the first forward mode.

Keywords Rotating machinery • Rotating damping • Experimental modal analysis • Damping matrix estimation • Decay rate

16.1 Introduction

High speed applications are trending in industry. These applications result in smaller component dimensions and lead to increased production. As a consequence, rotating speeds above the first resonant frequency are more often encountered. A major drawback of this supercritical rotation is instability. At a certain speed, the stability threshold is reached. This stability threshold speed is defined by the ratio between the damping in the nonrotating parts and the damping in the rotating parts [1]. The latter causes, together with the gyroscopic effect, skew-symmetric system matrices and speed dependency. Because of this asymmetry, classical modal analysis techniques are insufficient to describe these systems. Theoretically, decoupling can still be performed by solving the left and right eigenvalue problem [2]. In practice this implies that left and right eigenvectors have to be estimated from measurements. This means that both a row and a column frequency response function matrix have to be measured [3]. In practical applications, generally, not all measurement points can be accessed so an alternative has to be found.

The stability threshold speed of rotating systems is highly dependent on the damping, and damping is very difficult to model. Moreover, the calculation of the stability threshold speed depends on the chosen damping model [4]. Therefore, experimental validation is important. Ideally, an accurate model and some well chosen measurements in combination with a model updating technique can provide a new model that predicts the stability threshold speed. However, model updating proves to be very sufficient for mass and stiffness, but when damping is involved the complexity increases.

In this paper an experimental setup is presented, dedicated to analyze the stability problem and the influence of the rotating damping on the stability threshold speed. A linear speed dependent model is derived to perform the stability analysis. Experimental modal analysis is performed at 0 and 3,000 rpm. It is shown that by treating the system as nonrotating at both speeds still leads to an estimation of the rotating damping effect. As such, insight is gained in the physical phenomenon of rotating damping and its influence on the real part of the poles or the decay rate. In Sect. 16.2, experimental modal analysis of

B. Vervisch (✉)
Department of Industrial Systems and Product Design, Ghent University, Graaf Karel de Goedelaan 5, 8500 Kortrijk, Belgium

Department of Electrical Engineering, Systems and Automation, Ghent University, Technologiepark 914, 9000 Ghent, Belgium
e-mail: bramf.vervisch@ugent.be

K. Stockman
Department of Industrial Systems and Product Design, Ghent University, Graaf Karel de Goedelaan 5, 8500 Kortrijk, Belgium

M. Loccufier
Department of Electrical Engineering, Systems and Automation, Ghent University, Technologiepark 914, 9000 Ghent, Belgium

© The Society for Experimental Mechanics, Inc. 2015

R. Allemang (ed.), *Special Topics in Structural Dynamics, Volume 6*, Conference Proceedings of the Society for Experimental Mechanics Series, DOI 10.1007/978-3-319-15048-2_16

rotating systems is discussed and it is shown that because of the cyclic energy dissipation, the rotating damping can be treated as regular damping that is only speed dependent. In Sect. 16.3, the estimation of the damping matrix is discussed. In Sect. 16.4 the rotating damping setup is presented and a method to extract the rotating damping effect is proposed. Section 16.5 provides an experimental validation.

16.2 Experimental Modal Analysis of Rotating Systems

16.2.1 The Linear Speed Dependent Model

The model used in this paper includes speed dependency by means of rotating damping. The gyroscopic effect is neglected for a rotor symmetric in x and y direction the equations of motion can be written as

$$\begin{bmatrix} \mathbf{M} & \mathbf{0} \\ \mathbf{0} & \mathbf{M} \end{bmatrix} \begin{Bmatrix} \ddot{\mathbf{x}} \\ \ddot{\mathbf{y}} \end{Bmatrix} + \begin{bmatrix} \mathbf{C}_n + \mathbf{C}_r & \mathbf{0} \\ \mathbf{0} & \mathbf{C}_n + \mathbf{C}_r \end{bmatrix} \begin{Bmatrix} \dot{\mathbf{x}} \\ \dot{\mathbf{y}} \end{Bmatrix} + \left(\begin{bmatrix} \mathbf{K} & \mathbf{0} \\ \mathbf{0} & \mathbf{K} \end{bmatrix} + \Omega \begin{bmatrix} \mathbf{0} & \mathbf{C}_r \\ -\mathbf{C}_r & \mathbf{0} \end{bmatrix} \right) \begin{Bmatrix} \mathbf{x} \\ \mathbf{y} \end{Bmatrix} = \begin{Bmatrix} \mathbf{F}_x \\ \mathbf{F}_y \end{Bmatrix} \tag{16.1}$$

where $\mathbf{M}$ and $\mathbf{K}$ are the regular mass and stiffness matrices. $\mathbf{C}_n$ is the nonrotating damping or the damping that only influences the nonrotating frame. The matrix $\mathbf{C}_r$ includes the rotating damping or the damping that influences the rotating frame. $\mathbf{C}_r$ depends on the rotating speed Ω. The vector $\{\mathbf{x}\ \ \mathbf{y}\}^T$ represents the generalized coordinates in x- and y direction. As can be seen, the model not only deviates from a classical mass spring damper system by the speed dependency but both gyroscopic effect and rotating damping provide a coupling between x and y. Moreover, this coupling is skew-symmetric, which yields a difficulty in the use of classical modal analysis where symmetric matrices are assumed.

16.2.2 Cyclic Energy Dissipation and Stability

The appearance of skew-symmetric matrices have a particular effect on the energy dissipation and stability of the system as explained by Adams in [5] and [6]. Consider the equations of motion for a system with only two degrees of freedom

$$\begin{bmatrix} m & 0 \\ 0 & m \end{bmatrix} \begin{Bmatrix} \ddot{x} \\ \ddot{y} \end{Bmatrix} + \left(\begin{bmatrix} c_n + c_r & 0 \\ 0 & c_n + c_r \end{bmatrix} + \Omega \begin{bmatrix} 0 & g \\ -g & 0 \end{bmatrix} \right) \begin{Bmatrix} \dot{x} \\ \dot{y} \end{Bmatrix} + \left(\begin{bmatrix} k & 0 \\ 0 & k \end{bmatrix} + \Omega \begin{bmatrix} 0 & c_r \\ -c_r & 0 \end{bmatrix} \right) \begin{Bmatrix} x \\ y \end{Bmatrix} = \begin{Bmatrix} F_x \\ F_y \end{Bmatrix} \tag{16.2}$$

The work done by the displacement term or apparent stiffness term of the rotating damping can be expressed as

$$dw = -\Omega \begin{bmatrix} 0 & c_r \\ -c_r & 0 \end{bmatrix} \begin{Bmatrix} x \\ y \end{Bmatrix} \{dx\ dy\} = \Omega\left(-c_r y dx + c_r x dy\right) \equiv f_x dx + f_y dy$$

$$\therefore \frac{\partial f_x}{\partial y} = -\Omega c_r \quad and \quad \frac{\partial f_x}{\partial x} = \Omega c_r \tag{16.3}$$

Because $\partial f_x/\partial y \neq \partial f_y/\partial x$, dw is no exact differential, leading to a nonconservative force field. This means that the rotating damping appearing as a stiffness term actually dissipates energy. Together with the regular damping, the sum of the nonconservative forces of the rotating system can be written as

$$\begin{Bmatrix} P_x \\ P_y \end{Bmatrix} = -\begin{bmatrix} c_n + c_r & 0 \\ 0 & c_n + c_r \end{bmatrix} \begin{Bmatrix} \dot{x} \\ \dot{y} \end{Bmatrix} - \Omega \begin{bmatrix} 0 & c_r \\ -c_r 0 & \end{bmatrix} \begin{Bmatrix} x \\ y \end{Bmatrix} \tag{16.4}$$

Suppose that the rotor is operated at a certain speed Ω and whirling at ω. The whirling can be expressed as a harmonic motion in x and y direction, $x = Xsin(\omega t + \phi_x)$ and $y = Ysin(\omega t + \phi_y)$. The energy per cycle that is exchanged due to these forces can be expressed as

$$E_{cyc} = \oint P_x dx + P_y dy = \int_0^{2\pi/\omega} \left(P_x \dot{x} dt + P_y \dot{y} dt\right) \tag{16.5}$$

or

$$E_{cyc} = -\pi \left[\underbrace{\omega\left(c_n + c_r\right)\left(X^2 + Y^2\right)}_{\text{regular damping effect}} - \underbrace{2\Omega XYsin\left(\phi_x - \phi_y\right)}_{\text{speed dependent effect}} \right] \tag{16.6}$$

This is an important expression because it shows that the appearance of rotating damping can lead to a positive energy per cycle or instability. The first term is a regular damping effect and causes a cyclic decrease of the energy. The second term is not only speed dependent, but also depends on the phase difference between ϕ_x and ϕ_y. Whenever the whirling ω occurs in the same direction as the rotation Ω, which is called a forward mode, $sin(\phi_x - \phi_y) > 0$. As Ω increases the influence of the second term in (16.6) increases. When the energy per cycle becomes positive, instability occurs. For backward modes, $sin(\phi_x - \phi_y) < 0$, and the effect of the second term in (16.6) is always stabilizing.

16.2.3 Modeling of Rotating Damping

Although the gyroscopic effect plays an important role in rotating machinery, it is omitted in this paper since the focus lies on the rotating damping. When there is no gyroscopic effect, (16.1) can be written as

$$\mathbf{M}\ddot{\mathbf{q}} + \left(\mathbf{C}_n + \mathbf{C}_{rs}\right)\dot{\mathbf{q}} + \left(\mathbf{K} + \Omega\mathbf{C}_{rss}\right)\mathbf{q} = \mathbf{F} \tag{16.7}$$

where $\{\mathbf{x}\ \ \mathbf{y}\}^T$ is combined in $\mathbf{q}$ and therefore a distinction is made between the symmetric part of the rotating damping $\mathbf{C}_{rs}$ and the skew-symmetric part $\mathbf{C}_{rss}$. The appearance of the rotating damping as an apparent stiffness term has a relevant physical meaning [7] but leads to experimental difficulties. In order to decouple this system both left and right eigenvalue problem have to be solved. In a theoretical model this is possible, but in practice it implies that left and right eigenvectors have to be estimated from measurements. This is experimentally not always possible, thus this approach is difficult to apply. By slightly deviating from the exact physical model, an experimentally more interesting model is proposed

$$\mathbf{M}\ddot{\mathbf{q}} + \left(\mathbf{C} + \Omega\mathbf{C}_{re}\right)\dot{\mathbf{q}} + \mathbf{K}\mathbf{q} = \mathbf{F} \tag{16.8}$$

The physical relevance of this equation lies in (16.6) and the fact that all forms of energy dissipation are modeled in the damping matrix. The damping at standstill is combined in $\mathbf{C}$ and the difference in damping due to rotation is combined in $\mathbf{C}_{re}$. If both $\mathbf{C}$ and $\mathbf{C}_{re}$ are estimated from measurements, the effect of the rotating speed on the damping can be predicted.

16.3 Estimation of the Damping Matrix

16.3.1 Damping Matrix Estimation Methods

Identification of damping from measurements is an important topic in literature. Especially when it involves the identification of the damping matrix. Damping matrix identification methods are classified in matrix methods and modal methods [8]. Matrix methods are normally based upon the measurement of the whole frequency response function matrix and modal methods use modal parameters. The latter is preferred in this research because only one row of the frequency response function matrix is needed which highly limits the number of measurements. One of the modal methods is developed by Adhikari in a series of papers. Only the viscous damping identification method is used here [9]. The main advantage of this method is that it allows asymmetry of the matrices and that the reconstructed physical matrix has acceptable spatial distribution. It can also deal with errors due to modal truncation and noise.

16.3.2 Limitations of the Damping Estimation Method

16.3.2.1 Mode Shape Errors

Modal methods rely on an exact knowledge of the mode shapes and are extremely sensitive to errors [10–12]. However, several attempts have been made to optimize the damping matrix even with erroneous mode shapes. Pilkey [13] uses an iteration technique combined with Lancasters method to optimize the damping matrix. The main drawback is that the mass matrix has to be known. When symmetry of the damping matrix is a preliminary assumption, Adhikari proposes a symmetry preserving method, which forces experimental data to yield a symmetric matrix [14]. Balmès [15] uses Lancasters reconstruction method to define the orthogonality property between the eigenvectors, known as the properness condition. Subsequently an optimization procedure is used to find the eigenvectors that meet this properness condition.

The scaling of the mode shapes in Adhikari's method is unity modal mass scaling or

$$\mathbf{X}^{*}\mathbf{M}\mathbf{X} = \mathbf{I} \tag{16.9}$$

with $\mathbf{X}$ the matrix containing the mode shapes or eigenvectors, $\mathbf{I}$ is the identity matrix and $()^{*}$ denotes the conjugate transpose. So when the mass matrix is known, the mode shapes can be rescaled to unity modal mass yielding a better estimation. With finite elements it is possible to estimate mass matrices quite accurately. In order to improve the mass matrix even more, model updating techniques can be used.

16.3.2.2 Incomplete Data and Modal Truncation

When performing experimental modal analysis, the measured data is always incomplete. Due to the limited bandwidth of both excitation and response, modal truncation occurs. The poles and mode shapes that result from a measurement are limited by this bandwidth. Also, not all degrees of freedom that are assumed in the model can be measured, especially rotational degrees of freedom. Truncated poles and mode shapes can still be used in Adhikari's method but the limited degrees of freedom limit the dimension of the resulting damping matrix.

16.4 Extracting the Rotating Damping Effect from Measurements

16.4.1 The Rotating Damping Setup

The rotating damping setup is described in [16]. The purpose of the setup is a shaft that has clamped boundary conditions on both sides. The setup is shown schematically in Fig. 16.1. Five measurement locations are chosen equally divided along the shaft. The motion inside the bearings is difficult to measure so this is avoided. These five measurement locations are sufficient to identify the first five poles and mode shapes.

The shaft can be operated at a speed up to 6,160 rpm. For safety purposes, the range is kept between 0 and 4,500 rpm. As excitation, an impact hammer is used. Because impact testing requires an experienced operator, an automated impact is designed (Fig. 16.2). The responses are measured with eddy current probes. These probes typically generate a noisy signal and therefore every measurement is an average in frequency domain of ten impacts. The frequency response functions are estimated with an H1 estimator [17].

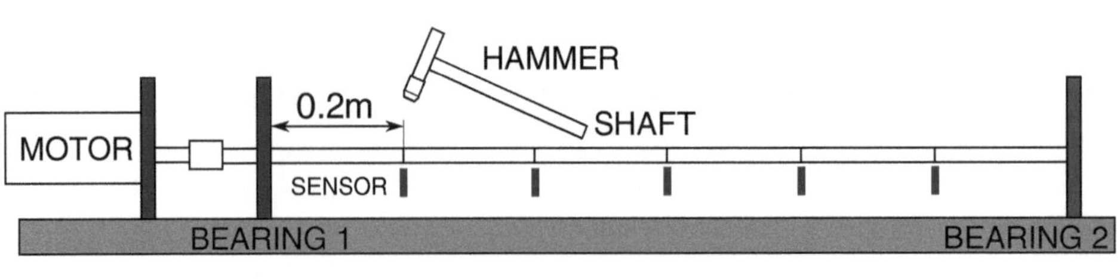

Fig. 16.1 The rotating damping setup is the equivalent of a clamped beam on both sides from which five measurement points are taken

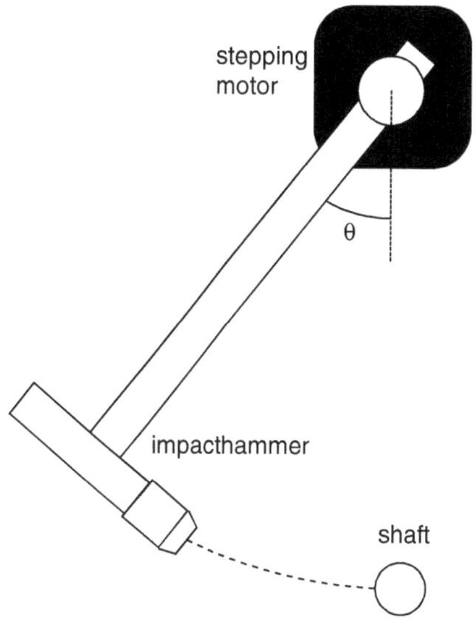

Fig. 16.2 The impact hammer is automated by means of a stepping motor that is controlling the hit in order to improve the reproducibility

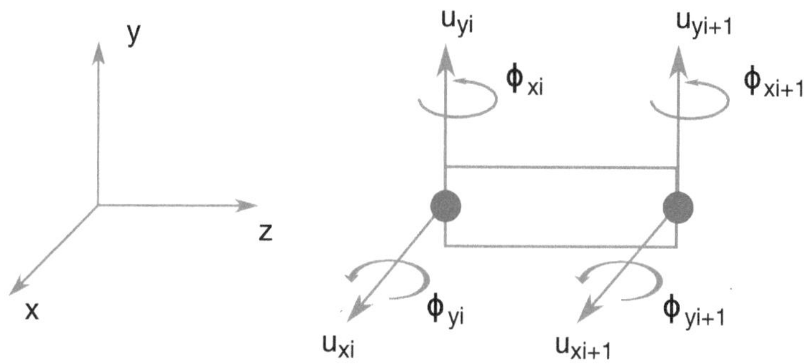

Fig. 16.3 Each element of the finite element model has eight degrees of freedom

16.4.2 Nonrotating Shaft

16.4.2.1 Modeling, Updating and Reducing of the Model

When the shaft is not rotating, there are no speed dependent effects, so a classical mass, spring damper system is assumed. The shaft is modeled with finite elements [18]. Six elements used, with each eight degrees of freedom (Fig. 16.3), so the resulting model has a dimension of 28×28. Only five frequency response functions are measured. The model updating method applied here [19] is based upon the frequency response functions and can deal with incomplete data by using an iterative procedure. The mass and the stiffness matrix of the model are updated and the dimension remains 28×28. Consequently, the (purely imaginary) poles and the corresponding (real) mode shapes of this undamped model are calculated. Because there are only five frequency response functions in the measurement, it is only possible to derive a damping matrix with dimension 5×5. Therefore, the modeled mass and stiffness matrices are reduced to a dimension of 5×5. This is done by truncating both calculated poles and mode shapes to the first five natural frequencies and by deleting the coordinates in the mode shapes that cannot be measured. By doing this the dimensions of the calculated poles and mode shapes correspond to the measurement. The resulting poles and mode shapes are used to reconstruct a new mass and stiffness matrix by transforming the modal matrices back to physical coordinates.

16.4.2.2 Identification of the Damping Matrix

The previous description results in updated mass and stiffness matrices with a dimension of 5×5. Next, the measured frequency response functions are used to extract poles and mode shapes with a least squares complex exponential method. The extracted mode shapes are unity modal mass scaled using the reduced mass matrix from the model. The newly scaled mode shapes are used together with the poles to estimate the damping matrix $\mathbf{C}_0$.

16.4.3 Rotating Shaft

When the shaft is rotating at a certain speed it behaves like the system in (16.1). However, if a system as (16.8) is assumed, then all the dissipation effects are combined in the damping matrix. This means that the 5×5 mass and stiffness matrix of the rotating system are not changing due to rotation. The measured frequency response functions at rotating speed Ω are then used to extract the poles and mode shapes. The resulting mode shapes are erroneous, so they are scaled to unity modal mass using the mass matrix derived at 0 speed. These scaled mode shapes are then used to estimate a damping matrix of the form $\mathbf{C}_{\Omega} = \mathbf{C} + \Omega \mathbf{C}_{re}$. The rotating damping effect is now estimated by

$$\mathbf{C}_{re} = \frac{C_{\Omega} - \mathbf{C}_0}{\Omega} \tag{16.10}$$

16.5 Experimental Validation

In the experiment, the shaft is operated at 0 and 3,000 rpm. At both speeds, the frequency response functions are measured. The identified damping matrices are given in Tables 16.1 and 16.2. The damping matrices are not symmetric and there are negative elements on the diagonal. However, together with the updated mass and stiffness matrix, they can be used to estimate the frequency response functions accurately. The measured and estimated frequency response functions at the driving point are given in Figs. 16.4 and 16.5. Because of the impact measurement, an exponential window has been used to estimate

Table 16.1 The identified damping matrix at 0 rpm

14.98719	−3.7457	17.42148	5.822773	10.20963
−4.36665	−9.92687	−15.0152	−13.9544	−11.8712
7.730844	6.800062	24.19674	11.88099	16.13112
−1.65739	−17.3205	−6.53153	−20.9999	−13.4505
16.69047	−14.5339	27.78173	−2.4798	13.98526

Table 16.2 The identified damping matrix at 3,000 rpm

−13.2302	−36.2871	−25.6383	−5.76573	−4.88674
24.98378	26.9026	26.24019	8.278528	5.051465
−25.6916	−7.65284	−34.3064	−12.0996	−4.71555
−9.68825	33.90984	9.023721	16.4902	−1.24859
−4.99275	−5.65304	−8.7208	−5.00681	4.067461

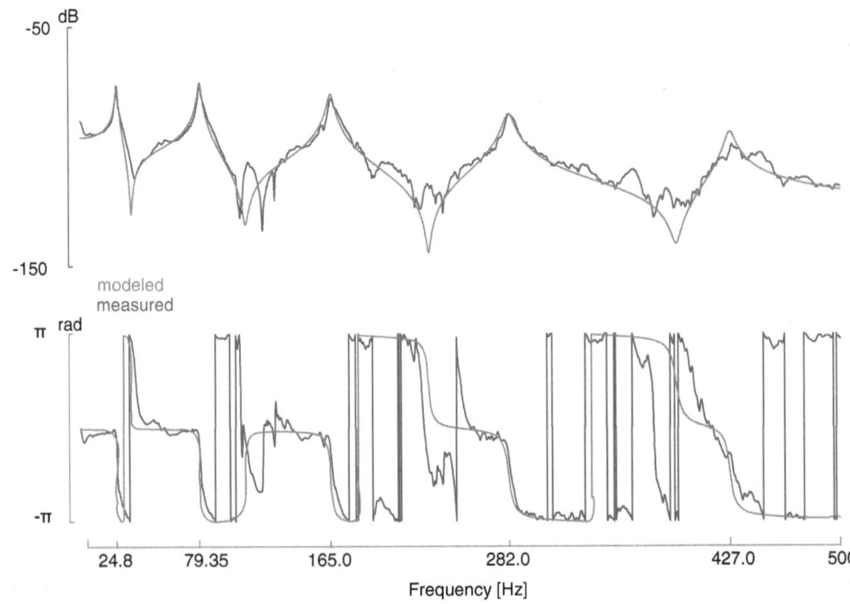

Fig. 16.4 The frequency response function taken at the driving point at 0 rpm can be reconstructed by using the identified damping matrix

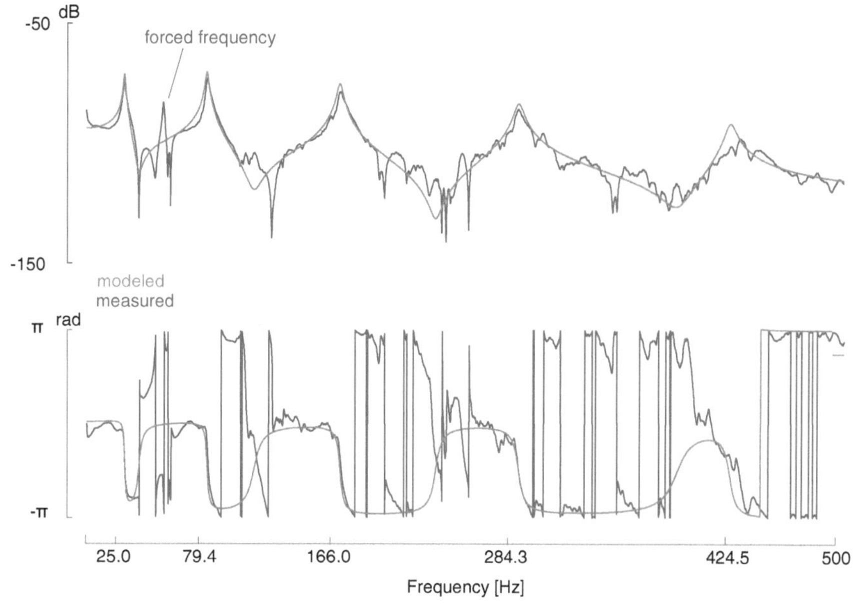

Fig. 16.5 The frequency response function taken at the driving point at 3,000 rpm can be reconstructed by using the identified damping matrix

the frequency response functions. After compensation of this window effect [17], the estimated damping is slightly lower. At 3,000 rpm, there is an unavoidable peak that is caused by the residual unbalance of the shaft.

The estimated mass, damping and stiffness matrices are used to predict the decay rate plot of the first forward mode (Fig. 16.6). In order to validate the prediction the system is operated at several speeds going from 0 to 4,500 rpm with a resolution of 500 rpm. The measurement at 1,500 rpm is not performed because it was too close to the first natural frequency leading to unwanted high amplitudes. At each speed, the real part of the poles is estimated and compared to the prediction. As can be seen, the prediction fits remarkably well. It should be noted that whereas the model predicts that the system will become instable. This prediction indicates that the first forward mode will not cause instability. At about 3,000 rpm, there seems to be some sort of stabilizing effect.

16.6 Conclusion

In this paper the phenomenon of rotating damping is investigated through experiments. Instead of approaching the system as skew-symmetric a different approach is proposed. A skew-symmetric model is experimentally not convenient. When the energy loss per cycle is observed, the rotating damping that appears in the displacement term or apparent stiffness term, actually behaves as regular damping term that is speed dependent. Therefore, a rotating system can be observed as a classical mass spring damper system that has just a speed dependency of the damping matrix.

A rotating damping setup is used to validate the proposed method. In combination with a finite element model and a model updating technique, the damping matrix is estimated both at 0 and 3,000 rpm. The difference between the two damping matrices is a measure for the rotating damping effect. By using the estimated damping matrices together with the updated mass and stiffness matrices, the decay rate plot of the first forward mode is predicted. The prediction is validated with single measurements at different speeds. It is important to notice that a linear speed dependent model with viscous damping predicts a certain stability threshold speed. The prediction in this paper however shows that there appears to be a stabilizing effect. The same effect is validated in the single measurements. In this research, both nonrotating and rotating damping are very low because of the configuration of the setup. In future, the influence of extra damping can be investigated together with the inclusion of the gyroscopic effect.

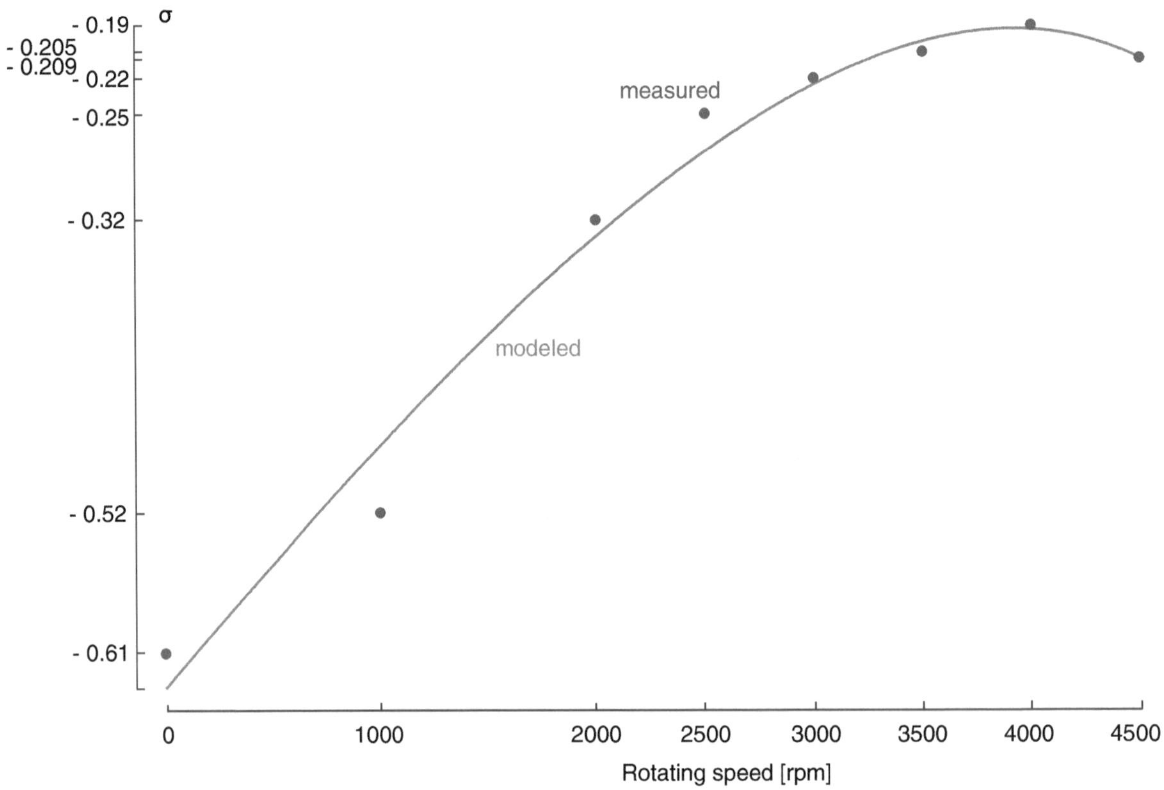

Fig. 16.6 The decay rate plot is constructed with the identified damping matrices at 0 and 3,000 rpm. The results are validated by individual measurements at different speeds

References

1. Vervisch B, Stockman K, Loccufier M (2014) Estimation of the damping matrix in rotating machinery for the calculation of the stability threshold speed. Int J Struct Stab Dyn 14:1450012
2. Tisseur F, Meerbergen K (2006) The quadratic eigenvalue problem. Soc Ind Appl Math 43(2):235–286
3. Bucher I, Ewins DJ (2001) Modal analysis and testing of rotating structures. Philos Trans R Soc A Math Phys Eng Sci 359(1778):61–96
4. Vervisch B, Stockman K, Loccufier M (2012) Sensitivity of the stability threshold in linearized rotordynamics. In: ISMA conference, Leuven
5. Adams ML (2009) Rotating machinery vibration: from analysis to troubleshooting. CRC Press/Taylor & Francis, Boca Raton
6. Adams ML, Padovan J (1981) Insights into linearized rotor dynamics. J Sound Vib 76(1):129–142
7. Genta G (2005) Dynamics of rotating systems, vol 1. Springer, New York
8. Srikantha Phani A, Woodhouse J (2007) Viscous damping identification in linear vibration. J Sound Vib 303(3–5):475–500
9. Adhikari S, Woodhouse J (2001) Identification of damping: part 1, viscous damping. J Sound Vib 243(1):43–61
10. Adhikari S, Woodhouse J (2002) Identification of damping: part 4, error analysis. J Sound Vib 251(3):491–504
11. Srikantha Phani A, Woodhouse J (2009) Experimental identification of viscous damping in linear vibration. J Sound Vib 319(3–5):832–849
12. Foltête E, Gladwell GML, Lallement G (2001) On the Reconstruction of a damped vibrating system from two complex spectra, part 2: experiment. J Sound Vib 240(2):219–240
13. Pilkey DF, Inman DJ (1997) An iterative approach to viscous damping matrix identification. In: IMAC XV proceedings, pp 1152–1157
14. Adhikari S, Woodhouse J (2002) Identification of damping: part 3, symmetry-preserving methods. J Sound Vib 251(3):477–490
15. Balmes E (1997) New results on the identification of normal modes from experimental complex modes. Mech Syst Signal Process 11(2):229–243
16. Vervisch B, Derammelaere S, Stockman K, Loccufier M (2014) Frequency response functions and modal parameters of a rotating system exhibiting rotating damping. In: ISMA conference 2014
17. Brandt A (2011) Noise and vibration analysis: signal analysis and experimental procedures. Wiley, Chichester
18. Forrai L (2000) A finite element model for stability analysis of symmetrical rotor systems with internal damping. JCAM 1(1):37–47
19. Lin RM, Ewins DJ (1994) Analytical model improvement using frequency response functions. Mech Syst Signal Process

Chapter 17
Estimation of Modal Damping for Structures with Localized Dissipation

M. Krifa, N. Bouhaddi, and S. Cogan

Abstract Damping plays an important role in bolted joints of assembled structures due to their significant capacity to dissipate energy. The underlying mechanisms of these dissipative phenomena are generally poorly understood and result from contact and friction effects within the joint interfaces. In order to provide useful virtual prototyping tools for reducing response levels, accurate model-based estimation of modal damping is required. The present study employs an energetic method to calculate the loss factor associated with the localized dissipative interfaces of a global linear structure. This method is based on the concept of the dissipated energy in the interfaces for which the closed-form expression of the loss factor is the ratio between dissipated energy and maximal potential energy, over a cycle of periodic vibration. The aim of this work is to investigate the advantages and drawbacks of this approach for particular conditions such as: modal projection, localized damping level and model density. Simulated academic examples, where accurate estimations of the exact solutions are available, will be used to illustrate the methodology and to explore the potential difficulties that may arise in more complex industrial applications.

Keywords Bolted joints • Loss factor • Proportional damping • Localized damping • Energetic method

17.1 Introduction

The study of dissipation in the joints has been the subject of many investigations [1]. However, the prediction of damping in assembled structures from the design phase started over the past decade [2, 3]. Constitutive modeling of mechanical joints is an important part of predictive dynamic modeling of joints structures because joints are often the dominant source of energy dissipation and vibration damping in those structures [2]. For very expensive components that are critical for safety, predictive dynamic modeling is necessary. In order to predict modal damping in an assembled structures two kinds of methods exist in the literature: experimental methods [4] and simulation methods [5]. In this article, the focus will be on the latter. Mechanical engineers often assume proportional damping for vibration calculations, even if the structure has localized damping effects. However, the modal damping assumption may not respect the physical properties of the dynamic system. This can lead to errors in vibration amplitudes and modal damping estimations with respect to a more detailed model [6]. The purpose of this paper is to investigate the limitations of the energetic method when predicting modal damping for localized structures. This paper is structured as follows: Sect. 17.2 presents the energetic method to estimate modal damping and Sect. 17.3 illustrates the prediction level of this methodology and its validity domain on academic examples.

17.2 Energetic Method

Modal damping with localized dissipation in bolted joints can be estimated, when the eigenfrequencies are well separated, with the state space formulation and the energetic method. In this work we will only describe the second method. But results of these two methods are taken into account to calculate modal damping of proposed academic examples.

M. Krifa (✉) • N. Bouhaddi • S. Cogan
Applied Mechanics Department, FEMTO-ST - UMR 6174, University of Franche-Comté-24, rue de l'Épitaphe, 25000 Besançon, France
e-mail: mohamed.krifa@femto-st.fr

© The Society for Experimental Mechanics, Inc. 2015

R. Allemang (ed.), *Special Topics in Structural Dynamics, Volume 6*, Conference Proceedings
of the Society for Experimental Mechanics Series, DOI 10.1007/978-3-319-15048-2_17

The present study employs this approach to calculate the loss factor associated with the localized linear interfaces of a globally linear structure. In practice, the response of the system is not calculated explicitly due to the high computational burden. When assuming a linear and dissipative problem of assembled structures, the discrete form of the damped vibration problem may be governed by the equation:

$$M\ddot{y}(t) + C\dot{y}(t) + Ky(t) = f(t) \tag{17.1}$$

where K, M and C are respectively the stiffness, mass and damping matrices, f is the vector of the external loads.

The objective of the energetic method is to determine the damping factor corresponding to each vibration mode of the structure. It is based on the concept of the dissipated energy in the interfaces for which the close form expression of the loss factor is the ratio between dissipated energy and maximal potential energy, over a cycle of periodic vibration. as shown in this relation:

$$\xi_\nu = \frac{1}{4\pi}\frac{E_\nu^{diss}}{E_\nu^{pot}}; \nu = 1, 2, \ldots, n \tag{17.2}$$

where E_ν^{diss} and E_ν^{pot} are respectively the dissipated energy and maximal potential energy. In the following we will present the necessary formulas for calculating these energies.

The estimation of the dissipated and maximal potential energies requires an accurate characterization of the response levels of the system and the latter remain an approximation since they depends a priori on a knowledge of the different dissipation mechanisms. So, dissipated energy is calculated by the following expression:

$$E_\nu^{diss} = \int_0^T \dot{y}(t)^T f_c(t)dt \tag{17.3}$$

Where

- $T = \frac{2\pi}{\omega_\nu}$ the cycle of periodic vibration
- $f_c(t) = C\dot{y}(t)$: dissipated force
- $\dot{y}(t) = \mathrm{Re}(j\omega_\nu y(\omega_\nu)e^{j\omega_\nu t})$ the velocity of harmonic response

Potential energy is calculated as follow:

$$E_\nu^{pot} = \frac{1}{2}\{y(w_\nu)\}^T K \{y(w_\nu)\} \tag{17.4}$$

where $y(w_\nu)$ is the frequency response of the system.

We distinguish two cases: proportional damping and localized damping. In the first case the frequency response of the system is equal to

$$y(w_\nu) = \sum_{\nu=1}^{n} \phi_\nu q_\nu \tag{17.5}$$

If the response is projected on a single mode of vibration then Eq. (17.5) becomes

$$y(\omega_\nu) = \phi_\nu q_\nu \tag{17.6}$$

where ϕ_ν and q_ν are respectively the eigenmode and modal amplitude corresponding to eigenfrequency ω_ν. The exact frequency response is equal to:

$$y(\omega) = (K + j\omega C - \omega^2 M)^{-1} \times f \tag{17.7}$$

In the case of proportional damping, assuming that the eigenmodes ϕ_ν are orthonormalized with respect to the mass matrix M, i.e. $\phi_k^{TM}\phi_l = \delta_{kl}$, Eqs. (17.4) and (17.5) allow the total strain energy can be derived in the form:

$$E_\nu^{pot} = \frac{1}{2}\sum_{\nu=1}^{n} \omega_\nu^2 q_\nu^2 \tag{17.8}$$

If the damping is assuming to be hysteretic, we have:

$$C = \frac{\eta}{\omega} K \tag{17.9}$$

where η is the unknown loss factor. It can readily be shown that at frequency ω_ν, the modal damping q_ν can be expressed in function of ξ_ν by:

$$q_\nu = -j \frac{\phi_\nu f}{\omega_\nu^2 \eta_\nu} = -j \frac{\phi_\nu f}{2\omega_\nu^2 \xi_\nu} \tag{17.10}$$

where: $\eta_\nu = 2 * \xi_\nu$. Equation (17.10) leads to the amplitude q_ν, from which one can determine the energy dissipated in each joint. Hence, the energy dissipated in all joints leads to the estimated value of the total damping factor:

$$\xi_\nu^{joints} = \frac{1}{4\pi} \frac{E^{diss}}{E^{pot}} \tag{17.11}$$

17.3 Advantages and Limitations of the Energetic Method

The energetic method proposed has been implemented on an academic example in MATLAB. Three numerical simulations have been performed in order to highlight the validity domain of energetic method. The goal of this part consists in studying the influence of the following parameters on the accuracy of this method: modal projection effect, localized damping level and modal density when predicting modal damping.

17.3.1 Influence of the Mode Projection

The system studied to demonstrate the effect of a projection on a user device is composed of two beams connected together by a localized dissipative interface (Fig. 17.1).

Each component is discretized into ten 2D beam finite elements (2 DOFs per node), so that the full model has 40 DOFs. Two lumped masses (m1,m2) are fixed at the free end of the beams. The excitation force is applied to node number 1 of the first beam. This model represents two structures assembled by a bolted joint. The simulation data are summarized in the Table 17.1. The joint is represented by a linear lumped model composed of spring k_i and a viscous damper c_i elements.

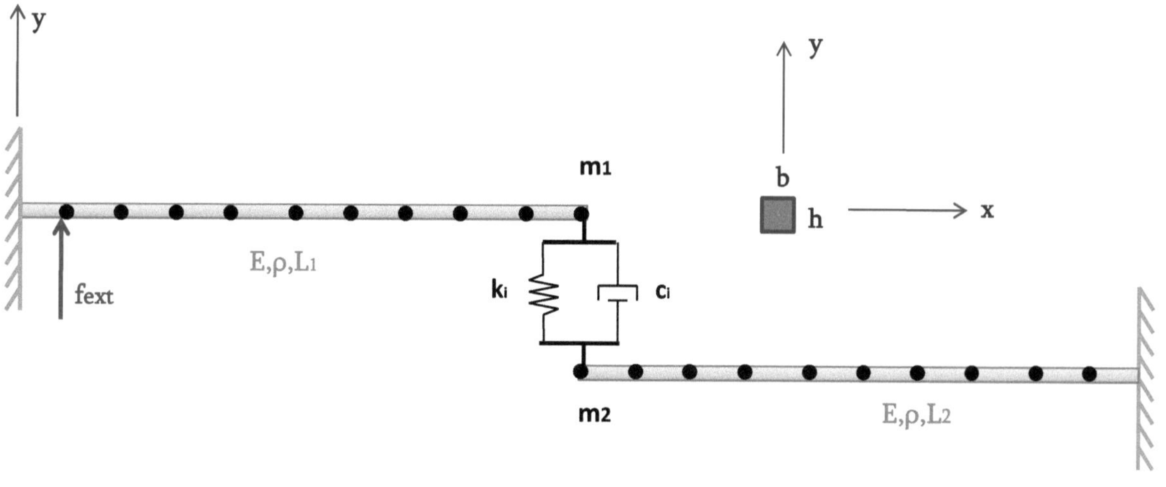

Fig. 17.1 Double-beam with simple bending

Table 17.1 Physical and geometric properties of the structure/interface parameters

E	Steel Young modulus (GPa)	210	k_i	Linear stiffness (N/m)	10^5
ρ_s	Steel density (kg/m^3)	7,800	c_i	Viscous damper (N/(m/s))	20
h	Beam thickness (m)	2 10^{-2}	m_1	Mass (Kg)	0.2
b	Beam width (m)	2 10^{-2}	m_2	Mass (Kg)	0.2
L_1	Beam first length (m)	0.6			
L_2	Beam second length (m)	0.5			
ν	Steel Poisson ratio	0.3			

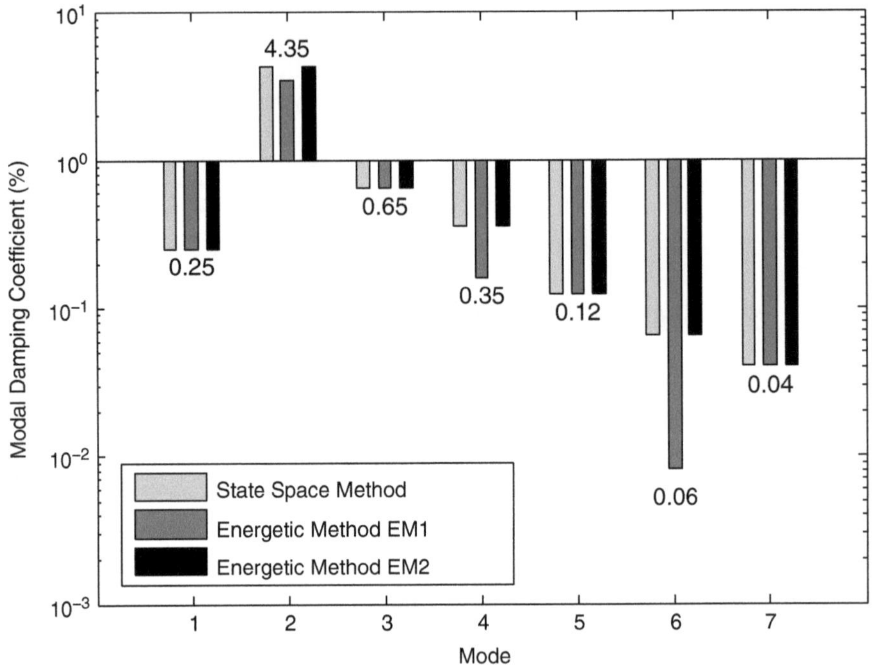

Fig. 17.2 Comparison of estimated modal damping

The dissipation force is expressed in terms of the velocity by the equation:

$$f_c(t, \dot{u}) = c_i \times \Delta\dot{u} \tag{17.12}$$

The modal damping results calculated with three methods are illustrated on Fig. 17.2: The reference method is, in this case, the state space method [7], the methods **EM1** and **EM2** are the variants of the **Energetic Method EM** respectively with mode projection on the truncated modal basis using Eq. (17.6) and with modal projection on a single mode of interest using Eq. (17.7). The method **EM1** predicts the modal damping with unacceptable errors: for mode 2 (19 %), 4 (55 %) and 6 (>80 %). Whereas for the method **EM2** these error levels drop significantly to reach 0.06 and 0.07 respectively for modes 2 and 4, (Fig. 17.3). This improvement is due, in this case, at the projection in a single mode of the forced response of the system.

The resulting damping errors found between the energetic method **EM1** and **EM2** versus the reference method (Fig. 17.3) are expected if one takes into account the contribution of neighboring terms before calculating the modal damping coefficients. The frequency response of the system (Fig. 17.1) when looking for a modal damping can be expressed by:

$$y(\omega) = \sum_{\nu=1}^{n} \phi_\nu \frac{\phi_\nu^T f^{ext}}{\omega_\nu^2 - \omega^2 + 2j\omega\omega_\nu\xi_\nu} = \sum_{\nu=1}^{n} \phi_\nu \times q_\nu \tag{17.13}$$

where q_ν is the contribution of mode ν. The coefficient $(\frac{q_i}{q_j})$ presents the ratio of the contribution of mode i relative to mode j. We can thus define a theoretical coefficient of the sum of neighboring modal contributions on a single mode j as follows:

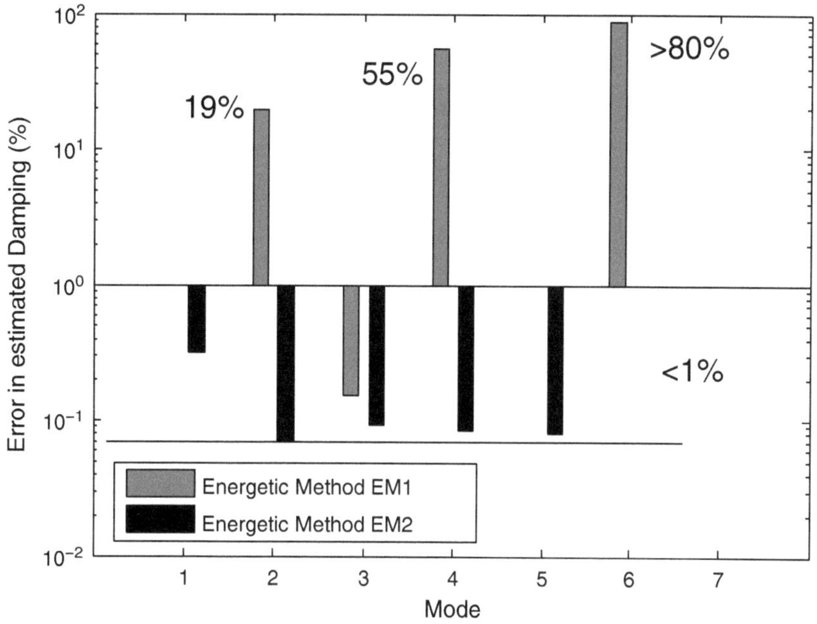

Fig. 17.3 Error in estimating modal damping

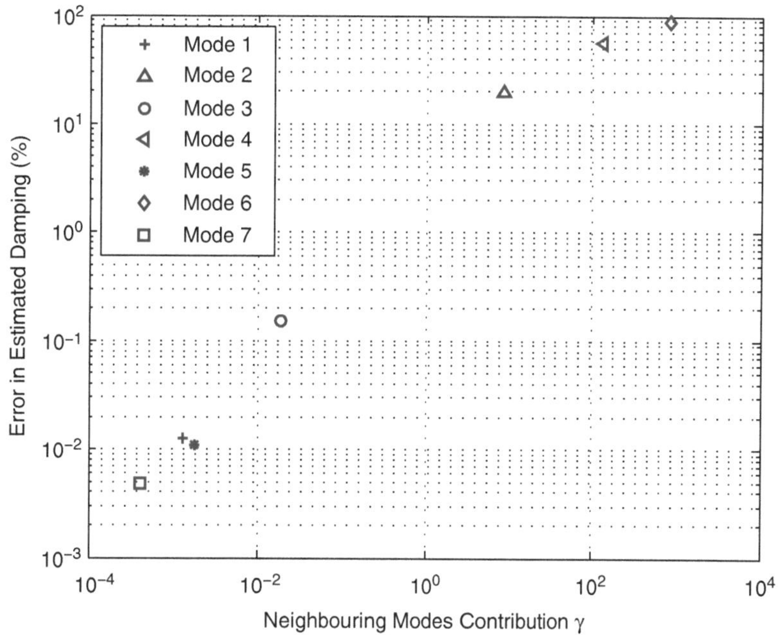

Fig. 17.4 Modal participation of energetic method EM1

$$\gamma_j = 100 \times \frac{\sum\limits_{i \neq j} {q_i}^2}{{q_j}^2} \tag{17.14}$$

Figure 17.4 presents the variation of damping errors with respect to the variation of the sum of neighboring modal contributions γ_j.

When γ increases for one mode, the damping error of this mode increases too. For this case modes 2,4 and 6 have the largest neighboring modal contributions and the largest estimated damping errors. In the reminder of this study, we will use the dynamic responses with projection on a single mode as this generates better results in the prediction of the modal damping coefficients.

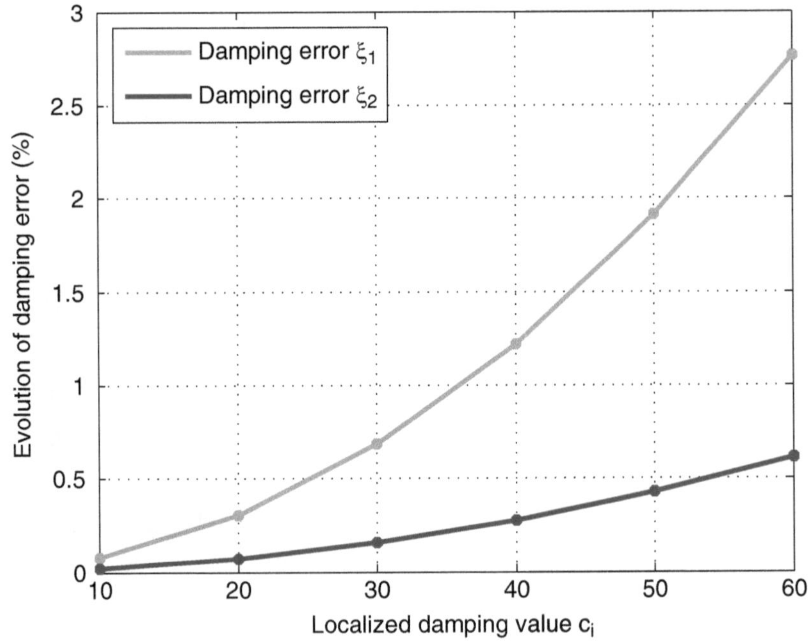

Fig. 17.5 Influence of the degree of dissipated energy at the interface

17.3.2 Effects of the Localized Damping Level

In this section, we are interested in studying the effect of the localized damping level (Fig. 17.1) on the damping prediction by the Energetic Method **EM2**. Six numerical simulations are performed with progressively higher viscous local damping ($c_i = 10, 20, 30, 40, 50$ and 60 Ns/m). When the localized damping value c_i increases the dissipated energy also increases. Figure 17.5 illustrates the evolution of relative error for modes 1 and 2, which is calculated between the energetic method **EM2** and the reference method. Figure 17.5 represents also the curves for modes 1 and 2 when localized damping value increases. For other modes we notice the same curve shape but with lower levels of error.

When c_i increases, for example when c_i = 60 N/(m/s), damping value ξ_2 goes up 10 % and attains a peak of 13 % (Fig. 17.6). Beyond this modal damping level, damping error increases to unacceptable values (Fig. 17.7). The reference method (state space method) may not give good results when c_i exceeds a specific modal damping level.

The raison is that state space method is based on the assumption described by the following equation

$$s_\nu = -\xi_\nu \omega_\nu + j\omega_\nu \sqrt{1 - \xi_\nu{}^2}, \xi_\nu << 1 \tag{17.15}$$

Where s_ν is the eigensolution of νth mode ; $j^2 = -1$. So to evaluate the quality of the estimated modal damping by the energetic method, and the previous assumption is valid only with low modal damping values. In the following, we compare the frequency responses functions FRFs of the four methods (energetic method **EM1**, energetic method **EM2**, state space and exact method). The exact method consists in the calculation of the FRFs by direct inversion of the full impedance matrix at each frequency (Eq. (17.7)). The FRFs computing by the energetic method (**EM1**, **EM2**) use a truncated basis of ten modes contained in the frequency band $(0, 2 \times f^{max})$, $f^{max} = 2{,}000$ Hz is the maximum frequency of interest.

To calculate FRFs level error in (%) between two responses y_1 and y_2 in the frequency band of interest [0, 2,000 Hz], we use the following expression:

$$\varepsilon_y = 100 \times \frac{\|y_2 - y_1\|}{\|y_1\|} \tag{17.16}$$

For the state space method, the FRF comparison with the exact method yields similar results (Fig. 17.8) when $c_i = 20$ Ns/m. However, the amplitude levels are different for high level of viscous damping, for example when $c_i = 60$ Ns/m, (Fig. 17.9). This is why the exact method will be used as a reference instead of state space approach, especially when dealing with high damping level.

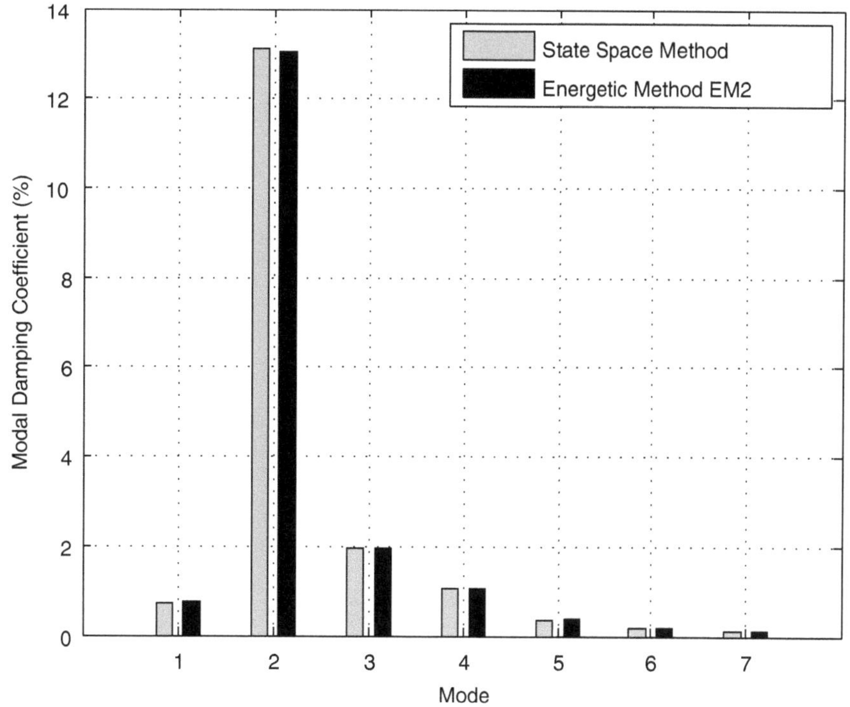

Fig. 17.6 Comparison of estimated modal damping

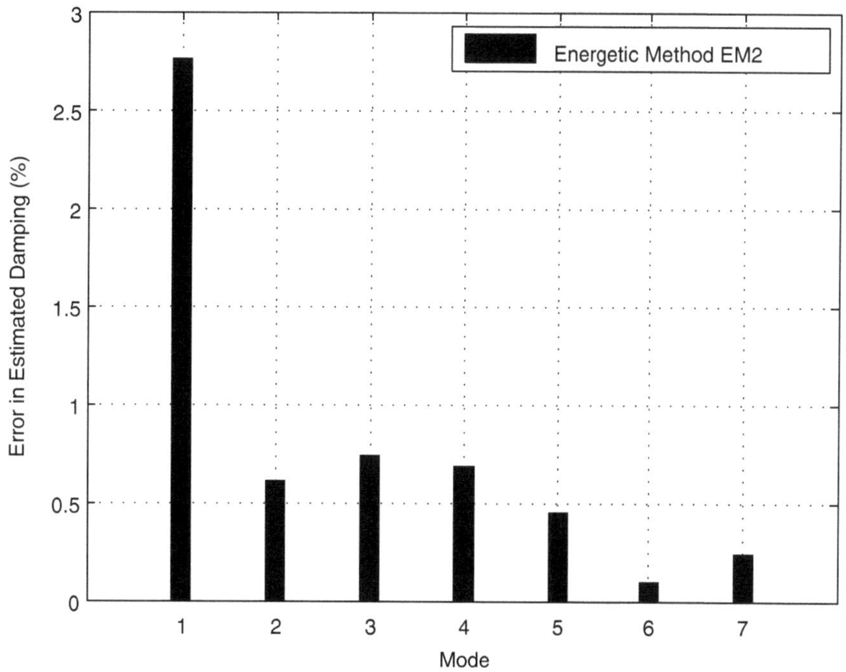

Fig. 17.7 Error in estimating modal damping

Figure 17.10 illustrates that the method **EM2** estimates the damping factor with good accuracy for low damping level. But, Fig. 17.11 shows that this method is not valid for high damping value.

As discussed in Sect. 17.3.1 the method **EM1** does not yield accurate results with respect to the exact method for both cases ($c_i = 20$ Ns/m and $c_i = 60$ Ns/m), (Figs. 17.12 and 17.13).

In the next section, we will investigate the limitations of the method **EM2** due to the influence of modal density.

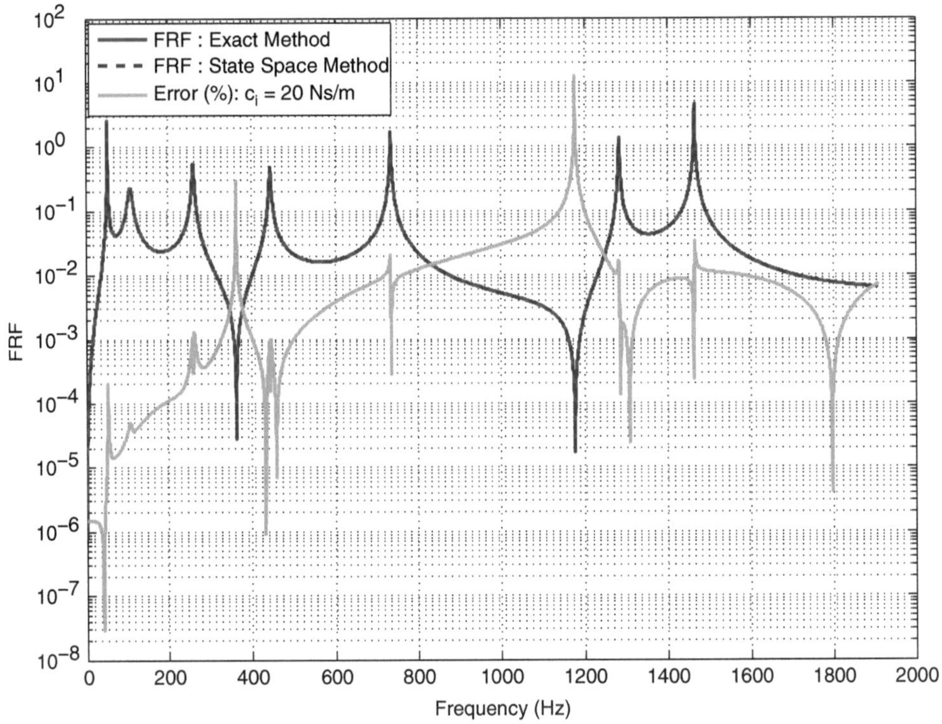

Fig. 17.8 FRFs error: state space vs exact method—$c_i = 20$ Ns/m

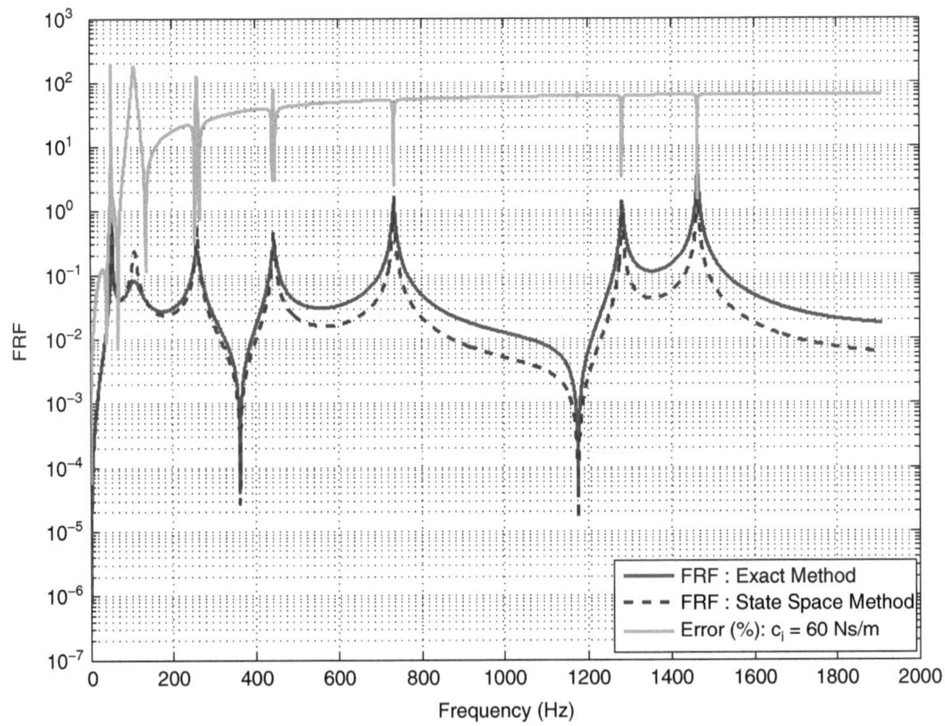

Fig. 17.9 FRFs error: state space vs exact method—$c_i = 60$ Ns/m

17.3.3 Influence of Modal Density

In this section, the system studied is adapted to investigate the effect of high modal density or the effect of neighboring modes is the triple beam as described on Fig. 17.14. The structure was chosen for its ability to generate neighboring modes by adjusting the parameters of the beam and those of linear interfaces. The triple beam structure is composed of three linear interfaces respectively between nodes 6 and 12, 12 and 18, and between the frame and node number 18, (Fig. 17.14).

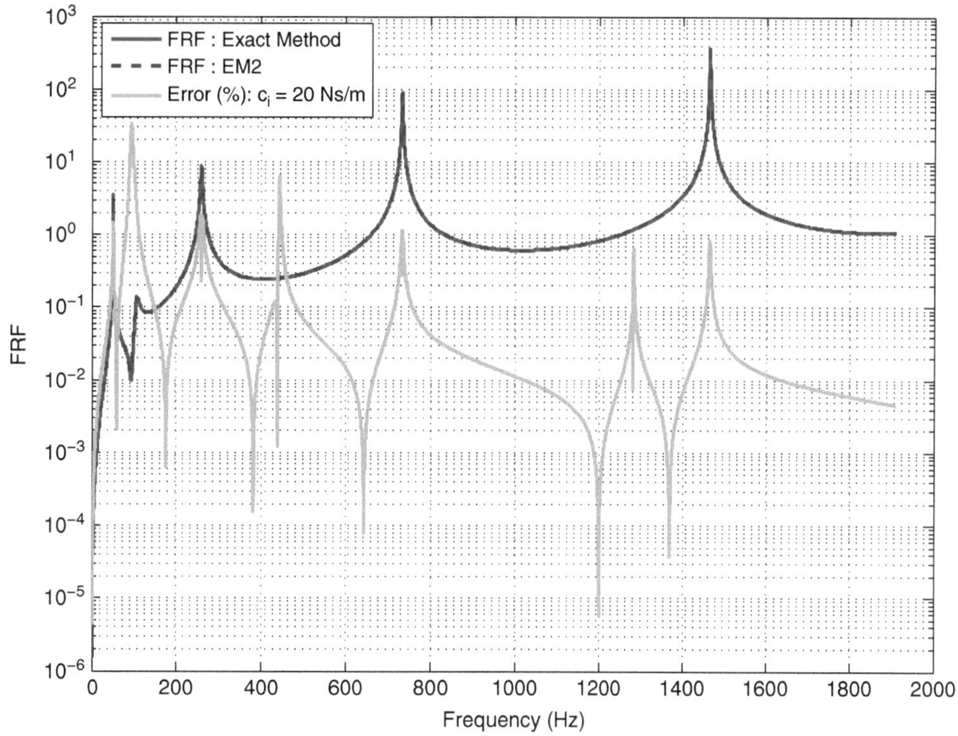

Fig. 17.10 FRFs error: **EM2** vs exact method—$c_i = 20\,\text{Ns/m}$

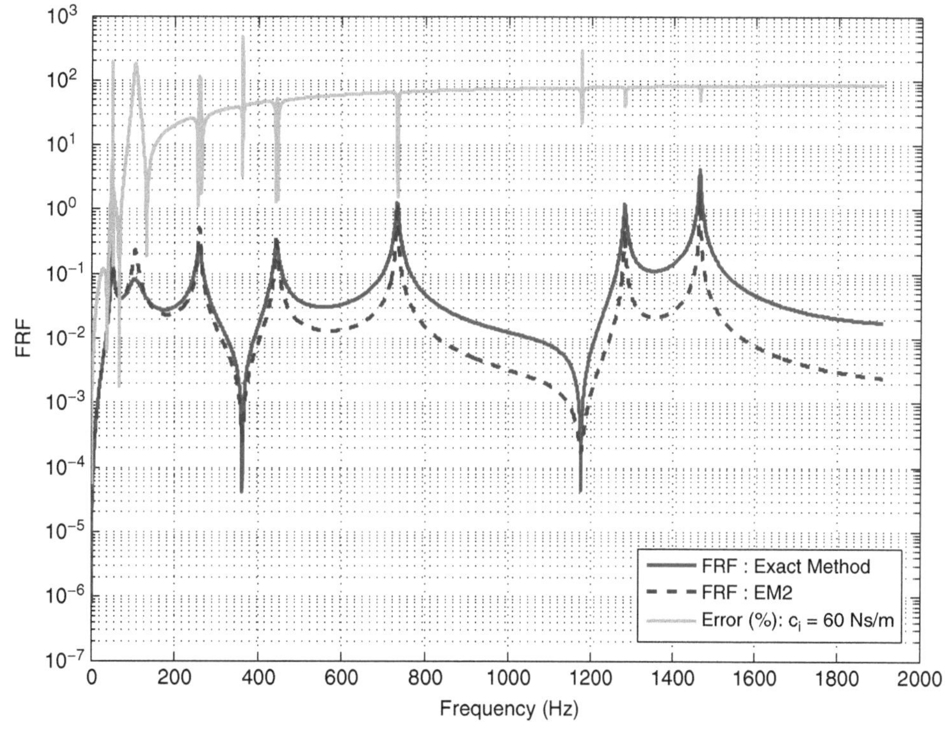

Fig. 17.11 FRFs error: **EM2** vs exact method—$c_i = 60\,\text{Ns/m}$

Each interface consists of a stiffness k_i and a linear damper c_i. The simple bending excitation is applied to the node number 1. The used parameters are performed in Table 17.2.

Hasselman [8] proposed the following condition for uncoupled modes:

$$\omega_j - \omega_i \gg 2 \times \xi_i \times \omega_i \tag{17.17}$$

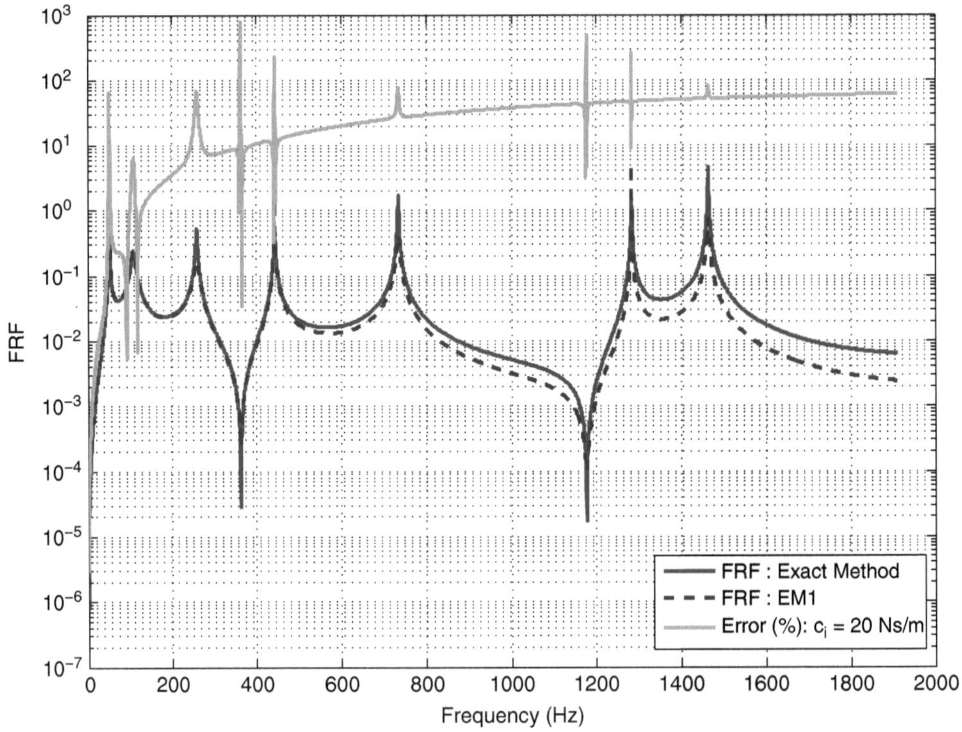

Fig. 17.12 FRFs error: **EM1** vs exact method—c_i = 20 Ns/m

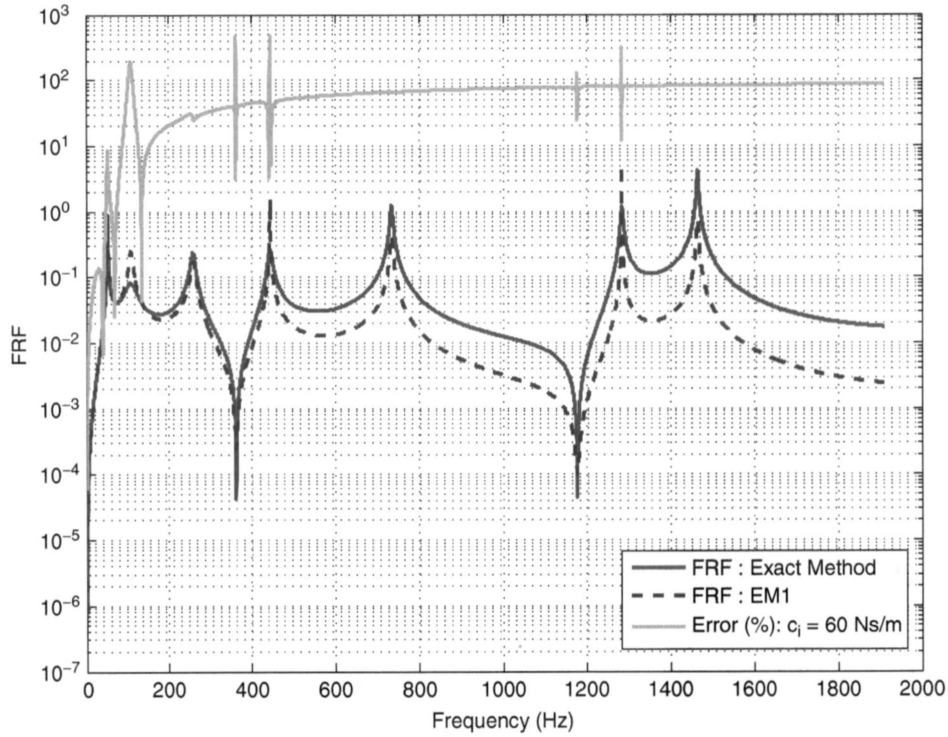

Fig. 17.13 FRFs error: **EM1** vs exact method—c_i = 60 Ns/m

Namely, two modes i and j are uncoupled if the pulsation difference between these two modes is much greater than the product of the angular frequency and the damping coefficient multiplied by two. We focus on the case of neighboring modes so eigenfrequencies ($\omega_i = 2 * \pi * f_i$) must respect the following inequality:

$$f_j - f_i \leq 2 \times \xi_i \times f_i \tag{17.18}$$

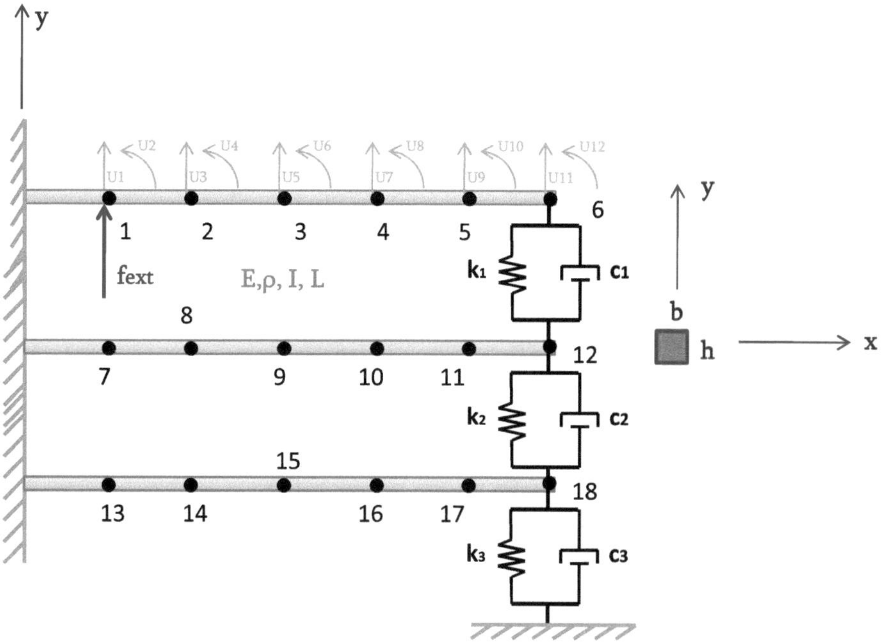

Fig. 17.14 Triple-beam with simple bending

Table 17.2 Physical and geometric properties of the triple beams

E	Steel Young modulus (GPa)	210
ρ_s	Steel density (kg/m^3)	7,800
h	Beam thickness (m)	$10\ 10^{-3}$
b	Beam width (m)	$50\ 10^{-3}$
L_1	Beam first length (m)	0.6
L_2	Beam second length (m)	0.5
L_3	Beam second length (m)	0.5
ν	Steel Poisson ratio	0.3

Table 17.3 Interfaces parameters evolution

Damping	Stiffness
Nonidentical and fixed damping c_i	Identical and evolutive stiffness k_i
$c_1 = 40$; $c_2 = 80$; $c_3 = 40$	case 1: $k_1 = k_2 = k_3 = 1\ 10^4$
...	case 2: $k_1 = k_2 = k_3 = 2\ 10^4$
	case 3: $k_1 = k_2 = k_3 = 4\ 10^4$
	...
	case N: $k_1 = k_2 = k_3 = 6\ 10^5$

We can thus define a theoretical limit frequency neighborhood as follows:

$$f_{j\ \mathrm{lim}} = (1 + 2 \times \xi_i) f_i \tag{17.19}$$

To investigate the effect of high modal density a series of simulations have been performed by changing the stiffness parameters while keeping the damping parameters constant. The aim of changing the stiffness of the interfaces is to obtain neighboring modes, so that we can study the importance of this phenomena when applying the energetic method to calculate damping. In this section, the used parameters are performed in Table 17.3.

For the first case the damping results are presented in Table 17.4. We obtain two triplets of neighboring modes, the first set is composed of modes $\{1, 2, 3\}$, and the second one is composed of modes $\{4, 5, 6\}$.

It is found that the error increases when the eigenmodes are closer together. The error can exceed 10 % when the frequency of neighbor modes becomes less than the theoretical limit frequency. For example the mode 4 depicts an error equal to 12.08 % with a frequency spacing of 0.35 Hz. ($f_1 = 920.59$ Hz, $f_2 = 920.93$ Hz, $f_{lim} = 923.13$ Hz and $f_2 \prec f_{\mathrm{lim}}$). Figure 17.15 shows the evolution of the damping error of mode 4 and mode 5 based on the difference between the frequencies of modes 4 and 5.

Table 17.4 Comparison of modal damping calculated using three methods

Modes (Hz)	1	2	3	4	5	6
Frequencies f_i	141.26	143.09	145.25	920.59	920.93	921.01
$f_{i+1} - f_i$	1.83	2.17		0.35	0.08	
f_{lim}		145.74	173.79		923.13	938.43
$\xi_\nu^{Ref}(\%)$	0.79	5.36	16.53	0.06	0.47	1.48
$\xi_\nu^{var2}(\%)$	0.88	5.64	16.14	0.07	0.50	1.45
Error (%)	11.74	5.15	2.32	12.08	5.24	2.06

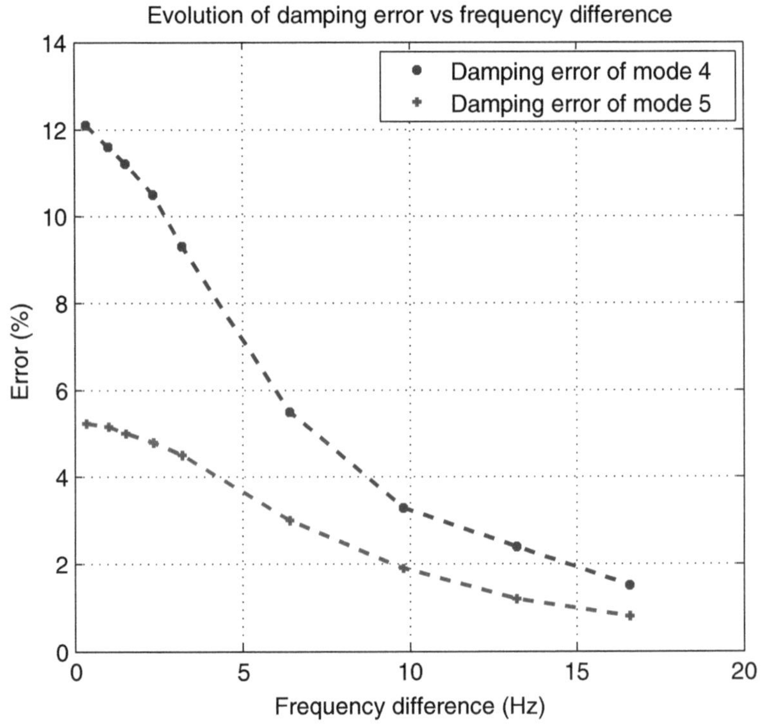

Fig. 17.15 Evolution of damping errors vs frequency difference

Each point of the two curves represents a test case. The frequency difference between modes 4 and 5, presented in Table 17.4 and equal to 0.35 Hz, shows the first point on the x-axis graph. The first point on the y-axis graph of the curve of the mode 5 shows the relative error (%) between damping calculated by the reference method and the energetic method **EM2**. The error is equal to 12.08 %. Similarly, for the second curve which represents the error for mode 5, the first point on the y-axes presents an error of 5.24 %. Figure 17.15 shows that when modes are close in frequency then the damping error increases. Indeed, in the case of high modal density it becomes difficult to isolate a single mode and in this case the generalized damping matrix presents large off-diagonal terms which are neglected under the proportional damping hypothesis.

17.4 Conclusion

In this paper, the energetic method is presented and investigated for specific conditions in linear interface joints in order to investigate its limitations. The method is based on the ratio between the dissipated energy in the interfaces, and potential energy of the global structure. Firstly, the authors compared the localized damping in interfaces calculated by three methods with the aim of demonstrating the importance of the effects of modal projection using a double beam system. Secondly, the effects of the localized damping level when localized viscous damping increases was investigated. Finally the influence of increasing modal density in the calculation of modal damping by the energetic method was studied using a triple beam system. In this case, significant prediction errors were found in the presence of close eigenfrequencies.

Acknowledgements The financial support of the "Agence Nationale de la Recherche ANR" (ANR-12-MONU-00016-01) is acknowledged.

References

1. Bograd S, Reuss P, Schmidt A, Gaul L, Mayer M (2011) Modeling the dynamics of mechanical joints. Mech Syst Signal Process 25(8):2801–2826
2. Gaul L, Roseira J, Becker J (2008) Structural damping with friction beams. Shock Vib 15(3–4):291–298
3. Ibrahim RA, Petit CL (2005) Uncertainties and dynamic problems of bolted joints and other fasteners. J Sound Vib 279:857–936
4. Dion JL, Chevallier G, Peyret N (2013) Improvement of measurement techniques for damping induced by micro-sliding. Mech Syst Signal Process 34:106–115
5. Caignot A, Ladeveze P, Neron D, Durand J.-F (2010) Virtual testing for the prediction of damping in joints. Eng Comput 5:621–644.15
6. Adhikari S (1999) Rates of change of eigenvalues ans eigenvectors in damped dynamic systems. AIAA J 37(11):1152–1158
7. Geradin M, Rixen D (1997) Mechanical vibrations. Wiley, New York
8. Hasselman TK (1997) Modal coupling in lightly damped structures. AIAA J 11(2):229–243

Chapter 18
Design of UAV for Surveillance Purposes

F. Cheli, F. Ripamonti, and D. Vendramelli

Abstract Two main algorithms are presented to use an UAV for surveillance purposes. A very fast algorithm for the scan of an unknown area has been implemented and tested: it permits to scan a domain for the definition of layout, boundaries, obstacles... Once that the area has been acquired, a second algorithm is used to monitor the regions of interest in an efficient way. A neural network has been built in order to choose the shortest path to reach a determined point, giving the drone the possibility to avoid unexpected obstacles. Finally these two algorithms has been tested to verify their accuracy and speed.

Keywords Quadcopter • Obstacle avoidance • Path following algorithm • Neural networks

18.1 Introduction

Over the last years, Unmanned Aerial Vehicles (UAV) have been applied in several interesting fields. The improvement and diffusion of control strategies both in terms of vibration suppression/stability [1] and path following [4] allowed to extend their application field. In this paper we propose to use a quadcopter for private, both domestic and industrial, surveillance purposes. The work begins with the implementation of an algorithm to scan an empty area: the UAV moves freely into given boundaries that have been identified from a sonar or laser camera, having the ability to avoid every kind of obstacle. Complete coverage path planning methods are many in literature. Already at the end of the 80s Lumelsky and Stepanov gave the idea of a complete sensor-based methods [5]. Neumann de Carvalho, Vidal, Viera, Ribeiro implemented a complete coverage path planning algorithm for a cleaning robot, but in their case the map of the area was a priori known [6]. In [7] Yang and Lao propose to scan the entire area by the implementation of a neural network.

In this work we propose a simple and fast algorithm using the information received by the sensors mounted on the drone to scan the unknown area: this is filled with sparse obstacles (fixed or moving), added randomly all over the feasible surface. Our algorithm is thus able to identify boundaries and obstacles, and to avoid them.

Moreover, pursuing the idea of security, the UAV should be forced to come back, after a given time, on regions that have been already scanned. Many articles has been written on this goal, achieving the result with genetic algorithms [8] and fuzzy logic approaches [9].

However we focused our attention on the neural network approach, which seems the best compromise between speed and complexity requests, and which is a strengthened method for that kind of problems. One of the first attempts to apply a neural network in an industrial environment was proposed by Rivals, Canvas, Peronnaz, Dreyfus in [10], solving a complex automotive control problem using dynamical neural network models. Then Glasius, Komoda, Gielen in [11] tried to face a problem of path planning using a topologically organized neural network of a Hopifield type with nonlinear analog neurons. In terms of obstacle avoidance, a neural network model learning from human driving data is used by Djugash, Hammer in [12].

Therefore, a pseudo-neural network was added to the main algorithm to determine the fastest path to return to "old scanned" areas, avoiding possible unknown obstacles.

Finally, in case a threat was identified, an engage algorithm gives to the drone the possibility to follow the movement of the object in the whole permitted area. For enhanced surveillance levels, the camera mounted on the UAV can be used for facial identification as well.

The article is organised as follows: Sect. 18.2 describes the first algorithm, which aim is to scan a whole unknown area and to get its features; in Sect. 18.3 the algorithm based on the neural network for surveillance purpose is presented; Sect. 18.4 reports the conclusion and the proposals for future developments of the project.

F. Cheli • F. Ripamonti (✉) • D. Vendramelli
Department of Mechanical Engineering, Politecnico di Milano, Via La Masa 1, Milan 20156, Italy
e-mail: francesco.ripamonti@polimi.it

© The Society for Experimental Mechanics, Inc. 2015

R. Allemang (ed.), *Special Topics in Structural Dynamics, Volume 6*, Conference Proceedings of the Society for Experimental Mechanics Series, DOI 10.1007/978-3-319-15048-2_18

18.2 Scan and Identification

The first request for a surveillance application is the scan of the area of interest. In this work we propose a very simple and fast algorithm permitting to an UAV to move in an unknown area in the respect of all possible boundaries and obstacles. This ability is reached with the support of a laser or sonar sensor which is able to transmit to the drone the presence of something that makes the current path inaccessible.

Even if the UAV is able to modify its altitude, at this stage we considered this parameter as fixed: therefore it is possible to assume the space which the UAV is moving as bidimensional (2D). Then, the outputs of the algorithm are the coordinates x and y of the moving space, and the θ angle, which is the yaw angle of the UAV referred to the coordinate system of the maps (or to the current path). Indeed, while the UAV is following its path, it is required to rotate left and right to scan in a more efficient way the area.

On the other hand, the only inputs are the initial position (x_0, y_0, θ_0) and the nominal visual depth. This parameter depends on the laser/sonar camera and it represents how far the UAV could follow the current path: a high value of the visual depth is safer in the obstacles avoidance but it forces the UAV to stay too much distant from the boundaries, not permitting to scan in a complete way the whole area. A compromise value, considering also the velocity of data transfer from laser camera to UAV CPU, is required.

The quadcopter motion and the information received from the sensor are memorized into a matrix through a discrete acquisition. A discrete representation is very convenient in terms of computational effort and memory storage but on the other side it is not possible to "see" all the places reached and occupied by the UAV. This issue has been partially avoided with a suitable discretization of the real scanned space.

Besides the quadcopter motion, also the view ability represents an important parameter. The "shape of the view" of the UAV depends on the type of sensor and on the visual depth: in this paper a triangular shape, which is typical for common sonar sensor and small distance, has been assumed (Fig. 18.1a). This triangle covers a determined number of pixel of scan matrix, and it is built in the manner that the height relative to hypotenuse is direct as θ. This dimension depends on the visual length (λ).

Strictly related to the "shape of the view" there is the problem of the shadow regions that don't permit to scan the whole area. If there is an obstacle on the vision line, everything which is behind the obstacle cannot be seen. The problem has been faced converting the visual triangle in a tree diagram, in which every branch is a "visual line" (Fig. 18.1b). If a pixel belonging to a branch is an obstacle-pixel, all the pixels of the branch behind that one are clouded.

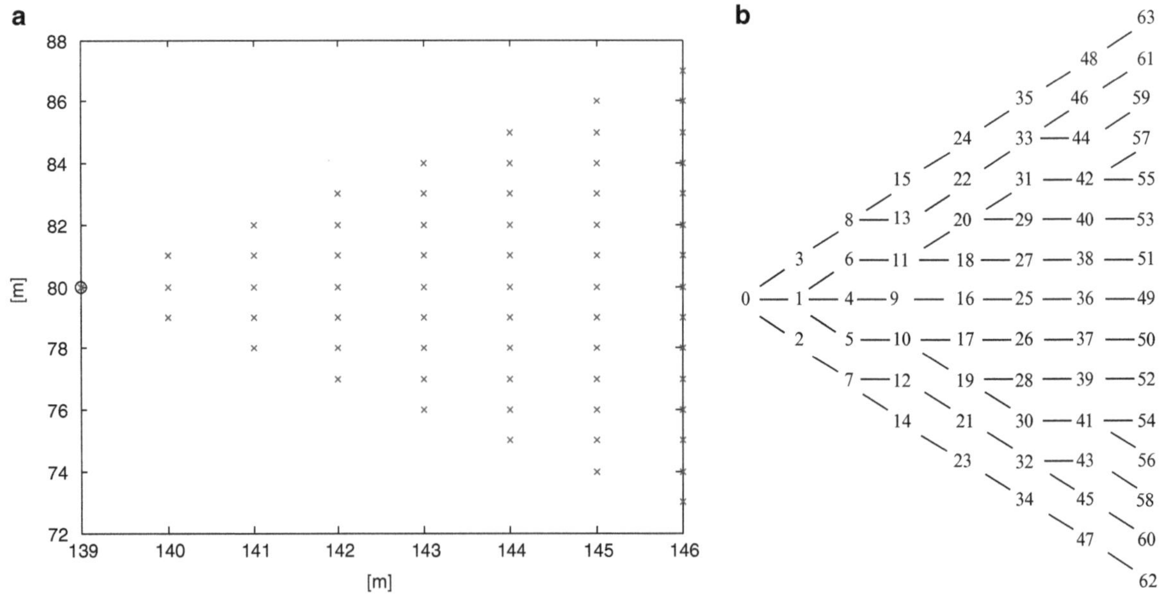

Fig. 18.1 UAV numeral simulations: the triangular "shape of view" (**a**) and the visual tree (**b**)

Fig. 18.2 The map used to test the algorithm

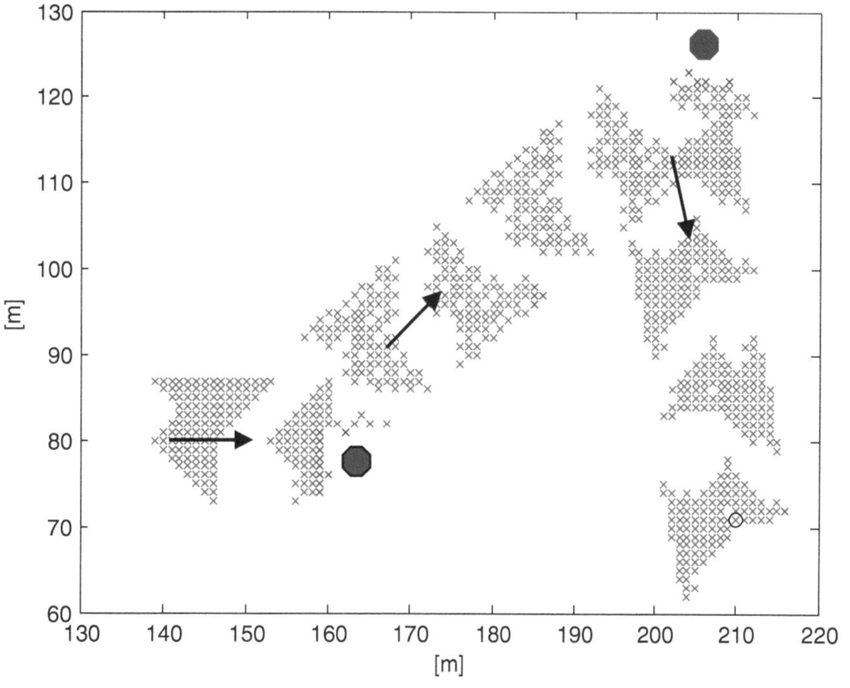

Fig. 18.3 The scanned unknown area after three steps

So, in the first part of the algorithm the quadcopter is programmed to move into the feasible area: his trajectory is defined only by one degree of freedom (dof), which gives the path direction. If in the whole visual depth of the sensor there are not obstacles, the UAV moves along its path for a distance equal to it. In the corresponding virtual matrix, the pixel associated to the real path are coloured in red.

While the UAV is moving along its path, it is forced to look ahead, left and right (hanging the yaw angle θ) for a more complete scan of the whole area. If during these recognitions an obstacle is seen at a certain distance, the corresponding pixels of the matrix are coloured in blue and the UAV continues the inspection. If the obstacle is found along the current path, a new trajectory is generated and the scan restarts (new step).

The algorithm has been tested on an area of about 80 m × 150 m with some sparse obstacles discretized with a resolution of 1 m^2 (Fig. 18.2). The UAV begins its exploration from a casually generated point (x = 132 m, y = 80 m) and a horizontal orientation. The value of the visual depth is set to 7 m.

Once the algorithm verified that the direction chosen is accessible, the quadcopter starts moving with a velocity of 1 m/s, following the current path, and looking ahead, left and right searching for possible obstacles and boundaries.

When an obstacle is identified, it's marked with the darker coloured cross. Note that if the obstacle is not directly on the path followed by the drone, the drone continues moving along the current trajectory. In Fig. 18.3 the several paths, are highlighted by the arrows, while the obstacle are represented by the octagons.

The discontinuities between the visual triangles are consequent to the discrete acquisition of the sonar sensor: in this case a very slow refresh rate has been assumed ($f_{SONAR} = 1/7Hz$).

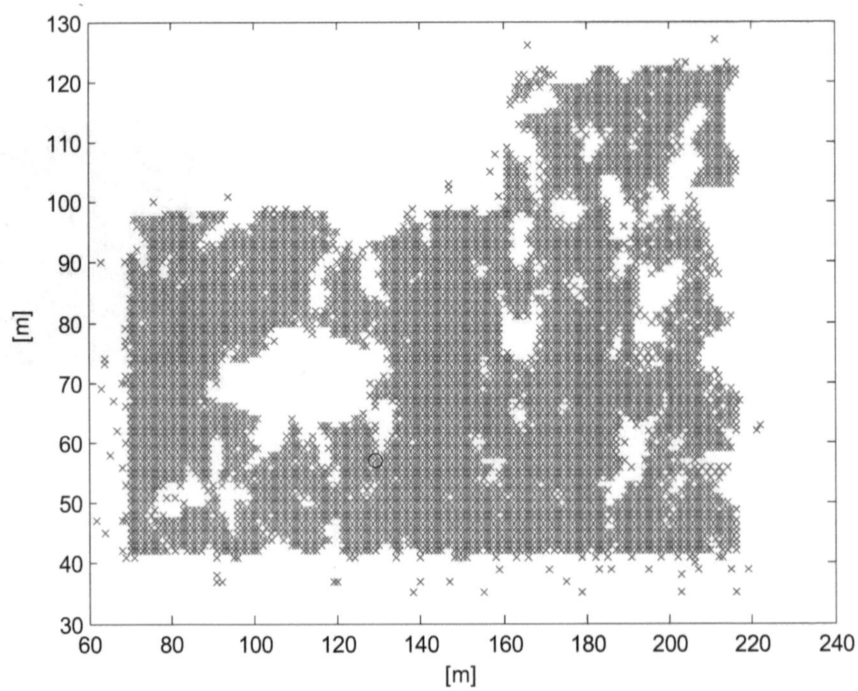

Fig. 18.4 The scanned unknown area after 150 steps

Moreover following this approach in many cases the contours of the area and of the obstacles are not completely defined (Fig. 18.4). However the algorithm is able to identify these regions and, under a certain tolerance, the open contours can be closed or otherwise the UAV is forced to a deep investigation.

Once that the UAV finished to scan the area, the obtained matrix is stored and it let the quadcopter to re-scan the known area in surveillance mode. The map extracted by this matrix is the starting point for the following algorithm.

18.3 Definition of the Path, Priority and Coming Back: The Neural Network

As previous stated, the second request for the surveillance application is to frequently control a whole known area. It means that the drone must check the surface under its supervision several times, having the possibility to remember "where and when" it passed, and then come back to the oldest monitored places for further surveillance check. It is important to remember that while the UAV is moving, it's possible to encounter not expected object on its track. Therefore it has to be able to avoid the obstacle and to reach the target point choosing the best path. These goals have been obtained with the implementation of a second algorithm.

It is based on a neural network (NN), and it allows to decide the shortest path to reach a particular point and to avoid possible unexpected obstacle. Artificial neural networks try to simulate the behaviour of a biological ones. This structure is composed by several processors, parallel, ordered in layer and linked each other (Fig. 18.5a). They are called neurons and they are characterised by a particular function defining their activity. Once that a neuron receive the information from a previous one, it can be activated by its defined function, it processes the input and it returns the corresponding output to the nearest neurons (Fig. 18.5b). Note that the value of the input processed by the single neuron is weighted, indeed not all the inputs have the same importance for all the neurons. The most important step for the implementation of a neural network is its training, which means to assign the values of the weight for every neuron. This is generally achieved with a set of data, called validation set, which is used to verify the convergence of the network with specific values of the weight.

Several neural networks have been described in the scientific literature. In this paper the starting point is the work of Glasius, Komoda, and Gielen [11], implementing a neural network of a Hopfield type for path planning and obstacle avoidance. During over the years their idea has been adopted in several cases, especially in industrial applications. Yang and Meng developed this idea initially for robot planning [13] and then for a path planning with enhanced obstacle avoidance [14]. In last years the algorithms based on [11] are evolved by the addition of penalty function [15] or a modified pulse-coupled neural network model [16].

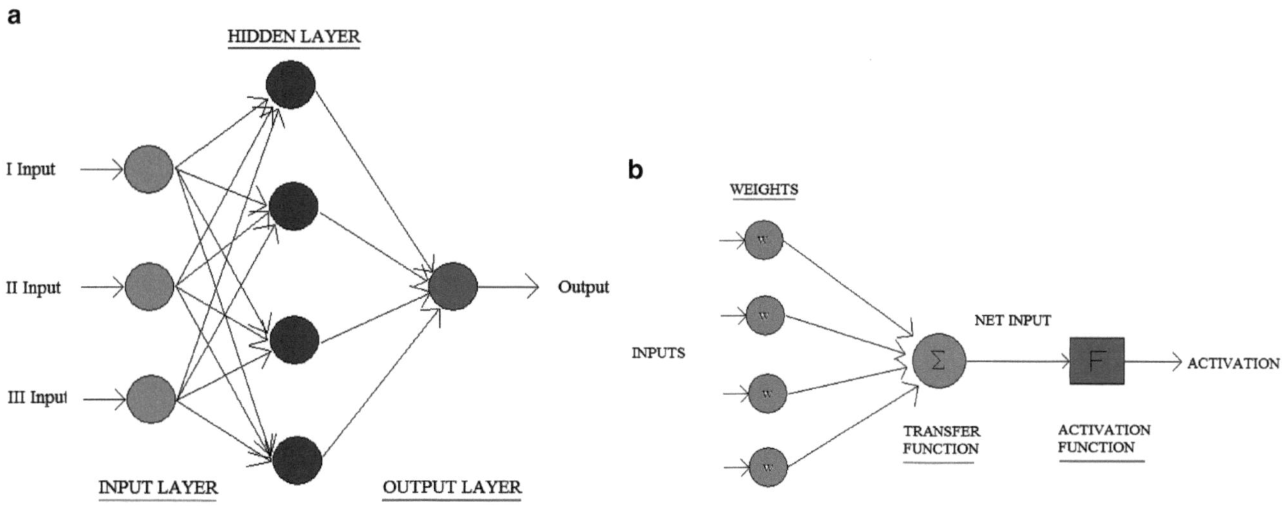

Fig. 18.5 The neural network structure (**a**) and the single neuron scheme (**b**)

In the present work, the total input u_i of the generic i -th neuron is

$$u_i(t) = \sum_{j}^{N} T_{ij}\sigma_j(t) + I_j \tag{18.1}$$

where the connections T_{ij} are excitatory ($T_{ij} > 0$), symmetric ($T_{ij} = T_{ji}$) and short range.

The connections are

$$\begin{cases} T_{ij} = 1, \text{ if } \rho(i,j) < r \\ T_{ij} = 0, \text{ otherwise} \end{cases} \tag{18.2}$$

where r is a positive number and $\rho(i,j)$ is the Euclidean distance between neuron i and j in N (number of neurons).

Defining the state variables of the neurons σ_i $(i = 1,\ldots,N)$ and the activation sigmoid function $g(x)$, Hopfield assumed that the activity of neurons changes each time the neurons receive an input

$$\sigma_i(t) = g(u_i(t)) \tag{18.3}$$

The network cannot display an oscillating behaviour and the final equilibrium states σ_j^* are the solution of the fixed point equations

$$\sigma_i^* = g\left(\sum_{i}^{N} T_{ij}\sigma_j^* + I_i\right) \tag{18.4}$$

This final equilibrium state is unique when the Liapunov function

$$L_{ij} = -T_{ij} + \delta_{ij}\frac{1}{g'(g^{-1}(\sigma_i))} < -T_{ij} + \frac{\delta_{ij}}{\beta} \tag{18.5}$$

is strictly convex.

In (18.5), β is the steepest slope of the function $g(x)$ and δ_{ij} is the Kronecker delta-function. It is possible to demonstrate that when all the eigenvalues of the matrix T_{ij} are negative, L_{ij} is a strictly convex function.

Based on these considerations the second algorithm consists in another random scan of the area. Also in this case, even if the area is known (Fig. 18.6a), the drone still moves its point of view to enhance the area monitoring.

The choice of the target point (required by the neural network to define the optimal path) is made in order to come back to the already visited region, giving priority to the oldest monitored ones.

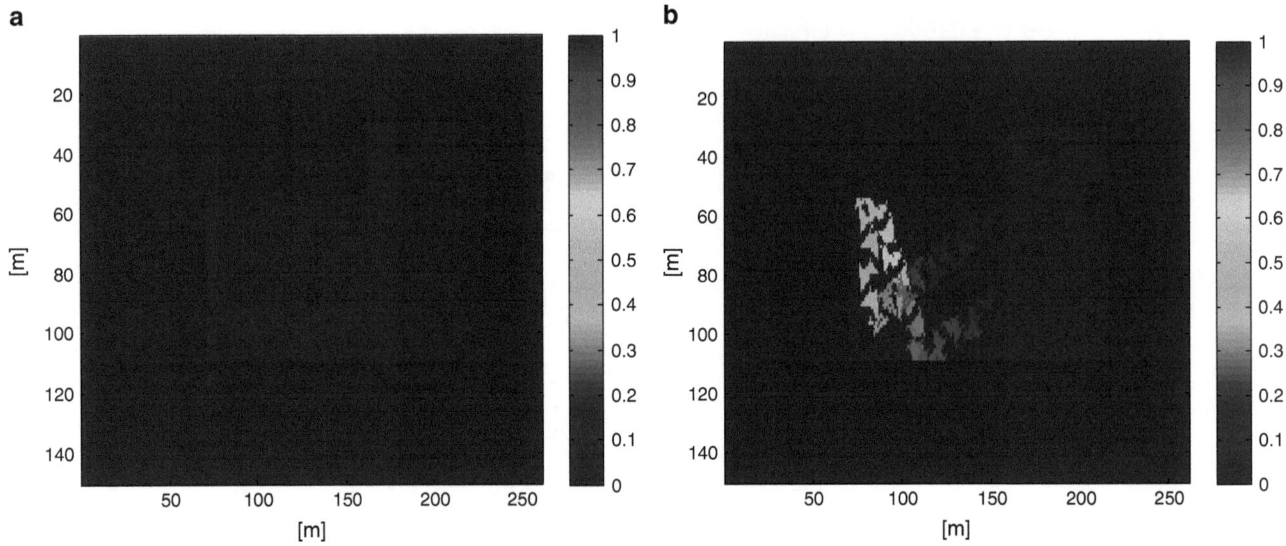

Fig. 18.6 The "visual age" at the beginning of the simulation (**a**) and a t = 70s (**b**)

The "visual age" of the different parts of the area is identified by the colours assigned to those regions. Indeed after a point has been visualised, it's coloured in the darkest tonality, which is the tonality of the obstacle too. It means that the region just checked is conceived by the drone as an obstacle, and so it avoids to review it. Then, while the time is elapsed, the region colour changes from the darkest to the lightest tones (Fig. 18.6b). Brighter the shade, higher the probability to choose that region as a target.

18.4 Conclusions

In the presented work two main algorithms are presented. The first one, focused on the scan of an unknown area, has been successfully tested simulating a quadcopter equipped with a sonar sensor. The surface has been correctly identified with a sequence of steps during which the position and orientation of the drone is modified along random paths. A development of this algorithm deals with the possibility of UAV to choose which zones check in a smarter way. Moreover if the drone gets into a space with several obstacles surrounding, it has difficulties to get out of it: an enhanced version of the algorithm should implement a suitable escape procedure.

The second algorithm, implementing a neural network to came back to critical zones (oldest monitored) in the fastest way and avoiding possible unexpected obstacles has been evaluated and tested too.

At the moment, to complete the application, a third algorithm for the object tracking is under investigation. Indeed for the aim of surveillance the drone has to be able to follow an eventual threat when identified. Moreover some algorithms for the diagnostics [17] and the failure prevention [18] may be included to improve the drone robustness.

References

1. Reitsma CL (2009) A novel approach to vibration isolation in small, unmanned aerial vehicles. Technologies for Practical Robot Applications, 2009, TePRA
2. Cola F, Resta F, Ripamonti F (2014) A negative derivative feedback design algorithm. Smart Mater Struct 23(8):art. no. 085008
3. Serra M, Resta F, Ripamonti F (2013) Dependent modal space control. Smart Mater Struct 22(10):art. no. 105004
4. Rysdyk R (2003) UAV path following for constant line-of-sight. In: 2nd AIAA "Unmanned Unlimited" system, technologies, and operations – Aerospace, San Diego, 15–18 Sept 2003
5. Lumelsky V, Stepanov A (1987) Path planning strategies for point mobile automation moving amidst unknown obstacles of arbitrary shape. Algorithmica 2:403–430
6. Neumann de Carvalho R, Vidal HA, Vieira P, Ribeiro MI (1997) Complete coverage path planning and guidance for cleaning robots. In: IEEE catalog number: 97TH8280 ISIE'97, Guimariies

7. Yang SX, Chaomin Luo (2004) A neural network approach to complete coverage path planning. IEEE Trans Syst Man Cybern Part B 34(1):718–725
8. Lin Lei Houjun Wang, Qinsong Wu (2006) Improved genetic algorithms based path planning of mobile robot under dynamic unknown environment. In: Proceedings of the 2006 IEEE international conference on mechatronics and automation, Luoyang, 25–28 Jun 2006
9. Meng Wang, Liu JNK (2005) Fuzzy logic based robot path planning in unknown environment. In: Proceedings of the fourth international conference on machine learning and cybernetics, Guangzhou, 18–21 Aug 2005
10. Rivals I, Canvas D, Peronnaz L, Dreyfus G (1994) Modeling and control of mobile robots and intelligent vehicles by neural networks. In: IEEE conference on intelligent vehicles, Paris, 24–28 Oct 1994
11. Glasius R, Komoda A, Gielen SCAM (1995) Neural network dynamics for path planning and obstacle avoidance. Neural Netw 8(1):125–133, Copyright © 1994 Elsevier Science Ltd
12. Djugash J, Hammer B Neural networks for obstacle avoidance. Informal Publication
13. Yang SX, Max Meng (2000) An efficient neural network approach to dynamic robot motion planning. Neural Netw 13:143–148
14. Yang SX, Max Meng (2003) Real-time collision-free motion planning of a mobile robot using a neural dynamics-based approach. Neural Netw 14(6):1541–1552
15. Willms AR, Yang SX (2008) Real-time robot path planning via a distance-propagating dynamic system with obstacle clearance. IEEE Trans Syst Man Cybern Part B 38(3):884–893
16. Hong Qu, Yang SX, Willms AR, Zhang Y (2009) Real-time robot path planning based on a modified pulse-coupled neural network model. IEEE Tran Neural Netw 20(11):1724
17. Cazzulani G, Moschini S, Resta F, Ripamonti F (2013) A diagnostic logic for preventing structural failure in concrete displacing booms. Autom Constr 35:499–506
18. Ambrosio P, Cazzulani G, Resta F, Ripamonti F (2014) An optimal vibration control logic for minimising fatigue damage in flexible structures. J Sound Vib 333(5):1269–1280

Chapter 19
An Innovative Solution for Carving Ski Based on Retractile Blades

F. Cheli, L. Colombo, and F. Ripamonti

Abstract This paper presents the design of a new ski to improve the performances focusing especially on steering conditions. In the proposed solution an additional retractile edge has been added modifying the lower surface of the ski. It consists of a thin stainless steel plate suitably reinforced with two longitudinal ribs welded along the plate borders to create the blades. Then this element is constrained to the ski along the longitudinal symmetry axis. In unload and straight conditions the gap between the ski and the edge is null while in curves the retractile blade is deformed and pushed out by the force applied by the skier, penetrating the snow. The mechanism for the force transmission is based on two revolution joints, allowing small rotations of the boot with respect to the ski and compensating the ski deformation, and several pistons, pushing on the edge. Two ski prototypes have been manufactured, instrumented and tested on a snow track. In particular, in order to show the differences in terms of performance, the modified skis have been compared to the classic version. Additional considerations about the variations on stiffness and friction behavior introduced by the new solution have been drawn too.

Keywords Sport engineering • Ski performance improvement • Stainless blades • Downhill tests • Tilting mechanism

19.1 Introduction

There has been a lack of significant research and innovation concerning skis and snowboards in the recent years. It was in 1988 when the last important innovation cycle started with the introduction of the carving, ending with its total implementation and success in 1997. The implementation of the carving ski has been important as it allowed to easily perform a smooth curve and to keep the direction of it, contrarily to the stemming. After that, the research in this field has been slow and it affected only marginal aspects like improvements in the materials adopted, in the shape, in the production process, etc. For instance, recent studies have been conducted about the adoption of a material with superhydrophobic properties to realize the ski base [1, 2]. Another research had the aim to determine the effect of the core material on the ski properties by making tests on ski prototypes with different core materials and thicknesses [3]. In 2012 Petrone N. built a static bench to help improving the design of the skis by evaluating their Edge Load Profile [4]. In 2014 Rajek M. et al. worked on the implementation of a passive shunt damping system realized with piezoelectric elements, to control the vibration behavior of the skis and improve their performance [5]. Other studies focused on the production process improvement by using techniques such as hydroforming to easily realize complicated shaped components, such as a pad for extreme carving [6], or the laser to change the appearence of the wooden cores by analyzing the effect of the thermal tool on the wood [7].

In this context, a new idea has been developed to open new possibilities of improvement and innovation for the winter sports based on the usage of skis and snowboards.

The concept is about the addition of a second edge realized with a steel plate applied on the lower surface of the ski and made retractile by an auxiliary mechanical system (as it will be described in detail in the following) deforming it thanks to the load applied by the skier during the curve. The aim of the second edge is to improve the steering maneuver by supporting the primary edge in the penetration into the snow. Moreover, the contact between the snow and a steel surface is another new element opening new doors to a possible enhancement of the skiing performance. Ski-snow contact is in fact one of the main factors affecting the skiing performance [8, 9], due to the friction between the two surfaces affecting the deformation of the snow.

F. Cheli • L. Colombo • F. Ripamonti (✉)
Department of Mechanical Engineering, Politecnico di Milano, Via La Masa 1, Milan 20156, Italy
e-mail: francesco.ripamonti@polimi.it

© The Society for Experimental Mechanics, Inc. 2015

R. Allemang (ed.), *Special Topics in Structural Dynamics, Volume 6*, Conference Proceedings of the Society for Experimental Mechanics Series, DOI 10.1007/978-3-319-15048-2_19

Although recently the development of new solutions in this field started being supported by computer simulation models, skis come from a history of trial and error tests made directly on the field. The presented project is characterized by some innovative elements not yet supported by previous studies or experiences. Because of this, the aim of the current work is to design and build a prototype of a pair of skis implementing the concept described above and to test it on the field to verify the behavior of the new solution and to compare it with a non-modified pair of skis of the same type.

19.2 The Prototype

A standard pair of skis 1.7 m long has been chosen to realize the prototype (Fig. 19.1). The lower surface has been modified to host the retractile edges. The latters are made by a 0.5 mm thick AISI 301 stainless steel plate, reinforced with two longitudinal ribs of the same material with a 5×5 mm section properly welded along the borders to create the blades. The plate is then constrained to the ski along the longitudinal symmetry axis through a set of M2×15 stainless steel screws, placed 20 cm one from each other (Fig. 19.2a). Two more of those screws have also been applied at the sides in the front part of the ski to prevent the possible infiltration of the snow in the tip region during skiing (Fig. 19.2b).

As stated above, a mechanical system has been designed to make null the gap between the retractile blade and the ski during unload and straight conditions, while during the steering maneuver the steel plate is deformed so that the blade comes out to a maximum of 5 mm, penetrating the snow (Fig. 19.3).

For that purpose, the bindings of the boot have been applied no more on the upper surface of the ski, but, as shown in Fig. 19.5, on an elevated plate, which can tilt respect to the ski thanks to two revolution joints (Fig. 19.4b). Moreover, two

Fig. 19.1 The prototype of one of the skis

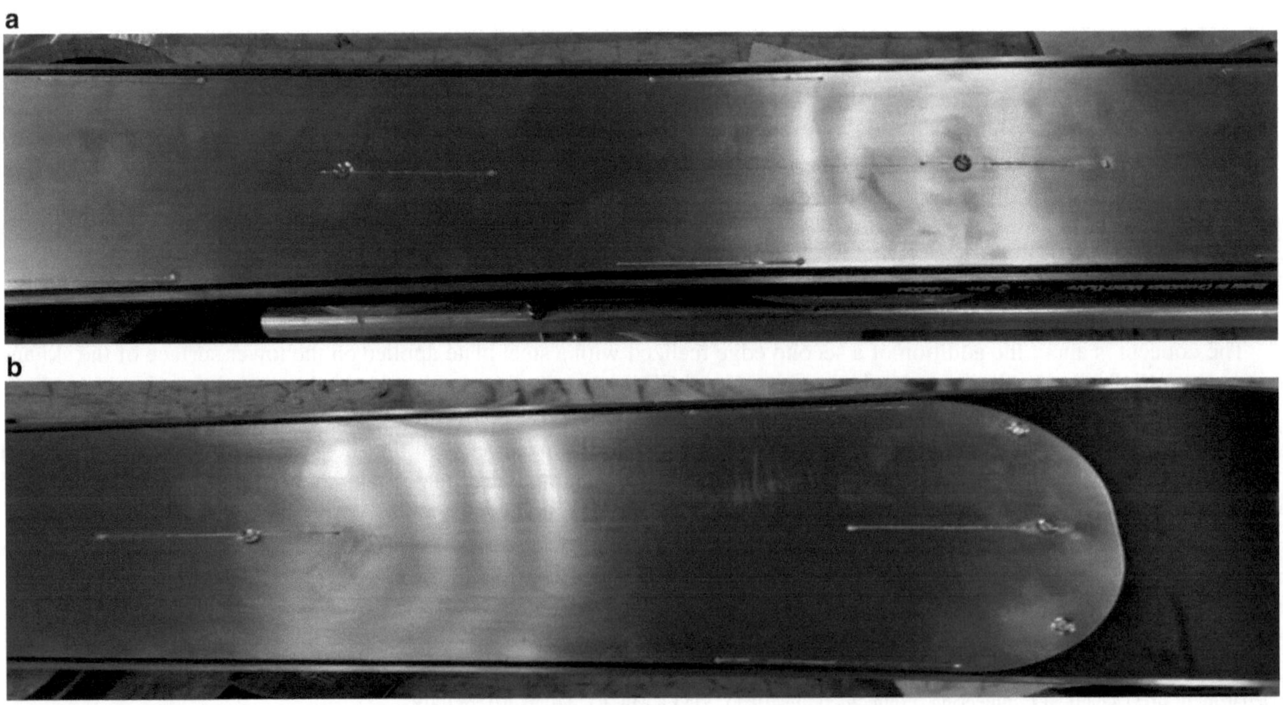

Fig. 19.2 The central part (**a**) and the front part (**b**) of the retractile steel plate which constitutes the retractile edges, constrained to the ski through a set of M2×15 screws

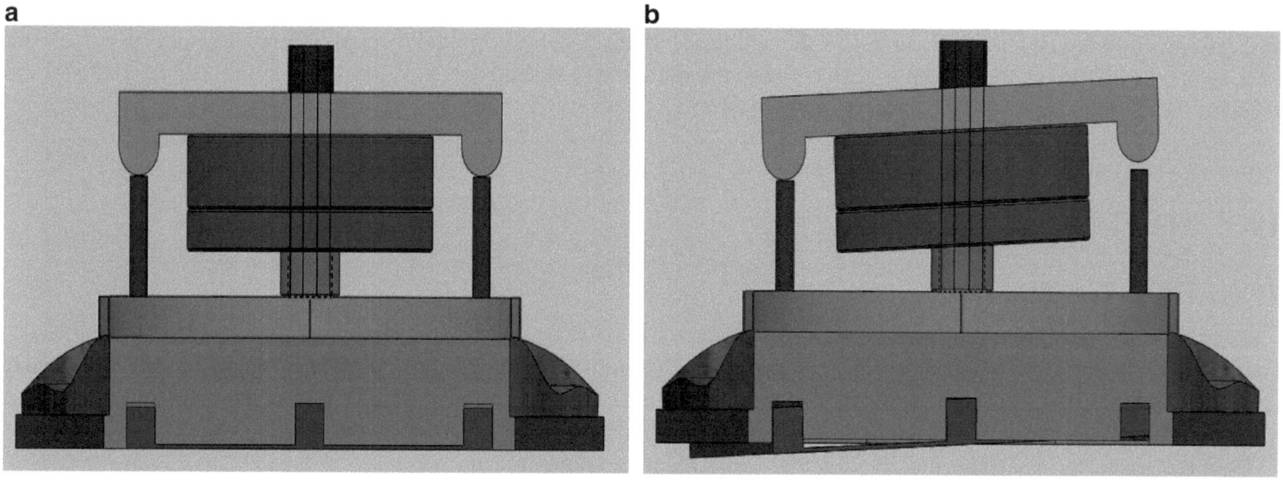

Fig. 19.3 The cross section of the prototype, showing how its mechanism functions in unload or straight condition (**a**) and in curves (**b**)

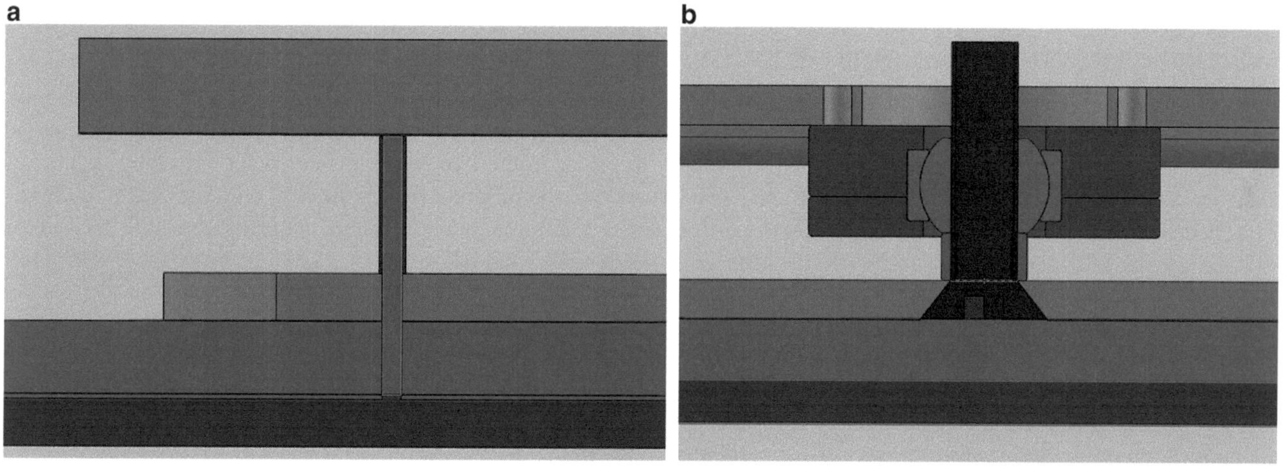

Fig. 19.4 The sections respectively of one of the pistons and the hole in which it is located (**a**) and of one of the revolution joints (**b**)

Fig. 19.5 Detail of the mechanism of the prototype

sets of four pistons have been applied at the sides, between the upper plate and the retractile blade (Fig. 19.4a). A series of holes have been created on the ski in order to let the pistons pass through, and to hold them in position at the same time. This way, when the skier bends to perform a curve, the upper plate tilts and the load applied by the skier pushes the pistons, which deform the retractile blade (Fig. 19.3b). When the curve is over the skier returns to the straight position, tilting the plate on the other direction; this way the pistons on the other side push the retractile blade, making it return to the starting position with no gap (Fig. 19.3a). Obviously, for a correct functioning of the mechanism, the tilting angle of the upper plate and consequently the maximum excursion of the additional blade have been limited by applying two bumpers per side. This way, as previously stated, the maximum gap between the retractile blade and the ski is limited to 5 mm. Figure 19.5 shows a zoom of the mechanism adopted on the prototype.

19.3 The Field Test

19.3.1 The Test Track

Experimental tests have been carried out at the ski track "Quarta Baita" in Foppolo, located in the province of Bergamo (Italy), in March 2014. The track can be divided into three main parts: a first downhill section, a second one almost in plane, and a final downhill section for a total altitude gap of 135 m and a track length of about 1,200 m.

Four test tracks have been accurately monitored with the measurement instruments:

1. Normal test with the modified skis;
2. Test performed with small radius curves with the modified skis;
3. Test performed with small radius and applying a high load on the skis with the modified skis;
4. Test performed with small radius curves with the classic skis.

In the following analysis, only the data regarding the tests 2 and 4 are going to be discussed, as they are the most meaningful and a direct comparison between the performances of the two pair of skis can be carried out. Figure 19.6b shows the satellite map of the ski track and the paths followed by the skier in the two tests considered for the analysis, tracked thanks to a GPS device. The paths of the tests 1 and 3 (Fig. 19.6a) are too different to be compared with the others as the test 1 has been executed with too large curves, while the path of test 3 is different than the others in the last downhill section.

19.3.2 Experimental Setup

The prototype pair of ski and an equivalent pair of non-modified skis have been used on the ski track by an expert skier after the application of measuring instruments [10], in particular:

- A GPS positioned on the athlete's back;
- A triaxial accelerometer K330 by ST Microelectronics positioned on the athlete's back, with:

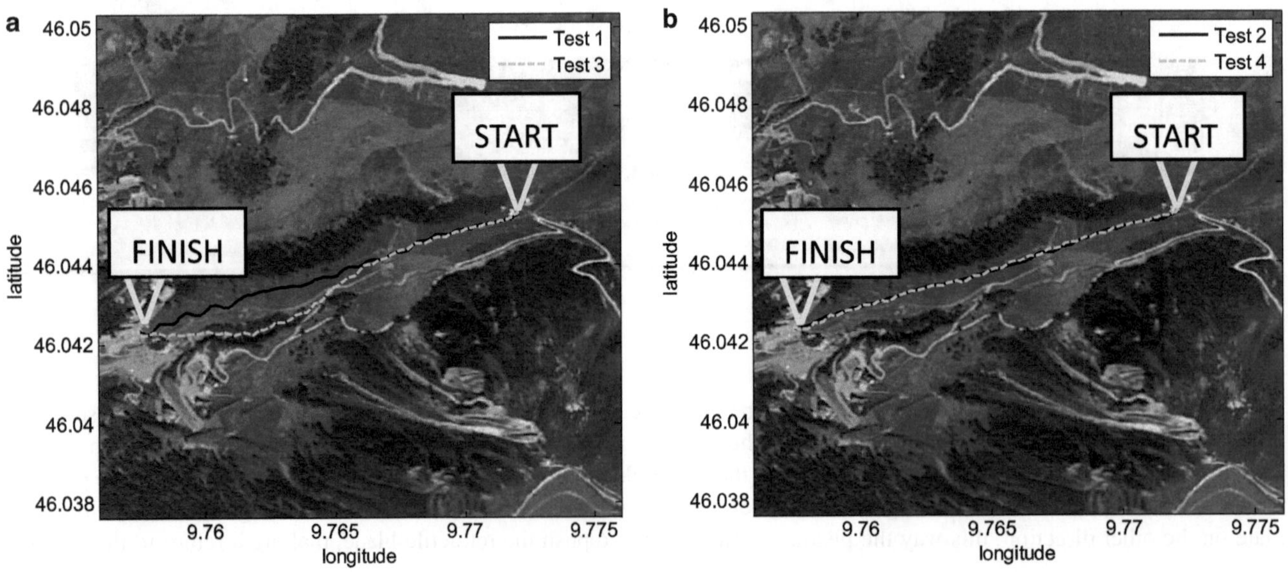

Fig. 19.6 The satellite map of the ski track with the paths of the tests 1 and 3 on the figure (**a**) and the paths of the tests 2 and 4 on the figure (**b**)

  - fs = 50 Hz
  - Range = 2 g

- An uniaxial accelerometer 333B30 by PCB Piezotronics positioned on the right ski, with:
  - fs = 2,048 Hz
  - Range = 50 g

- A triaxial accelerometer 356A02 by PCB Piezotronics positioned on the left ski, with:
  - fs = 2,048 Hz
  - Range = 500 g

The aim of these tests were to verify the behavior of the new solution on the snow as well as compare its performance to the classic skis.

19.3.3 The Experimental Results

The data gathered during the test have been analyzed and sorted in the following graphs, which show the performances of both the modified and classic skis measured by the GPS and the accelerometer on the athlete's back. The data gathered by the accelerometers positioned on the skis are not considered in this preliminary study and they will be taken into account in a following analysis. The results of the two tracks have been properly overlapped to easily compare the performances.

As already mentioned above, the track is 1,200 m long and has an altitude gap of 135 m.

The graphs in Fig. 19.7 bring out some similarities between the performance with the prototypes and with the classic skis. In both cases the track has been executed in about 100 s, and in terms of speed both the tests present an average speed of 43 km/h and a maximum speed of about 60 km/h. Moreover, both the graphics clearly show how the acceleration with the prototypes is lower in the first 20 % of the track (belonging to the first downhill).

The graphs of the accelerations (Fig. 19.8) confirm the similarity of performance between the modified and the classic skis. In the longitudinal and vertical acceleration graphs it is possible to see how, in both the tests, at the very start of the track the gravitational acceleration is applied to the vertical component, while after the first 100 m and until the end of the performance it's transferred to the longitudinal component because of the different position that the skier assumes while going down.

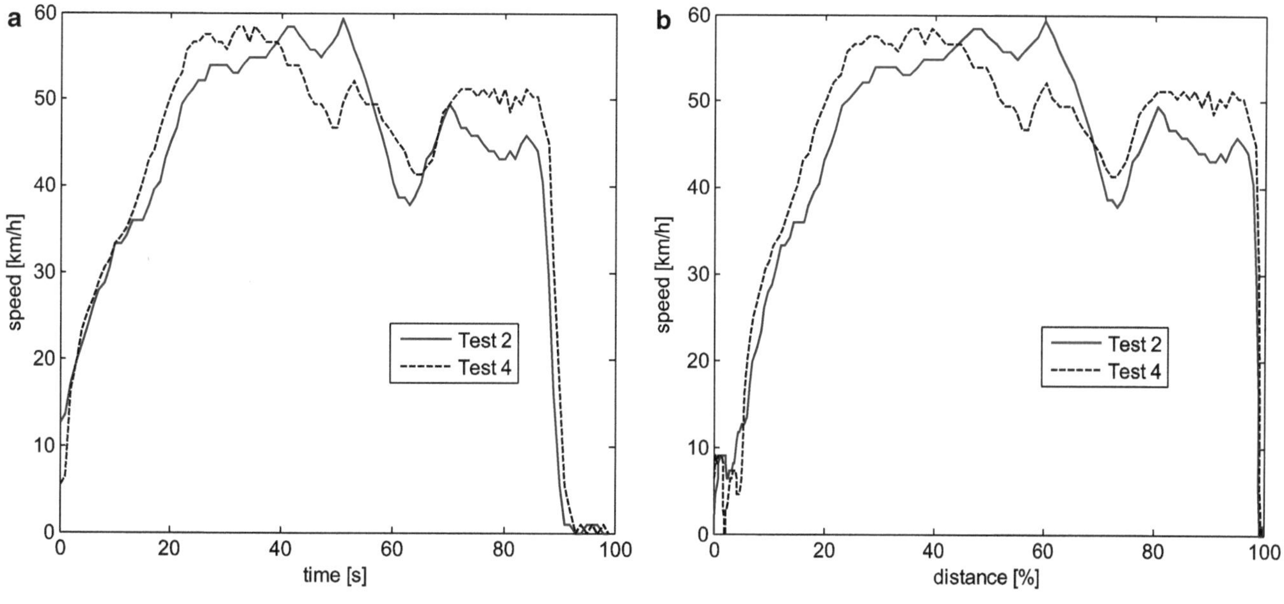

Fig. 19.7 The time histories (**a**) and the space in % histories (**b**) of the speed measured by the GPS. The *solid line* belongs to the test 2 and the *dashed line* belongs to the test 4, just as before

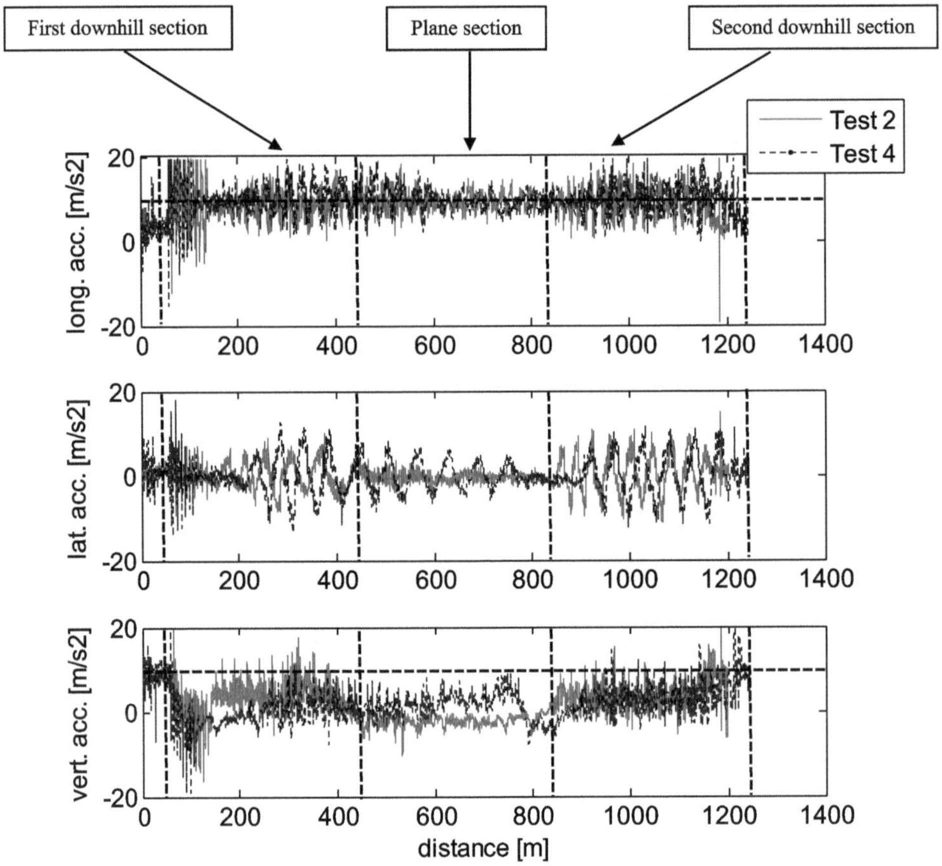

Fig. 19.8 The space histories of the longitudinal, lateral and vertical accelerations measured by the triaxial accelerometer positioned on the back of the athlete

The graph of the lateral acceleration clearly shows the number of curves and their frequency for both the performances in the first and in the second downhill sections:

- First downhill:
  - Modified skis: 5 curves, f = 0.27 Hz
  - Classic skis: 5 curves, f = 0.30 Hz
- Second downhill:
  - Modified skis: 8 curves, f = 0.33 Hz
  - Classic skis: 6 curves, f = 0.26 Hz

In the first part the athlete performed the same number of curves with the prototypes and with the classic skis, with only a slight difference in the frequency. In the last part of the track more curves have been performed with the modified skis, and at a higher frequency. As a direct consequence of this, the athlete has been slower with the prototypes in that section, as can be seen also from the speed graphs (Fig. 19.7).

19.4 Conclusion and Future Work

In this paper a pair of skis modified with an additional retractile blade with the intent to improve the performance is presented. The feedback gained from the test day and the data gathered demonstrated that the prototype works as intended, but at the current state the performances are similar to the classic pair of skis.

The next step is to carry out further field tests to collect more meaningful information. In particular, a higher number of tests on an accurately tracked and easily repeatable path is required to gather a good amount of actual, statistically valid and comparable data. Moreover, to further develop and improve the project, analysis and research on the behavior of the new surface (stainless steel) in contact with the snow is required, along with a study on the possible application of grooves (mode of application, shape, depth, effectiveness test, etc.) to avoid undesired ski-snow interaction effects and consequent issues on the control of the skis. The influence of the stiffness variation introduced by the prototype modifications will be considered and analyzed in a future study.

Acknowledgments The authors wish to thank Studio Polaris S.a.s. and Foppolo SKI for the financial support to the present research.

References

1. Nordin NA, Styring P (2014) Superhydrophobic ski bases for reduced water adhesion. Procedia Eng 72:605–610
2. Breitschädel F, Haaland N, Espallargas N (2014) A tribological study of UHMWPE ski base treated with nano ski wax and its effects and benefits on performance. Procedia Eng 72:267–272
3. Mastrogiuseppe P (2007) Skibuilders.com: The effects of core material and thickness on the performance and behaviour of a ski. McGill University, Montreal, Canada, accessed 23 Sept 2014 skibuilders.com/articles/EffectsofCore.pdf/
4. Petrone N (2012) The use of an Edge Load Profile static bench for the qualification of alpine skis. Procedia Eng 34:385–390
5. Rajek M, Lubieniecki M, Pieczonka U, Uhl T (2014) Piezoelectric shunt damping for alpine ski traction improvement. Diagnostyka 15(1):3–9
6. Krotak S, Masek B, Urbanek M (2011) Use the hydroforming by produced the pad for extreme carving. In: Annals of DAAAM and proceedings of the international DAAAM symposium, vienna, Austria, pp 1279–1280
7. Petutschnigg A, Stöckler M, Steinwendner F, Schnepps J, Gütler H, Blinzer J, Holzer H, Schnabel T (2013) Laser treatment of wood surfaces for ski cores: an experimental parameter study. Adv Mater Sci Eng 2013:1–7, art. no. 123085
8. Theile T, Szabo D, Luthi A, Rhyner H, Schneebeli M (2009) Mechanics of the ski-snow contact. Tribol Lett 36(3):223–231, Cited 11 times
9. Federolf P, JeanRichard F, Fauve M, Lüthi A, Rhyner H-U, Dual J (2006) Deformation of snow during a carved ski turn. Cold Reg Sci Technol 46(1):69–77, Cited 7 times
10. Bassetti M, Braghin F, Castelli Dezza F, Ripamonti F (2014) Design of an inertial measurement unit for enhanced training. In: Conference proceedings of the Society for Experimental Mechanics series, Miami, Florida, USA, vol 6, pp 275–285

FSC
www.fsc.org
MIX
Papier aus verantwortungsvollen Quellen
Paper from responsible sources
FSC® C105338

If you have any concerns about our products,
you can contact us on
ProductSafety@springernature.com

In case Publisher is established outside the EU,
the EU authorized representative is:
Springer Nature Customer Service Center GmbH
Europaplatz 3, 69115 Heidelberg, Germany

Printed by Libri Plureos GmbH
in Hamburg, Germany